개정증보판

집합론

집합론 (개정증보판)

지은이 You-Feng Lin · Shwu-Yeng T. Lin
옮긴이 이홍천
펴낸이 조경희
펴낸곳 경문사
펴낸날 2011년 9월 1일 개정증보판 1쇄
2025년 1월 6일 개정증보판 9쇄
등 록 1979년 11월 9일 제1979-000023호
주 소 04057, 서울특별시 마포구 와우산로 174
전 화 (02)332-2004 팩스 (02)336-5193
이메일 kyungmoon@kyungmoon.com

값 30,000원

ISBN 978-89-6105-377-8

개정증보판

집합론

Set Theory
An Intuitive Approach

You-Feng Lin · Shwu-Yeng T. Lin 지음

이흥천 옮김

KM 경문사

개정판에 즈음하여

한 마디로 번역은 "제 2의 창작" 이란 신념으로 이 개정판을 독자 앞에 내놓으면서 옮긴이는 다음과 같은 점을 강조하고 싶다.

첫째　독자 누구나 내용을 편하게 읽을 수 있도록 원문과 번역문의 글자체를 날씬하게 바꾸었다.

둘째　번역문 각주에 별표 *를 훨씬 많이 붙여 설명을 보충함으로써 독자의 학습효과를 올리도록 하였다.

셋째　처음부터 끝까지 거듭 수정하여 완벽한 번역이 되도록 온갖 힘을 썼다.

이 책의 수준도 수준이려니와 설명 또한 극히 쉬워 혼자 공부하는 사람에게도 알맞으리라 본다. 아무튼 이 개정판이 집합론을 학습하려는 초보자 누구에게나 큰 보탬이 되길 간절히 바란다.

끝으로 이와 같은 개정판이 나오기까지 남다른 노고가 컸었음을 여기에 덧붙인다.

우선 400쪽이 넘는 교정지를 열 번씩이나 묵묵히 처리해낸 경문사 편집부 이정순씨의 성의와 더불어 조판의 수정을 수도 없이 거듭해 온 (주) 동국문화 박순임씨의 적극적인 협조를 두고 하는 말이다. 이 두 분에게 새삼 고마움을 나타내마지 않는다.

2005년 8월
옮긴이

개정(증보)판에 즈음하여

한 마디로 번역은 "제 2의 창작"이란 신념으로 이 개정판을 독자 앞에 내놓으면서 옮긴이는 다음과 같은 점을 강조하고 싶다.

첫째 독자 누구나 내용을 편하게 읽을 수 있도록 원문과 번역문의 글자체를 날씬한 것으로 바꾸었다.

둘째 개정판보다 각주 *를 수 없이 늘려 원문의 부족함을 보충하였다.

셋째 개정판 연습문제 1.5, 2.2, 2.3, 2.6, 3.1, 3.2에 새로히 문제를 추가하였다.

넷째 독자가 이해하기 힘들 것으로 여겨지는 대목에 대하여 다른 저자의 책을 더 많이 소개함으로써 독자의 학습을 넓혀 나가도록 하였다.

그렇다고 일일히 구입할 필요는 없다고 생각한다. 다만 인근 대학도서관에서 복사를 떠서 공부하는 길도 매우 좋으리라 믿는다.

이 책의 수준도 수준이려니와 설명 또한 극히 쉬워 혼자 공부하려는 독자에게도 알맞으리라 본다. 아무튼 이 개정판이 집합론을 학습하려는 초보자 누구에게나 큰 보탬이 되길 간절히 바란다.

끝으로 이와 같은 개정(증판)판이 나오기까지 남다른 노고가 컸었음을 여기에 덧붙인다.

우선 400쪽이 넘는 교정지를 열 번씩이나 묵묵히 처리해낸 경문사 편집부의 성의와 더불어 조판의 수정을 수도 없이 거듭해 온 (주) 동국문화 박순임씨의 적극적인 협조를 두고 하는 말이다. 이분들에게 새삼 고마움을 나타내마지 않는다.

2011년 1월

옮긴이

옮긴이 머리말

You-Feng Lin, Shwu-Yeng T. Lin 두 저자가 공동으로 집필한 Set Theory는 이미 우리나라의 많은 대학교 학생들이 공부하여 왔던 수학교재로서 대학 1, 2학년 학생에게는 여러 모로 알맞기에 이 책을 번역하기에 이르렀다.

누구나 다 알다시피 집합론은 그 자체의 학문으로서도 중요할 뿐만 아니라 수학의 모든 분야를 비롯하여 논리적 사고를 필요로 하는 영역에 있어서도 기초가 되므로 그 이론을 조기에 익혀두는 것은 지극히 바람직하다. 옮긴이가 몸 담고 있는 대학교의 수학과와 이공계 2학년 학생의 교과서로서 다년간 사용하여 오던 이 책을 번역함에 있어서 특히 다음과 같은 점을 유의하였다.

1. 대체로 원문을 충실히 의역하도록 힘을 썼으며 증명이나 문제풀이 과정에서 독자의 이해를 돕고자 필요할 때에는 옮긴이가 직접 간단히 설명을 보충하였다. 한편 원문에 잘못 쓰여진 문자나 기호는 일일이 바로 고쳤다. 이를테면 제 1 장에서부터 다루어 온 명제 중 함의를 뜻하면서도 이것이 조건문 $p \to q$로 적혀 있는 데에는 기호 → 대신 ⟹를 써서 $\boldsymbol{p \Rightarrow q}$와 같이 바꿔 놓았다.

2. 원문에서는 모든 보기를 EXAMPLE로 열거하였으나 그 중 간단한 것은 **보기** 그대로, 증명이나 풀이가 뒤따른 것은 **예제**로 구분하였다.

3. 연습문제 4.4, 5.1에는 문제를, 또 뒤의 참고문헌(417쪽)에는 책 다섯 권을 옮긴이가 추가하였다.

4. 본문 각주 중 숫자로 된 것은 저자가, 별표 *가 붙여진 것은 전적으로 옮긴이가 보충해 놓은 것이다.

다년간 이 책을 원서로 지도해 본 경험에 비추어 보면 한 학기(주당 3시간 강의)용으로서 적절한 내용과 분량이 실려 있다고 생각한다. 다시 말하면 강의 중 학생들에게 가끔 연습문제를 풀게 함으로써 학습효과를 더욱 올릴 수 있을 만큼 여유있게 지도할 수 있다고 본다. 더구나 우리나라 대학생은 이미 중고등학교 시절에 집합의 기초를 습득한 터이므로 제 2 장 1~5절은 간단히 다루어도 무방할 것이다.

번역판을 1992년 2월에 발간하자마자 많은 독자들로부터 올바른 지적(칸토어-베른슈타인의 정리에 관한 역사적 기록 : 245쪽 참조)과 함께 분에 넘치는 호평을 받아왔기에 1993년 3월 이래 거듭 내고 있다.

그 동안 우리나라의 국제화 추세에 따라 강의도 영어로 진행하는 대학이 점차 늘어가고 있는 현실에 맞춰서 저자의 원문(영문)과 옮긴이의 번역문을 함께 실은 책도 별도로 내놓기에 이르렀다. 독자가 이 새로운 책의 영문과 그 번역문을 일일이 대조해가며 공부함으로써 집합론을 터득할 수 있을 뿐만 아니라 영어 실력도 저절로 쌓을 수 있을 것으로 믿는다.

만일, 이 책의 잘못된 점이 또 있으면 독자의 서슴없는 지적이나 냉철한 비평이 있기를 간곡히 바란다.

끝으로 이 책이 발간되기까지 적극 협력하여 주신 경문사 박문규 사장과 편집부 여러분에게 진심으로 감사를 표한다.

2004년 9월
옮긴이

Preface

This book has been written for the sophomore-junior level student who wishes to know more about the "new math." The prerequisites for this book include some contact with college mathematics but not necessarily great mathematical skill. Ideally, the reader will have taken one quarter or one semester of calculus.

There is currently a significant amount of repetition in post-calculus mathematics courses. It is not uncommon for a student taking four such courses to receive four separate sets of lectures on set theory. In the mathematics department of the University of South Florida, we have eliminated this duplication of effort by offering a sophomore-junior level course in set theory. This course, which is required of all mathematics majors, is also popular with upper level education majors who anticipate teaching mathematics. It is for these two groups of students that this book is primarily written, and the book is the outgrowth of many years of teaching this course.

Presenting set theory within the framework of an axiomatic system is, without doubt, more satisfactory when one intends to be rigorous. But there is no axiomatization of set theory that is simple and easy enough for the beginning student. Therefore we have utilized, cautiously we hope, an intuitive approach to the subject.

Also, we have endeavored to make the book self-contained. We have included the necessary basic material for the student who wishes to go on to such courses as topology, analysis, modern algebra, etc. But it is our hope that the material in this book would prove interesting to the student and teacher in its own right. It would be unfortunate to discover that a subject as interesting as set theory be regarded as useful only as a prerequisite course.

지은이 머리말

대학 1, 2학년 학생 중 "새 수학"을 알고자 하는 독자를 위하여 집필한 이 책을 공부하려면 대학수학을 미리 약간 알고 있어야 하지만 수학적 기능을 충분히 갖출 필요는 없다. 바람직한 것은 독자가 (두 학기 또는 세 학기로 나누어져 있는) 미적분학을 한 학기에 마치는 일이다.

현재 미적분학 이후에 반복되고 있는 강의의 분량이 제법 많다보니 학생이 집합론 강의를 네 과목이나 따로 수강하는 사례를 흔히 볼 수 있으나 South Florida 대학교 수학과에서는 대학 1, 2학년 수준의 집합론을 강의함으로써 이러한 중복을 피해왔었다. 모든 학생에게 필수인 이 과목은 장차 수학을 지도할 상급수준의 교육 전공 학생들에게도 보급되어 있다. 이런 두 집단의 학생들을 위하여 쓰여진 이 책은 집합론을 다년간 지도하여 얻은 산물이라 할 수 있다.

집합론을 공리계*의 테두리 안에서 다룬다면 틀림없이 엄밀성을 바라는 독자는 만족하겠지만 초보자를 위한 집합론을 단순하게 직관적으로 공리화할 수는 없다. 이러한 형편을 감안할 때 우리 두 저자는 이 과목에 대하여 직관적 접근을 조심스럽게 시도해왔노라고 자부하고 싶다. 이 책은 그 자체로 완벽하도록 힘을 썼으며 장차 위상수학, 해석학, 현대대수학의 과목을 공부하려는 학생들에게 필요한 기본적 내용을 싣기로 하였다. 그리하여 이 책이야말로 학생뿐만 아니라 지도할 사람에게도 흥미롭다는 것이 입증되기를 기대하고 있으나 한편으로는 집합론과 같은 흥미 있는 과목이 선행 과목으로만 유용한 것으로 인정되고 있어 안타깝다.

이 책의 모든 내용은(두 학기로 나누어져 있는) 한 학기 또는(세 학기로 나누어져 있는) 두 학기에 다룰 수 있도록 짜여져 있으나(세 학기로 나누어져 있는) 짧은 한 학기 과정으로서도 지도할 수 있다.

---

* 공리적 방법에 관심이 있는 독자라면 이 책의 제 2 장 집합 집합의 개념을 배운 뒤 71쪽 각주*에 소개된 책 중 Pinter 제 1 장 Historical Introduction(1~30쪽) 또는 이장우 2.8 집합론의 쩨르멜로・프랜켈공리(75~85쪽)를 읽어보기 바란다.

The entire book is designed to be covered in one semester or two quarters. For a short one quarter course, we recommend omission of the last three chapters and a de-emphasis of Chapter 1. The instructor presenting such an abbreviated course may also want to omit sections 5 to 8 of Chapter 2 and section 3 of Chapter 3.

Of course this book would never have been published without considerable assistance from many other people almost too numerous to mention. We are indebted, as always, to many students at the University of South Florida who have taken a course in set theory from us in recent years. Almost without fail they have happily endured the uncertainties of a course taught from mimeographed notes and manuscript.

We are indebted to our colleagues, Professors Marcus McWaters, George Michaelides, Leonard E. Soniat, and Frederic Zerla, all of whom taught a course from the mimeographed form of this book and made many helpful suggestions. We would particularly like to thank the following people : Professor A.W. Goodman for urging us to write this book and reading the first two chapters of the manuscript : Professor Arthur Lieberman for reading the entire manuscript and for many helpful discussions : Professors Klaus E. Eldridge of Ohio University, David Isles of Tufts University, Eugene M. Kleinberg of M.I.T., and Wayne H. Richter of the University of Minnesota, for their constructive criticisms and helpful suggestions which were incorporated into the manuscript : and Mr. William R. Zigler for reading portions of the manuscript.

Finally, we are grateful to Miss Wanda Balliet (now Mrs. Jimmie Alcus Evans) for her expert typing of the manuscript and to the staff of Houghton Mifflin Company who have turned a manuscript into a book.

Shwu-Yeng T. Lin

You-Feng Lin

Tampa, Florida

그럴 경우 마지막 세 장을 생략하고 제 1 장은 간략하게 다루거나 제 2 장의 5~8절과 제 3 장의 3절을 생략하여도 무방하리라고 본다.

여기서 언급하기에는 너무나 많은 분들의 상당한 협조가 없었더라면 물론 이 책은 출간되지 못하였을 것이다. 저자들이 늘 그러하듯, 최근에 우리의 집합론을 수강한 South Florida대학교의 많은 학생들에게 신세를 지고 있다. 아마도 대부분의 학생들이 복사판의 노트나 원고에 의지하여 지도받던 단점으로 말미암아 생긴 강의의 불확실한 점들을 용하게 이겨냈다고 본다.

우리는 동료교수들 Marcus McWaters, George Michaelides, Leonard E. Soniat 그리고 Frederic Zerla에게도 은혜를 입었으며, 이들의 제안은 복사판으로 이 책을 지도함에 있어서 많은 도움이 되었다. 특히 처음 1, 2장의 원고를 독해한 A. W. Goodman 교수, 모든 원고를 읽고 검토를 많이 해 준 Arthur Lieberman 교수 그리고 원고에 반영한 바 있는 건설적 비평과 함께 유익한 제안을 아끼지 않았던 Ohio대학교의 Klaus E. Eldridge, Tufts대학교의 David Isles, M.I.T.의 Eugene M. Klenberg, Minnesota대학교의 Wayne H. Richter 등 여러 교수들과 더불어 원고의 부분들을 독해한 William R. Zigler씨에게도 감사하고자 한다.

끝으로 원고를 전문가답게 타이프라이터로 쳐준 Wanda Balliet 양 (지금은 Jimmie Alcus Evans 부인)과 책이 되어 출간할 수 있도록 여러 모로 도와준 Houghton Mifflin 회사 직원에게도 깊은 사의를 표하는 바이다.

Shwu-Yeng T. Lin

You-Feng Lin

Tampa, Florida

Contents

Chapter 1 Elementary Logic

Chapter 2 The Concept of Sets

Chapter 3 Relations and Functions

차례

제 1 장 초등논리

제 2 장 집합의 개념

제 3 장 관계와 함수

Chapter 7 Ordinal Numbers and Ordinal Arithmetic

Appendix

Chapter 1
Elementary Logic

제 1 장
초등논리

1

Elementary Logic

In this chapter we introduce a minimal amount of logic, just enough to serve as a working tool for the remainder of this book.

1.1 Statements and Their Connectives

The study of logic is the study of the principles and methods used in distinguishing valid arguments from those that are not valid. The purpose of this preliminary chapter in logic is to help the reader to understand the principles and methods used in each step of a proof.

The starting point in logic is the term "statement," which is used in a technical sense. By a **statement** we mean a declarative sentence that is either true or false, but not both simultaneously. It is not necessary that we know whether the statement is true or false; the only qualification is that it should definitely be one or the other. Usually we can determine immediately whether a statement is true or false, but in some cases a little effort is required, and in other cases it may be impossible to reach a conclusion. The following examples should show what we mean.

1 초등논리

여기서는 이 책을 공부하는 데 꼭 알아두어야 할 논리의 기본을 소개한다.

1.1 명제와 결합자

논리의 공부란 것은 타당하지 않는 **논증**으로부터 타당한 논증을 구별하는 데 쓰이는 원리와 방법을 익히는 일이다. 이 예비 장에서의 논리 공부의 목적은 학생들로 하여금 논증의 증명의 각 단계에서 이용되고 있는 원리와 방법을 이해할 수 있도록 돕는 데 있다.

논리는 전문적으로 사용되고 있는 용어 **명제**로부터 시작한다. 명제는 참, 거짓 중 어느 한 경우이되 동시에 양쪽은 아닌 서술*(주장)을 뜻한다. 주어진 어느 명제가 참인지 거짓인지를 알려고 파고들 필요는 없겠지만 명제라면 참 거짓 중 꼭 어느 한 쪽이어야함을 분명히 가릴만한 조건이 갖추어져 있어야 한다. 흔히 명제의 참 거짓은 곧바로 정할 수 있으나 깊이 생각할 경우가 있고 결론을 내릴 수 없을 때도 있다. 다음 예를 본다면 여기서 언급한 내용을 누구나 이해할 수 있을 것이다.

---

* "미지수 x에 관한 일차방정식 $x+1=0$"은 $x=-1$일 때 참이지만 $x\neq-1$일 때에는 거짓이다. 이와 같은 서술을 **명제함수**라 하고 $p(x)$로 나타내기도 한다. 이 때의 $p(x)$는 함수에서 말하는 변수 x에 관한 명제로서 그 치역은 집합 {참, 거짓}이거나 이것의 부분집합(공집합 'ϕ'일 경우도 있음)이다(47쪽 중간구절 및 각주 ** 참조).

Example 1 Each of the following is a statement.

(a) Tampa is a city in the state of Florida.
(b)* 2+1 is equal to 5.
(c) The digit in the 105th decimal place in the decimal expansion of $\sqrt{3}$ is 7.
(d) The moon is made of blue cheese.
(e) There is no intelligent life on Mars.
(f) It is raining.

Clearly (a) is true while (b) and (d) are false. We are in doubt about the status (true or false) of (c) and (e), but this is due only to a defect in our knowledge. Thus (c) and (e) are also statements. The truth or falsity of sentence (f) depends on the weather at the *time* this statement is made.

Example 2 None of the following is a statement, because it makes no sense to ask if any of them is true or false.

(a) Come to our party!
(b) The sky is rich.
(c) How are you today?
(d) Good-bye, honey.

The statements quoted in Example 1 are all simple statements. A combination of two or more simple statements is a compound statement. For example,

"2+1 is equal to 5 and the dight in the 105th decimal place in the decimal expansion of $\sqrt{3}$ is 7"

is a compound statement.

We are familiar with the use of letters to represent numbers in algebra. In the study of logic we use letters such as p, q, r, $\cdots$ to represent statements. A letter such as p may represent either a simple statement or a compound statement. Unless otherwise stated, we shall use capital letters P, Q, R, $\cdots$ to represent compound statements.

예제 1 다음은 어느 것이나 명제이다.

(a) 대구는 경상북도의 도시이다.
(b)* 2+1은 5와 같다.
(c) $\sqrt{3}$을 십진법으로 전개할 때 소수점 아래 105째 자리수는 7이다.
(d) 달은 푸른 치즈로 만들어졌다.
(e) 화성에는 지능을 지닌 생명체가 존재하지 않는다.
(f)** 지금 서울 지역에 비가 내리고 있다.

위의 예제에서 (a)는 분명히 참이지만 (b)와 (d)는 거짓이다. 한편 (c)와 (e)의 참 거짓은 곧 바로 단언하기에는 망설이게 되지만 이것들도 또한 명제이다. (f)의 참 거짓은 그렇게 주장할 때의 날씨에 달려 있다.

예제 2 다음 어느 것에 대해서도 참 거짓을 묻는다는 자체가 무의미하므로 이들은 모두 명제가 아니다.

(a) 우리의 파티로 오너라. (b) 하늘은 청명하다.
(c) 안녕하십니까? (d) 여보, 잘 가요!

예제 1에서와 같은 명제를 모두 **단순명제**라 하고 둘 또는 그 이상의 단순명제들이 결합된 것을 **합성명제**라 한다. 이를테면

> "2+1은 5와 같다. 그리고 $\sqrt{3}$을 십진법으로 전개할 때 105째 자리수는 7이다."

는 두 개의 단순명제가 결합된 합성명제이다.

흔히 수학에서 수를 나타낼 때 문자를 사용하듯이 논리에서도 명제는 p, q, r, $\cdots$과 같은 문자를 사용한다. 그러할 때 p처럼 하나의 소문자는 단순명제를, 합성명제는 대문자 P, Q, R, $\cdots$로 나타내기로 한다.

---

* 이항 관계의 하나로 등식 2+1=5에서의 등호 "="를 감안한다면 원서의 문장 중 "is"는 "is equal to"라야 옳다.
** "비가 내리고 있다."는 저자의 주장에 지역이 빠져 있어 아쉽다.

There are many ways of connecting statements such as $p, q, r, \cdots$ to form compound statements, but only five are used frequently. These five common connectives are

(a) "not," symbolized by $\sim$; (b) "and," symbolized by $\wedge$; (c) "or," symbolized by $\vee$; (d) "if $\cdots$ then $\cdots$" symbolized by $\longrightarrow$; and (e) "$\cdots$ if and only if $\cdots$" symbolized by $\longleftrightarrow$.

In this section we discuss the connectives $\sim$ and $\wedge$, postponing the remaining connectives $\vee$, $\longrightarrow$, and $\longleftrightarrow$ until the next section.

Let p be a statement. Then the statement $\sim \boldsymbol{p}$, read "not p" or "the negation of p," is true whenever the statement p is false and is false whenever p is true. For example, let p be the statement

"This is an easy course."**

Then its negation $\sim p$ represents

"This is not an easy course."**

The truth of $\sim p$ depends upon the truth of p. It's convenient to record this dependency in a truth table:

Table 1

p	$\sim p$
T	F
F	T

where the letters T and F stand for "true" and "false," respectively. In the first column of Table 1 we list the two possible truth values for the statement p, namely T or F. Each line in a truth table represents a case that must be considered, and of course in this very simple situation, there are only two cases. Using the lines in Table 1, we see that if p is true then $\sim p$ is false, and if p is false, then $\sim p$ is true. Thus Table 1 tells us the truth value of $\sim p$ in every case.

명제 p, q, r, ⋯을 연결하여 합성명제를 구성하는 방법으로서 다음 다섯 가지 **결합자**들이 이용되고 있다.

(a) "아니다", 기호로는 ~
(b) "이고", 기호로는 ∧
(c) "또는", 기호로는 ∨
(d) "⋯ 이면 ⋯", 기호로는 ⋯→⋯
(e) "⋯ 이면 그리고 그때에만 ⋯", 기호로는 ⋯↔⋯

여기서는 두 결합자 ~, ∧를 다루고 나머지 세 결합자 ∨, →, ↔는 다음 절에서 다루기로 한다.

하나의 명제 p에 대하여 "p가 **아니다**." 또는 "p의 **부정**으로 불리는 ~$\boldsymbol{p}$*는 p가 거짓일 때 참이고, p가 참일 때 거짓인 명제이다. 이를테면**

p : 3은 홀수이다. ~p : 3은 홀수가 아니다.

인 경우 p가 참이므로 ~p는 거짓이다. 이처럼 명제 ~p의 참, 거짓이 명제 p의 참, 거짓에 달려 있음을 다음과 같은 **진리표**에 실어두면 편리하다.

표 1

p	~p
T	F
F	T

여기서 문자 T와 F는 각각 참, 거짓을 나타낸다. 표 1의 첫째 열에는 p에 대한 두 가지의 가능한 **진리값**, 즉 T 또는 F를 적어둔다. 한편 진리표의 각 행은 검토하려는 경우를 나타낸다. 이를테면 위와 같이 간단할 때 검토할 것은 단 두 가지뿐이다. 즉 표 1의 행을 확인함으로써 p가 참이면 ~p는 거짓이고 p가 거짓이면 ~p는 참인 것을 알 수 있다. 그러고 보면 표 1은 ~p의 진리값마저 나타내고 있다.

* 저자에 따라서는 기호 ~p 대신 $\bar{p}$ 또는 ¬p로도 나타낸다.

** 사실 원서(6쪽 본문 위로부터 12행)의 "이것은 쉬운 과목이다."라는 주장은 명제라고 볼 수 없다. 마찬가지로(14행) "이것은 쉬운 과목이 아니다."도 명제는 아니다.

Definition 1. The connective $\wedge$ may be placed between any two statements p and q to form the compound statement $\boldsymbol{p \wedge q}$ whose truth values are given in the following truth table.

Table 2

p	q	$p \wedge q$
T	T	T
T	F	F
F	T	F
F	F	F

The symbol $p \wedge q$ is read "p and q" or "conjuction of p and q." For example, let p be the statement "The sky is blue" and q be "The roses are red." Then the conjunction $p \wedge q$ represents "The sky is blue and the roses are red." In a compound statement such as $p \wedge q$, the individual statements p and q are called components. A component may be a simple statement or a compound statement. In a compound statement with two components, such as $p \wedge q$, there are at most $4(=2 \times 2)$ possibilities, called the logical possibilities, to be considered; namely:

(1) p is true and q is true; (2) p is true and q is false;

(3) p is false and q is true; (4) p is false and q is false.

Each one of these four possibilities is covered in the four rows of Table 2. The last column gives the truth values of $p \wedge q$. Inspection shows that $p \wedge q$ is true in only one case. That is, $p \wedge q$ is true when both components are true, and in the other three cases $p \wedge q$ is false. The thoughtful reader will realize that Table 2 reflects the way in which the conjunction "and" is used in everyday English.

Using Tables 1 and 2, we can find the truth values of complicated statements involving only the connectives $\sim$ and $\wedge$.

정의 1 임의의 명제 p, q 사이에 결합자 $\wedge$를 붙여서 합성명제 $\boldsymbol{p \wedge q}$를 구성한다. 이 때 $p \wedge q$의 진리표는 다음과 같다.

표 2

p	q	$p \wedge q$
T	T	T
T	F	F
F	T	F
F	F	F

기호 $p \wedge q$는 "p**이고** q" 혹은 "p와 q의 **논리곱**"이라고 읽는다. 이를테면 "하늘은 푸르다."를 p, "장미는 붉다."를 q라고 하면 논리곱 $p \wedge q$는

"하늘은 푸르고 장미는 붉다."

를 나타내는 명제이다. $p \wedge q$와 같은 합성명제에 대하여 각각의 명제 p, q를 그 **성분**이라고 한다. 합성명제의 성분은 단순명제이거나 그 자신이 합성명제일 수도 있다. $p \wedge q$와 같이 두 성분으로 이루어진 합성명제에 있어서 검토할 모든 가능성 곧 **논리적 가능성**은 다음과 같이 4(=2×2)가지 있다.

(1) p는 참이고 q는 참이다. (2) p는 참이고 q는 거짓이다.

(3) p는 거짓이고 q는 참이다. (4) p는 거짓이고 q는 거짓이다.

이러한 네 가지 기능성의 각각은 표 2의 각 행에 실려 있으며 그 표의 마지막 열에는 $p \wedge q$의 진리값이 적혀 있다. 지금 그 열을 보면 단 한 경우에만 $p \wedge q$가 참인 것을 알 수 있다. 즉 성분 p, q가 모두 참일 때에만 $p \wedge q$가 참이고 나머지 세 경우는 모두 거짓이다. 앞으로 "p와 q"이거나 "p 그리고 q" 같은 표현을 모두 표 2에서의 논리곱 $p \wedge q$으로 받아 들이기로 한다*.

다음에서 볼 수 있듯이 두 결합자 ~, $\wedge$만이 포함되어 있는 복잡한 합성명제의 진리값도 표 1과 표 2를 이용하여 찾을 수 있다.

* 영어에서 "and", "or"가 혼동되듯이 우리의 일상 용어에서도 "와", "그리고", "또는"(15쪽) 모두는 비슷한 뜻으로 사용되고 있다. 그러나 논리에 관점에서 표 1, 표 4(17쪽)에 따라 이들 세 용어를 분명히 구분짓도록 한다.

Example 3 Construct the truth table for the compound statement

$$\sim[(\sim p)\wedge(\sim q)]$$

[Solution]

Table 3

p	q	$\sim p$	$\sim q$	$(\sim p)\wedge(\sim q)$	$\sim[(\sim p)\wedge(\sim q)]$
T	T	F	F	F	T
T	F	F	T	F	T
F	T	T	F	F	T
F	F	T	T	T	F
Step		1	1	2	3

If the method used in constructing Table 3 is not obvious, a word of explanation may help. The headings are selected so that the compound statement (last column) is gradually built from its various components. The first two columns merely record all cases for the truth values of p and q.

We then use Table 1 to obtain the entries in the third and fourth columns, the corresponding truth values for $\sim p$ and $\sim q$. In the next step we use the entries in the third and fourth columns and Table 2 to obtain the entries in the fifth column.

Finally the entries in the fifth column and Table 1 give the entries in the sixth column — the truth values of $\sim[(\sim p)\wedge(\sim q)]$. The serious student should now copy this last compound statement, close the book, and try to reproduce Table 3.

The statement in the above example, $\sim[(\sim p)\wedge(\sim q)]$, uses parentheses and brackets to indicate the order in which the connectives apply. Often an expression can be simplified if we can drop some of the parentheses or brackets. The usual convention is to agree that $\sim$ binds more strongly than $\wedge$, i.e., the connectives $\sim$ should be applied first. Thus, for example, the expression $(\sim p)\wedge(\sim q)$ is simplified as $\sim p\wedge\sim q$.

예제 3 다음 합성명제의 진리표를 작성하여라.

$$\sim[(\sim p)\wedge(\sim q)] \qquad \cdots\cdots ①$$

[풀이]

표 3

p	q	$\sim p$	$\sim q$	$(\sim p)\wedge(\sim q)$	$\sim[(\sim p)\wedge(\sim q)]$
T	T	F	F	F	T
T	F	F	T	F	T
F	T	T	F	F	T
F	F	T	T	T	F
단계		1	1	2	3

먼저 표 3에 대한 설명을 다음과 같이 보충하도록 한다.

우선 머리 부분에 해당하는 행에 주어진 합성명제(마지막 열)를 꾸밀 수 있도록 그 성분들을 골라서 실은 다음, 처음 두 열에는 p, q의 진리값에 대한 모든 가능성을 적는다. 그리고 표 1을 이용하여 제3열과 4열에 실릴 진리값 즉 $\sim p$, $\sim q$의 진리값을 얻는다. 다음으로 제3열과 4열의 진리값과 표 2를 이용하여 제5열의 진리값을 얻어낸다. 마침내 제5열의 진리값과 표 1을 이용하여 제6열의 합성명제 곧 위 ①의 진리값을 얻게 된다. 착실한 학생이라면 이 마지막 합성명제를 지금 옮겨 적은 뒤 바로 책을 덮고 나서 표 3을 다시 작성하도록 힘써봄 직하다.

지금까지 언급한 명제 ①에서 소괄호와 대괄호를 씀으로써 결합자 $\sim$, $\wedge$를 적용할 순서를 나타낸 것이지만 이들 소괄호나 대괄호를 생략함으로써 주어진 명제를 간단히 나타낼 수 있다. 그러려면 두 결합자 $\sim$, $\wedge$ 중 $\sim$이 $\wedge$보다 더욱 강하게 명제를 묶는 것으로 약속한다. 즉 결합자 $\sim$, $\wedge$ 중 $\sim$를 먼저 적용시켜야만 한다.

이러한 뜻에서 합성명제 $(\sim p)\wedge(\sim q)$를 $\sim p\wedge\sim q$처럼 괄호를 없애고 나타낼 수 있다.

Exercise 1.1

In Problems 1 through 10 an English sentence is given. In each case determine whether the sentence is a statement (S), or not (N).

1. On January 7, 1442, snow fell somewhere in Florida.

2. Aristotle had flat feet.

3. Socialism is wrong.

4. The richest man in the world is Mr. Hunt of Texas.

5. Jack and Jill are good.

6. How much is this car worth?

7. Keep off the grass.

8. Always fasten your seat belt.

9. The sum $2^{987654321}+37$ is a prime number.

10. Bethoven wrote some of Chopin's music.

11. Among the statements given in Problems 1 through 10, indicate those that you feel must be true (T), and those for which the status may be difficult to determine (D).

In Problems 12 through 19 find the truth values for the given statements. Use the format of Table 1 or Table 2 for the two or four cases, respectively.

12. $\sim(\sim p)$ **13.** $\sim[\sim(\sim p)]$ **14.** $p\wedge p$ **15.** $\sim(p\wedge\sim p)$

16. $p\wedge\sim q$ **17.** $\sim p\wedge q$ **18.** $(p\wedge q)\wedge\sim p$ **19.** $\sim(p\wedge q)$

20. In a compound statement involving three distinct components p, q, and r, how many cases are required to cover every logical possibility? How many cases are necessary if there are four distinct components? How many cases are necessary if there are n distinct components?

연습문제 1.1

다음 1~10의 각 문장이 명제(S)인지 아닌지(N)를 가려라.

1. 1442년 1월 7일 경상남도 부산에 눈이 내렸다.

2. Aristotels는 평발이었다.

3. 사회주의는 나쁘다.

4. 미국 Texas주의 Hunt 씨는 세계에서 제일가는 부자이다.

5. 철수와 선희는 착하다.

6. 이 자동차는 얼마입니까?

7. 잔디밭에 들어가지 마시오.

8. 자동차를 운전할 때 안전띠를 늘 매시오.

9. 두 수의 합 $2^{987654321}+37$은 하나의 소수(그 뜻은?)와 같다.

10. Bethoven은 Chopin의 음악 중 몇 가지를 썼다.

11.* 위 명제 1−10 중 진리값을 곧바로 정할 수 없는 것(D)을 지적하여라.

표 1과 표 2를 이용하여 아래 12~19의 각 명제의 진리표를 작성하여라.

12. $\sim(\sim p)$ **13.** $\sim[\sim(\sim p)]$ **14.** $p\wedge p$ **15.** $\sim(p\wedge\sim p)$

16. $p\wedge\sim q$ **17.** $\sim p\wedge q$ **18.** $(p\wedge q)\wedge\sim p$ **19.** $\sim(p\wedge q)$

20. 다음 성분으로 이루어진 합성명제의 논리적 가능성은 차례로 모두 몇 가지인가?

3개의 p, q, r 4개의 p, q, r, s n** 개의 $p_1, p_2, p_3, \cdots, p_n$

---

* 참인 명제를 곧바로 가려내라 하지만 어느 것 하나 가려내기 쉽지 않다.

** $n=2, 3, 4, \cdots, 10, \cdots$에 대하여 $2^n=4, 8, 16, \cdots, 1024, \cdots$
그러므로 합성명제의 진리값을 진리표로 가늠하기에는 너무 벅찬 작업이 뒤 따른다. 이런 어려움을 피하고자 우리는 연역적 추론(47쪽 참조)을 도입하고 있다.

21. The following is an attempt to arrange all the cases in a truth table for a statement involving three components p, q and r. Complete this unfinished work.

p	q	r
T	T	
T	T	F
T		T
T		F
	T	T
	T	F
F	F	
	F	F

In Problems 22 through 25, find the truth tables for the given statements. Use the pattern developed in Problem 21 for the various cases.

22. $(p \wedge q) \wedge r$ **23.** $p \wedge (q \wedge r)$ **24.** $(p \wedge \sim q) \wedge r$ **25.** $\sim q \wedge (r \wedge p)$

1.2 Three more Connectives

In the English language there is an ambiguity involved in the use of "or." The statement "I will get a Master's degree or a Ph. D." indicates that the speaker may get both the master's degree and the Ph. D. But in another statement, "I will marry Linda or Lucy," the word "or" means that only one of the two girls will be chosen. In mathematics and logic, we cannot allow ambiguity. Hence we must decide on the meaning of the word "or."

Definition 2 The connective $\vee$ may be placed between any two statements p and q to form the compound statement $\boldsymbol{p \vee q}$. The truth values of $p \vee q$ are defined by Table 4. Thus $\vee$ is defined to be the inclusive "or" as used in the first statement above.

21. 다음은 명제 p, q, r의 진리값에 대한 표의 일부이다. 빈 부분을 채워라.

p	q	r
T	T	
T	T	F
T		T
T		F
	T	T
	T	F
F	F	
	F	F

다음 22-25의 진리표를 작성하되 필요하면 문제 21의 본보기를 이용하여라.

22. $(p\wedge q)\wedge r$ **23.** $p\wedge(q\wedge r)$ **24.** $(p\wedge \sim q)\wedge r$ **25.** $\sim q\wedge(r\wedge p)$

1.2 나머지 세 결합자

우리의 일상 언어에서 "또는"을 애매하게 쓰는 경우가 흔하다. 이를테면

"나는 석사학위 또는 박사학위를 받는다." ……①

에 있어서 이것은 말하는 사람이 석사학위와 박사학위를 모두 받을 걸로 여기는 명제이다. 한편 명제

"나는 유미 또는 영숙이와 결혼할거다."

에서 인용된 "또는"은 물론 두 여자 중 한 사람만을 택한다는 뜻으로 사용되고 있다. 수학이나 논리에서는 이 같은 애매한 점을 피하기 위하여 용어 "또는"의 뜻을 분명히 정할 필요가 있다.

정의 2 임의의 두 명제 p, q 사이에 결합자 $\vee$를 붙여 합성명제 $\boldsymbol{p}\vee\boldsymbol{q}$를 구성할 때 그 것의 진리값은 다음 표 4에 따라 정의한다. 즉 결합자 $\vee$는 ①에서 언급된 명제에서와 같이 두 경우를 포함하는 뜻에서의 "또는"으로 받아들인다.

Table 4

p	q	$p \vee q$
T	T	T
T	F	T
F	T	T
F	F	F

The symbol $p \vee q$ is read "p *or* q" or "the disjunction of p and q." Notice that the conjunction of p and q is true when and only when the two components are both true (Table 2), whereas the disjunction is false when and only when the two components are false (Table 4).

Let us compare the truth tables for $p \vee q$ and $\sim(\sim p \wedge \sim q)$ in Tables 3 and 4. We find that in each case the last column is $TTTF$ so that these two statements have the same truth values in each of the four logical possibilities. Showing that certain statements have the same truth values in each case is an important part of logic. In fact, logic treats two such statements as the same.

Definition 3. When two statements P and Q, simple or compound, have the same truth values in each of all the logical possibilities, then P is said to be logically equivalent or simply equivalent to Q, and we write $\boldsymbol{P \equiv Q}$.

In short, two statements are logically equivalent provided that they have the same truth table. Thus, we have

$$p \vee q \equiv \sim(\sim p \wedge \sim q).$$

Although two logically equivalent statements are regarded as the same, as far as logic is concerned, we prefer the simple statement "p or q" to its more complicated equivalent statement, "It is not true that neither p nor q."

표 4

p	q	$p \vee q$
T	T	T
T	F	T
F	T	T
F	F	F

기호 $p \vee q$는 p **또는** q 혹은 p, q의 **논리합**이라고 읽는다. 우리는 논리곱과 논리합 사이의 차이점을 주의해야 한다. 즉 p, q가 모두 참일 때에만 $p \wedge q$는 참인데 반하여 p, q가 모두 거짓일 때에만 $p \vee q$는 거짓이다.

이제 $p \vee q$와 $\sim(\sim p \vee \sim q)$의 진리표, 즉 표 3과 표 4를 비교하면 어느 표에서나 마지막 열은 $T\ T\ T\ F$이므로 이 두 명제는 네 가지 논리적 가능성의 각각에 대한 진리값이 같다. 이와 같이 두 명제에 대한 각 논리적 가능성마다 같은 진리값을 갖는다는 사실을 알아보는 일은 논리에서 중요한 자리를 차지하거니와 논리에서는 그런 두 명제를 같은 류의 것으로 생각하고 다음과 같은 정의를 내린다.

정의 3 단순명제 p, q이거나 합성명제인 P, Q에 대한 모든 논리적 가능성의 각각의 경우 진리값이 같으면 P와 Q는 **논리적 동치** 또는 간단히 **동치***라 하고 이것을 $\boldsymbol{P \equiv Q}$와 같이 나타낸다.

정의 3을 간단히 말하면 두 명제에 대하여 진리표가 같으면 동치이다. 이를테면 $p \vee q$와 $\sim(\sim p \wedge \sim q)$의 진리표는 같으므로 이 두 명제는 동치이다.

즉

$$p \vee q \equiv \sim(\sim p \wedge \sim q)$$

이렇듯 논리에 관한 한 동치인 두 명제

"p 또는 q", "p가 아니고 q도 아님은 참이 아니다."

중 우리로서는 복잡한 후자보다 마땅히 간단한 전자를 선호하기 마련이다.

* 여기서 당장 동치인 두 명제의 보기(책의 것은 피해가며)를 들어라.

Definition 4. The connective $\rightarrow$ is called the conditional and may be placed between any two statements p and q to form the compound statement $\boldsymbol{p \longrightarrow q}$ (read: "if p, then q"). By definition the statement $p \longrightarrow q$ is equivalent to the statement $\sim(p \wedge \sim q)$, and the truth values of $p \longrightarrow q$ are given in Table 5.

Table 5

Case	p	q	$\sim q$	$p \wedge \sim q$	$p \rightarrow q [\equiv \sim(p \wedge \sim q)]$
1	T	T	F	F	T
2	T	F	T	T	F
3	F	T	F	F	T
4	F	F	T	F	T

A motivation of Definition 4 is in order. Let p be the statement "The sun is shining" and let q be the statement "I am playing tennis." Then the compound statement $p \longrightarrow q$ is "If the sun is shining, then I am playing tennis." Now when do we consider such a statement false? Clearly $p \longrightarrow q$ is false if the sun is shining and I am not playing tennis, and only in this case. In other words $p \longrightarrow q$ is false if $p \wedge \sim q$ is true, and only in this case. Consequently $p \longrightarrow q$ is true if and only if $\sim(p \wedge \sim q)$ is true. But this is precisely Definition 4. We now study the truth table of $p \longrightarrow q$, that is, of $\sim(p \wedge \sim q)$.

According to Definition 4, the meaning of the conditional statement $p \longrightarrow q$ departs radically from our ordinary usage of "If p, then q." In our ordinary language,[1)] a sentence of the form "If p, then q" is taken to mean that q is true whenever p is true. Therefore the cases in which p is false need not be considered.

For example, the statement, "If Lincoln had shot Grant, then Jefferson would have been the first president" is regarded as nonsense, since both of the components are false.

Consequently in ordinary usage one does not inquire whether such a compound statement is true. In creating a formal language, the logician wishes to assign a truth value to $p \longrightarrow q$ for each of the four logical possibilities, even though two of the cases appear to be nonsense in our ordinary language. For

[1)] As opposed to "ordinary language," logic is called a formal language.

정의 4 임의의 명제 p, q 사이에 조건부라고 불리는 결합자 →를 붙여 조건문 $\boldsymbol{p \to q}$(p이면 q라고 읽음)를 구성한다. 여기서 $p \to q$는 표 5에서와 같이 $\sim(p \wedge \sim q)$와 동치인 명제로 정의한다.

표 5

경우	p	q	$\sim q$	$p \wedge \sim q$	$p \to q [\equiv \sim(p \wedge \sim q)]$
1	T	T	F	F	T
2	T	F	T	T	F
3	F	T	F	F	T
4	F	F	T	F	T

여기서 정의 4를 도입한 것은 시기 적절하다고 본다. 지금

"햇빛이 빛나고 있다.", "나는 테니스를 치고 있다."

를 각각 p, q라고 하면 $p \to q$는

"햇빛이 빛나고 있으면 나는 테니스를 치고 있다."

를 나타내는 명제이다. 이러한 조건문의 진리값은 어느 경우에 거짓일까?

분명히 햇빛이 빛나고 있으나 내가 테니스를 치고 있지 않을 때, $p \to q$는 거짓이다. 다시 말하면 $p \wedge \sim q$가 참일 때에만 $p \to q$는 거짓이다. 따라서 $\sim(p \wedge \sim q)$가 참일 때에만 $p \to q$는 참이다. 독자는 이러한 배경을 이해함으로써 정의 4를 익힐 수 있다.

이제 $p \to q$, 즉 $\sim(p \wedge \sim q)$의 진리표를 살펴보자. 정의 4에 따른 조건문 $p \to q$의 뜻은 일상 용법에서의 "p이면 q"와는 현격한 차이가 있다. 일상 언어[1)]에서 "p이면 q"와 같은 꼴의 주장은 p가 참이기만 하면 q는 저절로 참인 것으로 받아들이며 p가 거짓인 경우는 생각조차 아니한다. 이를테면

"Lincoln이 Grant를 쏘았다라면 Jefferson은 초대 대통령이었을 것이다."

라는 주장은 무의미한 것에 지나지 않는다. 왜냐하면 이 명제의 성분은 모두 거짓이기 때문이다. 대체로 우리는 일상 용법에 있어서 그런 합성명제가 참인지 거짓인지 관심을 두지 않는다. 그러나 비록 두 경우는 무의미하더라도 $p \to q$의 모든 논리적 가능성(여기서는 4가지)에 대한 진리값을 표 5에서와 같이 매기도록 한다.

[1)] 일상 언어에 반하여 논리를 형식적 언어라고 부른다.

a variety of reasons, which will appear in due time, logicians have settled on the definition adopted here. Thus in our formal language, $p \rightarrow q$ is true in every case except case 2 (see Table 5). As a consequence of this agreement, we will be able to prove some very simple and useful theorems, without such an agreement the proofs of which would be awkward or very difficult.[2)]

We now introduce the last of the five most common connectives, one that appears frequently in the statements of mathematical theorems.

Definition 5. The connective $\longleftrightarrow$ is called the biconditional and may be placed between any two statements p and q to form the compound statement $\boldsymbol{p \longleftrightarrow q}$(read: "$p$ if and only if q"). The statement $p \longleftrightarrow q$ is defined to be equivalent to the compound statement$(p \rightarrow q) \wedge (q \rightarrow p)$, and the truth values of $p \longleftrightarrow q$ are given in Table 6.

Example 4 Find the truth table for $p \longleftrightarrow q$.

[Solution] Following the method described earlier, we obtain Table 6.

Table 6

Case	p	q	$p \rightarrow q$	$q \rightarrow p$	$p \longleftrightarrow q[\equiv(p \rightarrow q) \wedge (q \rightarrow p)]$
1	T	T	T	T	T
2	T	F	F	T	F
3	F	T	T	F	F
4	F	F	T	T	T
Step			1	1	2

From the above truth table, we observe that $p \longleftrightarrow q$ is true if both components are true or if both components are false. In any other case (case 2 and 3) the statement $p \longleftrightarrow q$ is false.

[2)] See, for instance, Theorems 1 and 7 in Chapter 2.

앞으로 적당한 시기에 더욱 설명하겠지만 논리학자는 여러 이론상 이유로 여기서 내린 정의를 받아들인다. 그러고 보면 형식적 언어에 있어서 $p \to q$는 경우 2(표 5 참조)만을 제외한 모든 경우에 참이다. 이러한 약속하에 몇 가지 중요한 정리들을 증명하여 볼 수 있으나 그러한 약속 없이는 증명이 몹시 번거로울 것이다.[2)]

이제 다섯 결합자 중 마지막으로 소개될 것은 수학에서 빈번히 등장한다.

정의 5 임의의 명제 p, q 사이에 **쌍조건부**라고 부르는 결합자 $\leftrightarrow$를 붙여 합성명제 $\boldsymbol{p \leftrightarrow q}$*($p$이면 그리고 그때에만 q라고 읽음 : 7쪽 (e)참조)를 구성하되 그것은 표 6의 합성명제

$$(p \to q) \wedge (q \to p)$$

와 동치인 것으로 정의한다.

예제 4 $p \leftrightarrow q$의 진리표를 작성하여라.

[풀이] 지금까지의 방법에 따라 다음 표 6을 얻는다. 즉 $p \leftrightarrow q$의 진리값은 p, q가 모두 참이거나 모두 거짓이면 참이되 그 이외(경우 2와 3)일 때에는 거짓이다.

표 6

경우	p	q	$p \to q$	$q \to p$	$p \leftrightarrow q\,[\equiv (p \to q) \wedge (q \to p)]$
1	T	T	T	T	T
2	T	F	F	T	F
3	F	T	T	F	F
4	F	F	T	T	T
단계			1	1	2

[2)] 이를테면 제2장 정리 1(77쪽)과 정리 7(115쪽)을 보아라.

* $p \leftrightarrow q$를 **쌍조건문**이라고 한다.

Exercise 1.2

In problems 1 through 12 construct the truth tables for the given statements.

1. $p \vee \sim q$

2. $\sim(p \vee \sim q)$

3. $\sim(\sim p \vee \sim q)$

4. $\sim p \vee q$

5. $(\sim q) \longrightarrow (\sim p)$

6. $q \longleftrightarrow p$

7. $p \wedge (q \vee r)$

8. $(p \wedge q) \vee (p \wedge r)$

9. $p \vee (q \wedge r)$

10. $(p \vee q) \wedge (p \vee r)$

11. $(p \vee q) \vee r$

12. $p \vee (q \vee r)$

13. Is the statement $(\sim q) \longrightarrow (\sim p)$ (Problem 5) logically equivalent to the statement $p \longrightarrow q$?

14. Is the statement $\sim p \vee q$ (Problem 4) logically equivalent to the statement $p \rightarrow q$?

15. From the statements in Problems 1 through 12, find those pairs of statement that are logically equivalent.

16. In each of the following, translate the given compound statement into a symbolic form using the suggested symbols.

(a) It is not the case that I am not friendly to you. (F)

(b) If she is an angel, then she has two wings. (A, W)*

(c) The price of meat goes up if and only if the supply does not meet the demand of meat. (P, S)

(d) Either the farmers will reduce the prices or the government will step in. (F, G)

(e) If meat exports increase or more livestock is not raised, then prices will rise. (E, R, P)

연습문제 1.2

다음 명제 1~12의 진리표를 작성하여라.

1. $p \vee \sim q$

2. $\sim(p \vee \sim q)$

3. $\sim(\sim p \vee \sim q)$

4. $\sim p \vee q$

5. $(\sim q) \rightarrow (\sim p)$*

6. $q \leftrightarrow p$

7. $p \wedge (q \vee r)$

8. $(p \wedge q) \vee (p \wedge r)$

9. $p \vee (q \wedge r)$

10. $(p \vee q) \wedge (p \vee r)$

11. $(p \vee q) \vee r$

12. $p \vee (q \vee r)$

13. 문제 5의 조건문은 조건문 $p \rightarrow q$와 동치인가?

14. 문제 4의 명제는 조건문 $p \rightarrow q$와 동치인가?

15. 문제 1~12의 명제 중 동치인 명제들의 쌍을 구하여라.

16. 다음 합성명제를 결합자와 ( ) 안의 문자를 이용하여 나타내어라.

(a)** 내 동생에게 음식을 나눠주지 않는다는 것은 사실이 아니다(F).

(b)** 제주도의 김효숙이 천사라면 그 여자에게는 두 날개가 있다(A, W).

(c) 올해 우리나라 육류의 수요를 공급이 채우지 못하면 그리고 그 때에만 육류 값이 오른다(P, S).

(d) 올해 우리나라 농부들이 곡물 가격을 내리거나 그렇지 않을 경우 정부가 개입할 것이다(F, G).

(e) 내년 우리나라 가축업계가 육류의 수출을 늘리거나 가축을 더 많이 사육하지 않으으면 육류 가격은 오를 것이다(E, R, P).

---

* 문제 5의 조건문 $(\sim q) \rightarrow (\sim p)$는 조건문 $p \rightarrow q$의 대우라고 한다[29쪽 정리 2(d) 참조].

** 사실 원서의 문제 16(a), (b) 그대로라면 모두 명제라고 볼 수 없어 위에서와 같이 바꿔놓았다. 앞으로도 이같은 사례가 종종 있을 것이다.

1.3 Tautology, Implication and Equivalence

Let us examine the truth table for the statement $p \vee \sim q$:

Table 7

p	$\sim p$	$p \vee \sim p$
T	F	T
F	T	T

We notice that the statement $p \vee \sim p$ is true in every case, that is, in all logical possibilities. Such an important type of statement deserves a special name.

> ***Definition 6.*** A statement is said to be a tautology provided that it is true in each of all logical possibilities.

Let P and Q be two statements, compound or simple. If the conditional statement $P \longrightarrow Q$ is a tautology, it is called an implication and is denoted by $\boldsymbol{P} \Longrightarrow \boldsymbol{Q}$ (read : P implies Q). Thus, the following conditional statements are tautologies :

(1) $p \rightarrow p$

(2) $p \wedge q \rightarrow q \wedge p$

(3) $p \rightarrow p \wedge p$

(4) $p \wedge q \rightarrow q$[3]

In logic or mathematics, "theorems" are meant to be true statements, and a "proof" (of a theorem) is a justification of the theorem.

[3] We shall consider $\vee$ and $\wedge$ to bind more strongly than $\rightarrow$ and $\leftrightarrow$, and shall write $p \rightarrow p \vee q$ for $p \rightarrow (p \vee q)$, etc. See also the last paragraph of Section 1.

1.3 항진, 함의, 동치

다음은 합성명제 $p\vee\sim q$의 진리표이다.

표 7

p	$\sim p$	$p\vee\sim p$
T	F	T
F	T	T

이 표를 보면 주어진 명제는 모든 논리적 가능성에 대하여 참인 것을 알 수 있다. 이와 같은 명제에는 특별한 이름을 붙일만 하다.

정의 6 모든 논리적 가능성에 대하여 참인 명제를 항진 또는 항진 명제라고 한다.

특히 단순명제이거나 합성명제인 P, Q에 대한 조건문 $P\rightarrow Q$가 항진일 때 이것을 함의 또는 함의명제라 하고

$$\boldsymbol{P\Longrightarrow Q}\text{('}P\text{는 }Q\text{를 함의한다.'라고 읽음)}$$

와 같이 나타낸다. 다음 조건문들은 모두 함의이다(과연 그러한지 일일이 진리표로 알아보아라!).

(1) $p\rightarrow p$ (2) $p\wedge q\rightarrow q\wedge p$

(3) $p\rightarrow p\wedge p$ (4) $p\wedge q\rightarrow q$[3]

다음 정리 1에서 알 수 있듯이 수학이나 논리의 명제 중 참인 것을 정리라 하며 정리의 타당성(53쪽) 여부를 밝히는 과정을 그것의 증명이라고 한다.

[3] 두 결합자 $\vee$, $\wedge$은 결합자 $\rightarrow$, $\leftrightarrow$보다도 더 강하게 명제를 묶는 것으로 약속하다 보면 $p\rightarrow(p\vee q)$를 $p\rightarrow p\vee q$와 같이 나타낼 수 있다(11쪽 본문 하단 문장 참조).

Theorem 1. Let p and q be any two statements. Then

(a) Law of addition (Add.) : $p \Rightarrow p \vee q$

(b) Laws of simplification (Simp.) : $p \wedge q \Rightarrow p$, $p \wedge q \Rightarrow q$

(c) Disjunctive Syllogism (D.S.) : $(p \vee q) \wedge \sim p \Rightarrow q$

[Proof] We leave the proofs for (a) and (b) to the reader as exercises. The following is a simplified truth table for $(p \vee q) \wedge \sim p \rightarrow q$:

Table 8

	$(p$	$\vee$	$q)$	$\wedge$	$\sim p$	$\rightarrow$	q
	T	T	T	F	F	T	T
	T	T	F	F	F	T	F
	F	T	T	T	T	T	T
	F	F	F	F	T	T	F
Step	1	2	1	3	2	4	1

Let us take a moment to explain the construction of simplified truth table : The truth values in Table 8 are assigned, column by column, in the order indicated by the numerals appearing in the bottom row of the table. In a simplified truth table, we write the truth values directly first under each component and then under the connectives. This saves space and time.

Now, returning to the proof of the theorem, since the final step (step 4) in Table 8 consists of all T's, the conditional statement $(p \vee q) \wedge \sim p \rightarrow q$ is indeed an implication.

If the biconditional statement $P \longleftrightarrow Q$ happens to be a tautology, it is called an equivalence and is denoted by $\boldsymbol{P} \Longleftrightarrow \boldsymbol{Q}$ (read : P is equivalent to Q).

정리 1 임의의 명제 p, q에 대하여 다음이 성립한다.

(a) **합의 법칙** $p \Rightarrow p \vee q$

(b) **단순화법칙** $p \wedge q \Rightarrow p$, $p \wedge q \Rightarrow q$

(c) **논리합의 삼단논법** $(p \vee q) \wedge \sim p \Rightarrow q$

[증명]* (a)와 (b)에 대해서는 연습문제로 돌리고 여기서는 명제 (c)만의 진리표를 다음과 같이 간단한 꼴로 나타낸다.

표 8

	$(p$	$\vee$	$q)$	$\wedge$	$\sim p$	$\rightarrow$	q
	T	T	T	F	F	T	T
	T	T	F	F	F	T	F
	F	T	T	T	T	T	T
	F	F	F	F	T	T	F
단계	1	2	1	3	2	4	1

여기서 이 표의 작성법을 설명하여 보자. 즉 표 8의 진리값은 그 맨 끝 행의 숫자의 순서에 따라 열에 매기기로 하되 먼저 각 성분 밑에 그 진리값을 적고 결합자 밑에도 해당되는 진리값을 적는다. 사실 이러한 방법으로 우리는 지면과 시간을 절약할 수 있다.

이제 정리 1의 증명을 염두에 둘 때 표 8의 마지막 단계(단계 4)의 진리값은 모두 T이므로 과연 주어진 조건문은 합의임을 알 수 있다.

특히 쌍조건문 $P \leftrightarrow Q$가 항진일 때 P, Q는 **동치****라 하고 이것을

$$\boldsymbol{P \Leftrightarrow Q^{**}}$$

와 같이 나타낸다.

다시 말해 P, Q에 대한 모든 논리적 가능성의 각 경우마다 이 두 명제의 진리값이 같으면 그리고 그때에만 $P \Leftrightarrow Q$.

---

* 여기서는 주어진 명제에 대한 진리표로 증명을 하고 있다. 그러나 증명에 관한 논의는 '1.5 연역적 추론(41쪽)' 및 '1.7 타당성의 증명(53쪽)'에서 본격적으로 다룬다.

** 그런 것의 보기를 어느 독자나 새로이 찾아 보아라. 한편 17쪽 정의 3에서도 동치 $P \equiv Q$를 다룬바 있다.

From Definition 5 and Table 6, $P \Leftrightarrow Q$ provided that P and Q have the same truth values in each of all logical possibilities. and conversely, P and Q have the same truth values in each of all logical possibilities provided $P \Leftrightarrow Q$. Thus, by Definition 3, $P \Leftrightarrow Q$ and $P \equiv Q$ have the same meaning, and hence we may use $\Leftrightarrow$ and $\equiv$ interchangeably.

Theorem 2. Let p and q be any two statements. Then

(a) Law of Double Negation (D.N.) : $\sim(\sim p) \equiv p$
(b) Commutative Laws (Com.) : $p \wedge q \equiv q \wedge p, \quad p \vee q \equiv q \vee p$
(c) Laws of Idempotency (Idemp.) : $p \wedge p \equiv p, \quad p \vee p \equiv p$
(d) Contrapositive Law (Contrap.) : $(p \rightarrow q) \equiv (\sim q \rightarrow \sim p)$

[Proof] We leave proofs for parts (a), (b), and (c) to the reader as exercises, but we give an outline of the proof of (d).

We have the following simplified truth table for the biconditional statement $(p \rightarrow q) \leftrightarrow (\sim q \rightarrow \sim p)$:

Table 9

	$(p$	$\rightarrow$	$q)$	$\leftrightarrow$	$(\sim q$	$\rightarrow$	$\sim p)$
	T	T	T	T	F	T	F
	T	F	F	T	T	F	F
	F	T	T	T	F	T	T
	F	T	F	T	T	T	T
Step	1	2	1	4	2	3	2

Thus, Table 9 shows that $p \rightarrow q$ is equivalent to $\sim q \rightarrow \sim p$.

The following theorem credited to Augustus De Morgan(1806-1871) is one of the most convenient tools in logic.

Theorem 3. De Morgan's Laws (De M.) Let p and q be any two statements. Then

$$\sim(p \wedge q) \equiv \sim p \vee \sim q, \quad \text{and} \quad (p \vee q) \equiv \sim p \wedge \sim q$$

따라서 정의 3을 감안할 때 $P \Longleftrightarrow Q$와 $P \equiv Q$는 같은 뜻을 지니므로 우리는 두 기호 $\Longleftrightarrow$, $\equiv$를 편의에 따라 번갈아 이용하기로 한다.

정리 2 임의의 명제 p, q에 대하여 다음이 성립한다.

(a) **이중부정법칙*** $\sim(\sim p) \equiv p$

(b) **교환법칙** $p \wedge q \equiv q \wedge p$, $p \vee q \equiv q \vee p$

(c) **멱등법칙** $p \wedge p \equiv p$, $p \vee p \equiv p$

(d) **대우법칙*** $(p \rightarrow q) \equiv (\sim q \rightarrow \sim p)$

[증명] (a), (b), (c)의 확인은 모두 독자의 연습문제로 돌리고 여기서는 (d)의 경우 다음과 같이 그 골자만을 제시하기로 한다.

실제로 쌍조건문 $(p \rightarrow q) \leftrightarrow (\sim q \rightarrow \sim p)$의 진리표 9(단계 4)를 보면 $p \rightarrow q$, $\sim q \rightarrow \sim p$는 동치임을 곧바로 알 수 있다.

표 9

	$(p$	$\rightarrow$	$q)$	$\leftrightarrow$	$(\sim q$	$\rightarrow$	$\sim p)$
	T	T	T	T	F	T	F
	T	F	F	T	T	F	F
	F	T	T	T	F	T	T
	F	T	F	T	T	T	T
단계	1	2	1	4	2	3	2

다음은 드 모르간(1806~1871)의 공로에 따른 것으로 논리에서 많이 인용되는 정리 중의 하나이다.

정리 3 **드 모르간의 법칙**** 임의의 명제 p, q에 대하여 다음이 성립한다.

$$\sim(p \wedge q) \equiv \sim p \vee \sim q, \qquad \sim(p \vee q) \equiv \sim p \wedge \sim q$$

---

* 이중부정법칙, 대우법칙을 각각 이중부정, 대우라고도 한다.

** 정리 6 드 모르간의 정리(99쪽) 참조.

[Proof] We shall prove the first part of this theorem and leave the other part to the reader as an exercise. We construct a simplified truth table for the biconditional $\sim(p\wedge q)\leftrightarrow(\sim p\vee\sim q)$:

Table 10

	$\sim$	$(p$	$\wedge$	$q)$	$\leftrightarrow$	$(\sim p$	$\vee$	$\sim q)$
	F	T	T	T	T	F	F	F
	T	T	F	F	T	F	T	T
	T	F	F	T	T	T	T	F
	T	F	F	F	T	T	T	T
Step	3	1	2	1	4	2	3	2

The above truth table shows that $\sim(p\wedge q)$ is equivalent to $\sim p\vee\sim q$.

Theorem 4. Let p, q and r be any statements. Then

(a) Associative Laws (Assoc.): $(p\wedge q)\wedge r\equiv p\wedge(q\wedge r)$,
$(p\vee q)\vee r\equiv p\vee(q\vee r)$

(b) Distributive Laws (dist.): $p\wedge(q\vee r)\equiv(p\wedge q)\vee(p\wedge r)$,
$p\vee(q\wedge r)\equiv(p\vee q)\wedge(p\wedge r)$

(c) Transitive Law (Trans.): $(p\rightarrow q)\wedge(q\rightarrow r)\Rightarrow(p\rightarrow r)$

[Proof] We leave the proofs for the Associative Laws and the second Distributive Law to the reader as exercises.

Let us prove that $p\wedge(q\vee r)\equiv(p\wedge q)\vee(p\wedge r)$. Since this involves three components, there are $2^3=8$ logical possibilities to consider. The following truth table shows that $p\wedge(q\vee r)$ and $(p\wedge q)\vee(p\wedge r)$ have the same truth values in each of all eight logical posibilities. Therefore, $p\wedge(q\vee r)$ and $(p\wedge q)\vee(p\wedge r)$ are equivalent.

[증명] 이 정리의 앞부분만을 증명하고 뒷부분의 증명은 독자의 연습문제로 돌린다.

우리는 다음과 같이 쌍조건문 $\sim(p\wedge q)\leftrightarrow(\sim p\vee\sim q)$의 진리표를 작성할 수 있다.

표 10

	$\sim$	$(p$	$\wedge$	$q)$	$\leftrightarrow$	$(\sim p$	$\vee$	$\sim q)$
	F	T	T	T	T	F	F	F
	T	T	F	F	T	F	T	T
	T	F	F	T	T	T	T	F
	T	F	F	F	T	T	T	T
단계	3	1	2	1	4	2	3	2

위 표 10(좌우 단계 3)에 따라 $\sim(p\wedge q)$, $\sim p\vee\sim q$는 동치이다.

정리 4 임의의 명제 p, q, r에 대하여 다음이 성립한다.

(a) 결합법칙 $(p\wedge q)\wedge r\equiv p\wedge(q\wedge r)$,
$(p\vee q)\vee r\equiv p\vee(q\vee r)$

(b) 분배법칙 $p\wedge(q\vee r)\equiv(p\wedge q)\vee(p\wedge r)$,
$p\vee(q\wedge r)\equiv(p\vee q)\wedge(p\vee r)$

(c) 추이법칙* $(p\rightarrow q)\wedge(q\rightarrow r)\Rightarrow(p\rightarrow r)$

[증명] (b)의 첫 번째와 (c)만을 각각 진리표로 밝히고 나머지 것은 모두 독자의 연습문제로 돌린다.

세 성분 p, q, r에 대한 논리적 가능성은 모두 $2^3(=8)$가지 있다. 다음 표 11에 따르면 그 8가지 논리적 가능성의 각 경우마다

$$p\wedge(q\vee r),\ (p\wedge q)\vee(p\wedge r)$$

의 진리값이 같다. 따라서 이 두 합성명제는 동치이다.

* 추이법칙을 추이율이라고도 한다.

Table 11

p	q	r	$q \vee r$	$p \wedge q$	$p \wedge r$	$p \wedge (q \vee r)$	$(p \wedge q) \vee (p \wedge r)$
T	T	T	T	T	T	T	T
T	T	F	T	T	F	T	T
T	F	T	T	F	T	T	T
T	F	F	F	F	F	F	F
F	T	T	T	F	F	F	F
F	T	F	T	F	F	F	F
F	F	T	T	F	F	F	F
F	F	F	F	F	F	F	F

For simplicity and space saving, we construct a simplified truth table as introduced in Table 8 for

$$(p \rightarrow q) \wedge (q \rightarrow r) \rightarrow (p \rightarrow r)$$

Table 12

	$(p$	$\rightarrow$	$q)$	$\wedge$	$(q$	$\rightarrow$	$r)$	$\rightarrow$	p	$\rightarrow$	r
	T	T	T	T	T	T	T	T	T	T	T
	T	T	T	F	T	F	F	T	T	F	F
	T	F	F	F	F	T	T	T	T	T	T
	T	T	T	T	T	T	F	T	T	F	F
	F	T	T	T	T	T	T	T	F	T	T
	F	T	F	T	F	T	T	T	F	T	T
	F	T	F	T	F	T	F	T	F	T	F
Step	1	2	1	3	1	2	1	4	1	2	1

Since the final step (step 4) consists entirely of T values, the Transitive Law is proved.

Because of the Associative Laws, the brackets in $(p \wedge q) \wedge r \equiv p \wedge (q \wedge r)$ and $(p \vee q) \vee r \equiv p \vee (q \vee r)$ become unnecessary, and the expressions $p \wedge q \wedge r$ and $p \vee q \vee r$ now have definite meanings, and similarly for $p_1 \wedge p_2 \wedge \cdots \wedge p_n$ and $p_1 \vee p_2 \cdots \vee p_n$.

표 11

p	q	r	$q\vee r$	$p\wedge q$	$p\wedge r$	$p\wedge(q\vee r)$	$(p\wedge q)\vee(p\wedge r)$
T	T	T	T	T	T	T	T
T	T	F	T	T	F	T	T
T	F	T	T	F	T	T	T
T	F	F	F	F	F	F	F
F	T	T	T	F	F	F	F
F	T	F	T	F	F	F	F
F	F	T	T	F	F	F	F
F	F	F	F	F	F	F	F

다음 표 12에서 볼 수 있듯이 조건문

$$(p\to q)\wedge(q\to r)\to(p\to r) \qquad \cdots\cdots ①^{*}$$

의 진리표는 지면을 아끼기 위하여 표 8에서와 같은 방법으로 간단히 작성 할 수 있다.

표 12

	$(p$	$\to$	$q)$	$\wedge$	$(q$	$\to$	$r)$	$\to$	p	$\to$	r
	T	T	T	T	T	T	T	T	T	T	T
	T	T	T	F	T	F	F	T	T	F	F
	T	F	F	F	F	T	T	T	T	T	T
	T	T	T	T	T	T	F	T	T	F	F
	F	T	T	T	T	T	T	T	F	T	T
	F	T	F	T	F	T	T	T	F	T	T
	F	T	F	T	F	T	F	T	F	T	F
단계	1	2	1	3	1	2	1	4	1	2	1

이 표의 마지막 단계(단계 4)는 T값만으로 이루어져 있으므로 추이율이 성립한다.

결합법칙에 따라 $(p\wedge q)\wedge r\equiv p\wedge(q\wedge r)$의 양변의 합성명제를 각각 괄호 없이 $p\wedge q\wedge r$, $p\vee q\vee r$로 나타낸다.

나아가 자연수 $n(n\geqq 4)$에 대한 $p_1\wedge p_2\wedge\cdots\wedge p_n$, $p_1\vee p_2\vee\cdots\vee p_n$에서도 괄호들이 모두 생략된 것으로서 받아들인다.

* 조건문 ①에서도 항진임에 주목하자. 사실 ①은 지금까지 여러분이 알아온 삼단논법을 말한다.

Theorem 5. Let p, q, r and s be any statements. Then

(a) Constructive Dilemmas (C. D.) :

$$(p \to q) \wedge (r \to s) \Rightarrow (p \vee r \to q \vee s),$$
$$(p \to q) \wedge (r \to s) \Rightarrow (p \wedge r \to q \wedge s)$$

(b) Destructive Dilemmas (D. D.) :

$$(p \to q) \wedge (r \to s) \Rightarrow (\sim q \vee \sim s \to \sim p \vee \sim r),$$
$$(p \to q) \wedge (r \to s) \Rightarrow (\sim q \wedge \sim s \to \sim p \wedge \sim r)$$

[Proof] The proof of Theorem 5 is left to the reader as an exercise.

Theorem 6. Let p and q be statements. Then

(a) Modus Ponens (M. P.) : $(p \to q) \wedge p \Rightarrow q$

(b) Modus Tollens (M. T.) : $(p \to q) \wedge \sim q \Rightarrow \sim p$

(c)* Reductio ad Absurdum (R. A.) : $(p \to q) \Leftrightarrow (p \wedge \sim q \to q \wedge \sim q)$

[Proof] Exercise.

Exercise 1.3

Prove the following statements 1 through 7.

1. parts (a) and (b) of Theorem 1.

2. parts (a), (b), and (c) of Theorem 2.

3. $\sim(p \vee q) \equiv \sim p \wedge \sim q$.

4. (a) of Theorem 4.

5. $p \vee (q \wedge r) \equiv (p \vee q) \wedge (p \vee r)$.

6. $(p \to q) \Rightarrow (p \wedge r \to q \wedge r)$.

7. $(p \leftrightarrow q) \equiv (p \wedge q) \vee (\sim p \wedge \sim q)$.

8. Using De Morgan's Law, write in ordinary language the negation of the statement " The function $f(x)$ has a derivative or I am stupid."

정리 5 임의의 명제 p, q, r, s에 대하여 다음이 성립한다.

(a) **구성적 양도논법** $(p\to q)\wedge(r\to s)\Rightarrow(p\vee r\to q\vee s)$,
$(p\to q)\wedge(r\to s)\Rightarrow(p\wedge r\to q\wedge s)$

(b) **파괴적 양도논법** $(p\to q)\wedge(r\to s)\Rightarrow(\sim q\vee\sim s\to\sim p\vee\sim r)$,
$(p\to q)\wedge(r\to s)\Rightarrow(\sim q\wedge\sim s\to\sim p\wedge\sim r)$

[증명] 연습문제 1.3의 12로 돌린다.

정리 6 임의의 명제 p, q에 대하여 다음이 성립하다.

(a) **긍정식 삼단논법** $(p\to q)\wedge p\Rightarrow q$

(b) **부정식 삼단논법** $(p\to q)\wedge\sim q\Rightarrow\sim p$

(c) **귀류법*** $(p\to q)\Leftrightarrow(p\wedge\sim q\to q\wedge\sim q)$

[증명] 연습문제 1.3의 13으로 돌린다.

연습문제 1.3

다음 1–7을 진리표로 밝혀라.

1. 정리 1의 (a), (b)

2. 정리 2의 (a), (b), (c)

3. $\sim(p\vee q)\equiv\sim p\wedge\sim q$

4. 정리 4의 (a)

5. $p\vee(q\wedge r)\equiv(p\vee q)\wedge(p\vee r)$

6. $(p\to q)\Rightarrow(p\wedge r\to q\wedge r)$

7. $(p\leftrightarrow q)\equiv(p\wedge q)\vee(\sim p\wedge\sim q)$

8. 드 모르간의 법칙을 이용하여 명제
"함수 $f(x)$의 도함수가 존재하거나 나의 키는 175cm이다."
의 부정을 일상 언어로 적어라.

* 귀류법을 배리법이라고도 한다.

9. Prove the following De Morgan's Laws for the three components.

(a) $\sim(p \wedge q \wedge r) \equiv \sim p \vee \sim q \vee \sim r$ (b) $\sim(p \vee q \vee r) \equiv \sim p \wedge \sim q \wedge \sim r$

10. Can you generalize, without proof, De Morgan's Laws for n components? See Problem 9 for $n=3$.

11. Prove the following Absorption Laws.

(a) $p \wedge (p \vee r) \equiv p$ (b) $p \vee (p \wedge q) \equiv p$

12. Prove Theorem 5.

13. Prove Theorem 6.

1.4 Contradiction

In contrast to tautologies, there are statements whose truth values are all false for each of the logical possibilities. Such statements are called contradictions. For example, $p \wedge \sim p$ is a contradiction.

It is obvious that if t is a tautology, then $\sim t$ is a contradiction ; conversely, if c is a contradiction, then $\sim c$ is a tautology.

Theorem 7. Let t, c and p be a tautology, a contradiction, and an arbitrary statement, respectively. Then

(a) $p \wedge t \Longleftrightarrow p$, $p \vee t \Longleftrightarrow t$

(b) $p \vee c \Longleftrightarrow p$, $p \wedge c \Longleftrightarrow c$

(c) $c \Longrightarrow p$, $p \Longrightarrow t$

[Proof] (a) The following truth table for $p \wedge t \longleftrightarrow p$ shows that $p \wedge t$ is equivalent to p.

Table 13

	p	$\wedge$	t	$\longleftrightarrow$	p
	T	T	T	T	T
	F	F	T	T	F
Step	1	2	1	3	1

9. 다음은 세 성분 p, q, r에 대한 드 모르간의 법칙이다. 이것을 밝혀라.

(a) $\sim(p\wedge q\wedge r)\equiv\sim p\vee\sim q\vee\sim r$ (b) $\sim(p\vee q\vee r)\equiv\sim p\wedge\sim q\wedge\sim r$

10.* $n(n\geq 4)$개의 성분에 대한 드 모르간의 법칙을 증명없이 적어보아라.

11. 다음 **흡수법칙**을 증명하여라.

(a) $p\wedge(p\vee r)\equiv p$ (b) $p\vee(p\wedge q)\equiv p$

다음을 진리표로 확인하여라.

12. 정리 5 **13.** 정리 6

1.4 모순

항진에 반하여 모든 논리적 가능성의 각 경우마다 진리값이 거짓인 명제를 **모순** 또는 **항위명제**라고 한다. 이를테면 $p\wedge\sim p$는 하나의 모순이다.

분명히 항진 t**의 부정 $\sim t$는 모순이고, 모순 c**의 부정 $\sim c$는 항진 t이다(연습문제 1.4의 2).

정리 7 항진 t, 모순 c, 임의의 명제 p에 대하여 다음이 성립한다.

(a) $p\wedge t\Leftrightarrow p$, $p\vee t\Leftrightarrow t$ (b) $p\vee c\Leftrightarrow p$, $p\wedge c\Leftrightarrow c$ (c) $c\Rightarrow p$, $p\Rightarrow t$

[증명] (a) 다음과 같은 $p\wedge t\leftrightarrow p$의 진리표에 따라 $p\wedge t$, p는 동치인 것을 알 수 있다.

표 13

	p	$\wedge$	t	$\leftrightarrow$	p
	T	T	T	T	T
	F	F	T	T	F
단계	1	2	1	3	1

* 이것을 연습문제 1.8의 8(68쪽)에서는 수학적 귀납법으로 증명한다.

** 앞으로 기호 t, c는 각각 항진, 모순을 나타내는 것으로 본다(원서 38쪽 Table 15 아래 문장 참조).

The other equivalence $p \vee t \Longleftrightarrow t$ may be proved similarly.

(b) From the following truth table we find that the biconditional statement $p \vee c \longleftrightarrow p$ is a tautology and hence, $p \vee p \Longleftrightarrow p$.

Table 14

	p	$\vee$	c	$\longleftrightarrow$	p
	T	T	F	T	T
	F	F	F	T	F
Step	1	2	1	3	1

The proof for $p \wedge c \Longleftrightarrow c$ is similar.

(c) The truth tables of $c \rightarrow p$ and $p \rightarrow t$ show that $c \rightarrow p$ and $p \rightarrow t$ are tautologies. Therefore, $c \Longrightarrow p$ and $p \Longrightarrow t$.

Table 15

c	$\rightarrow$	p
F	T	T
F	T	F

p	$\rightarrow$	t
T	T	T
F	T	T

For the remainder of this book, the symbol c with or without a subscript will stand for a contradiction; and the symbol t with or without a subscript will denote a tautology.

Exercise 1.4

Prove the following statements 1 through 5.

1. $p \vee t \Longleftrightarrow t$ and $p \wedge c \Longleftrightarrow c$. **2.** $\sim t \Longleftrightarrow c$ and $\sim c \Longleftrightarrow t$.

3. Reductio ad Absurdum : $(p \rightarrow q) \Longleftrightarrow (p \wedge \sim q \rightarrow c)$

4. $p \wedge (p \rightarrow q) \wedge (p \rightarrow \sim q) \Longleftrightarrow c$.

5. $(p \rightarrow q) \Longrightarrow (p \vee r \rightarrow q \vee r)$ for any statement r.

같은 방법으로 동치 $p \vee t \Longleftrightarrow t$도 밝힐 수 있다.

(b) 다음 진리표(단계 3)에 따라 쌍조건문 $p \vee c \longleftrightarrow p$는 항진이다.

표 14

	p	$\vee$	c	$\longleftrightarrow$	p
	T	T	F	T	T
	F	F	F	T	F
단계	1	2	1	3	1

즉 $$p \vee c \Longleftrightarrow p.$$

마찬가지로 $p \wedge c \Longleftrightarrow c$인 것도 알 수 있다(과연?).

(c) 조건문 $c \rightarrow p$, $p \rightarrow t$의 진리표(표 15)를 보면 이 두 명제는 모두 항진이므로 $c \Longrightarrow p$, $p \Longrightarrow t$.

표 15

c	$\rightarrow$	p
F	T	T
F	T	F

p	$\rightarrow$	t
T	T	T
F	T	T

연습문제 1.4

임의의 명제 p, q, r에 대한 다음 1−5를 진리표로 밝혀라.

1. $p \vee t \Longleftrightarrow t$, $p \wedge c \Longleftrightarrow c$ **2.** $\sim t \Longleftrightarrow c$, $\sim c \Longleftrightarrow t$

3.* 배리법 $(p \rightarrow q) \Longleftrightarrow (p \wedge \sim q \rightarrow c)$

4. $p \wedge (p \rightarrow q) \wedge (p \rightarrow \sim q) \Longleftrightarrow c$

5. $(p \rightarrow q) \Rightarrow (p \vee r \rightarrow q \vee r)$

* 여기서의 "$p \wedge \sim q \rightarrow c$"와 "$p \wedge \sim q \rightarrow q \wedge \sim q$"[정리 6(c) 참조] 사이의 차이점에 주목하여라. 수학이나 논리에서 조건문 $p \rightarrow q$를 직접 증명하기 곤란할 때 흔히 배리법이 활용된다. 그럴 경우 두 조건문 중 $p \wedge \sim q \rightarrow c$가 인용된다[예제 11(59쪽) 참조].

1.5 Deductive Reasoning

The 17 laws summarized in Theorems 1 through 6 are very useful tools for justifying logical equivalences and implications, as illustrated in Examples 5 through 7. We shall call these 17 laws the rules of inference. It should be noted that these rules are selected just for convenient reference and are not intended to be independent of each other. For instance, the Contrapositive Law can be established "deductively" by using other laws and relevant definitions, as the next example shows.

Example 5 Prove the Contrapositive Law, $p \to q \equiv \sim q \to \sim p$, by using relevant definitions and other rules of inference.

[Solution]

$p \to q \equiv \sim(p \wedge \sim q)$	Def. 4
$\equiv \sim(\sim q \wedge p)$	Com
$\equiv \sim[\sim q \wedge \sim(\sim p)]$	D.N.
$\equiv \sim q \to \sim p$	Def. 4

Therefore, $p \to q \equiv \sim q \to \sim p$, by the Transitive Law.

The method of proof used in Example 5 is called deductive reasoning or the deductive method, which differs from the method of proof by truth tables.

In general, in deductive reasoning any previously stated axioms, definitions, and theorems and the rules of inference may be used.

Example 6 Prove the disjunctive Syllogism by deductive reasoning.

[Solution]

$(p \vee q) \wedge \sim p \equiv \sim p \wedge (p \vee q)$	Com.
$\equiv (\sim p \wedge p) \vee (\sim p \wedge q)$	Dist.
$\equiv c \vee (\sim p \wedge q)$	$\sim p \wedge p \equiv c$
$\equiv (\sim p \wedge q) \vee c$	Com.
$\equiv \sim p \wedge q$	Th. 7(b)
$\Rightarrow q$	Simp.

Finally, by the Transitive Law, $(p \vee q) \wedge \sim p \Rightarrow q$.

1.5 연역적 추론

정리 1~6에 요약해 놓은 17개의 법칙들은 다음 예제 5-7에서와 같이 논리적 동치와 함의에 대한 정당성을 밝히는 데 매우 유용하게 활용되고 있다. 곧 **추론규칙**이라고 하는 이들 법칙은 단지 편리하게 활용하려고 택한 것일 뿐 그들 규칙상호간의 독립성(그 뜻은 비켜간다)은 염두에 두지 않고 있다. 이를테면 다음 예제 5에서 보듯 대우법칙을 유도할 때 그 것과는 다른 법칙과 거기에 관련된 정의를 써서 **연역적***으로 다룰 수 있다.

예제 5 대우 $p \to q \equiv \sim q \to \sim p$를 증명하여라.

[풀이]

$p \to q \equiv \sim(p \wedge \sim q)$	정의 4
$\equiv \sim(\sim q \wedge p)$	교환법칙
$\equiv \sim[\sim q \wedge \sim(\sim p)]$	이중부정
$\equiv \sim q \to \sim p$	정의 4

그러므로 ≡에 관한 추이율에 따라 $p \to q \equiv \sim q \to \sim p$

예제 5에서의 증명법은 진리표에 따른 증명법과 다른 이른바 **연역적 추론** 또는 **연역적방법**이다. 대체로 연역적 추론에서는 그 때 그 때 알려진 공리, 정의, 정리, 추론규칙 등을 활용한다.

예제 6 논리합의 삼단논법(정리 1(c))을 증명하여라.

[풀이]

$(p \vee q) \wedge \sim p \equiv \sim p \wedge (p \vee q)$	교환법칙
$\equiv (\sim p \wedge p) \vee (\sim p \wedge q)$	분배법칙
$\equiv c \vee (\sim p \wedge q)$	$\sim p \wedge p \equiv c$
$\equiv (\sim p \wedge q) \vee c$	교환법칙
$\equiv \sim p \wedge q$	정리 7(b)
$\Rightarrow q$	단순화법칙

그러므로 ≡에 대한 추이율에 따라 $(p \vee q) \wedge \sim p \equiv \sim p \wedge q$

여기서 $\sim p \wedge q \Rightarrow q$. 따라서 $(p \vee q) \wedge \sim p \Rightarrow q$(왜?)

* "연역적"이란 "귀납적"[1.8 수학적 귀납법(63쪽)]에 대등하는 용어이다.

Example 7 Prove the following *Exportation Law*:

$$p \wedge q \rightarrow r \equiv p \rightarrow (q \rightarrow r)$$

by deductive reasoning.

[Solution]

$$\begin{aligned} p \rightarrow (q \rightarrow r) &\equiv p \rightarrow \sim(q \wedge \sim r) && \text{Def. 4} \\ &\equiv \sim[p \wedge (q \wedge \sim r)] && \text{Def. 4, D.N.} \\ &\equiv \sim[(p \wedge q) \wedge \sim r] && \text{Assoc.} \\ &\equiv p \wedge q \rightarrow r && \text{Def. 4} \end{aligned}$$

Hence, $p \wedge q \rightarrow r \equiv p \rightarrow (q \rightarrow r)$.

Example 8 Prove that $(p \rightarrow r) \vee (q \rightarrow s) \equiv (p \wedge q \rightarrow r \vee s)$ by deductive reasoning.

[Solution]

$$\begin{aligned} (p \rightarrow r) \vee (q \rightarrow s) &\equiv \sim(p \wedge \sim r) \vee \sim(q \wedge \sim s) && \text{Def. 4} \\ &\equiv (\sim p \vee r) \vee (\sim q \vee s) && \text{De M., D.N.} \\ &\equiv (\sim p \vee \sim q) \vee (r \vee s) && \text{Com., Assoc.} \\ &\equiv \sim[(p \wedge q) \wedge \sim(r \vee s)] && \text{De M., D.N.} \\ &\equiv p \wedge q \rightarrow r \vee s && \text{Def. 4} \end{aligned}$$

Why we want to use deductive reasoning as opposed to truth tables may be seen from the following comparison : To verify the equivalence in Example 8 by the method of truth tables, we would have to construct a huge truth table with $16(=2^4)$ cases(see Problem 20 of Exercise 1.1 or Problem 12 of Exercise 1.3) : on the other hand, in the solution of Example 8 above, we established that equivalence in only five short steps.

예제 7 다음 **이출법칙**을 증명하여라.

$$p\wedge q\rightarrow r\equiv p\rightarrow(q\rightarrow r)$$

[풀이]

	$p\rightarrow(q\rightarrow r)\equiv p\rightarrow\sim(q\wedge\sim r)$	정의 4
	$\equiv\sim[p\wedge(q\wedge\sim r)]$	정의 4, 이중부정
	$\equiv\sim[(p\wedge q)\wedge\sim r]$	결합법칙
	$\equiv p\wedge q\rightarrow r$	정의 4
그러므로	$p\rightarrow(q\rightarrow r)\equiv p\wedge q\rightarrow r$	$\equiv$에 관한 추이율
따라서	$p\wedge q\rightarrow r\equiv p\rightarrow(q\rightarrow r)$	$\equiv$에 관한 대칭율*

예제 8 동치 $(p\rightarrow r)\vee(q\rightarrow s)\equiv(p\wedge q\rightarrow r\vee s)$를 증명하여라.

[풀이]

$(p\rightarrow r)\vee(q\rightarrow s)\equiv\sim(p\wedge\sim r)\vee\sim(q\wedge\sim s)$	정의 4
$\equiv(\sim p\vee r)\vee(\sim q\vee s)$	드 모르간의 법칙, 이중부정
$\equiv(\sim p\vee\sim q)\vee(r\vee s)$	교환법칙, 결합법칙,
$\equiv\sim[(p\wedge q)\wedge\sim(r\vee s)]$	드 모르간의 법칙, 이중부정
$\equiv p\wedge q\rightarrow r\vee s$	정의 4

그러므로

$(p\rightarrow r)\vee(q\rightarrow s)\equiv(p\wedge q\rightarrow r\vee s)$ $\equiv$에 관한 추이율

여기서 다루어본 몇몇 증명에 있어서 진리표보다도 연역적 추론을 선호한 이유는 다음과 같은 비교를 통하여 쉽게 알 수 있다. 이를테면 예제 8에서의 동치를 진리표로 밝히려면 무려 $16(=2^4)$ 경우(연습문제 1.1의 20 또는 연습문제 1.3의 12)에 달하는 벅찬 작업을 거쳐야 한다. 그러나 예제 8에서는 단지 다섯 과정만으로 동치의 증명을 끝낼 수 있는 장점이 있다.

---

* 임의의 명제 P, Q에 대하여 대칭율

$$P\equiv Q\Rightarrow Q\equiv P$$

가 성립한다. 이 대칭율은 임의의 실수 a, b의 등식 "="에 대한 대칭율, 즉 $a=b\Rightarrow b=a$와 유사하다.

Exercise 1.5

Prove the following tautologies by the deductive method.

1. Modus Ponens : $p \wedge (p \rightarrow q) \Rightarrow q$

2. Modus Tollens : $\sim q \wedge (p \rightarrow q) \Rightarrow \sim p$

3. Reductio ad Absurdum : $(p \rightarrow q) \Leftrightarrow (p \wedge \sim q \rightarrow c)$

4. Disjunctive Syllogism : $(p \vee q) \wedge \sim p \Rightarrow q$

5. Theorem 7(c) : $c \Rightarrow p$

6. $(p \rightarrow q) \Leftrightarrow (p \rightarrow p \wedge q)$

7. $(p \rightarrow q) \Leftrightarrow (p \vee q \rightarrow q)$

8. $(p \rightarrow q) \Leftrightarrow \sim p \vee q$

9. $(p \rightarrow r) \wedge (q \rightarrow r) \Leftrightarrow (p \vee q \rightarrow r)$

10. $(p \rightarrow q) \wedge (p \rightarrow r) \Leftrightarrow (p \rightarrow q \wedge r)$

11. $(p \rightarrow q) \wedge (p \rightarrow \sim q) \Leftrightarrow \sim p$

12. $(p \rightarrow q) \vee (p \rightarrow r) \Leftrightarrow (p \rightarrow q \vee r)$

13. $(p \rightarrow r) \vee (q \rightarrow r) \Leftrightarrow (p \wedge q \rightarrow r)$

연습문제 1.5

다음 1∼13을 연역적방법으로 증명하여라.

1. 긍정식 삼단논법 $p \wedge (p \to q) \Rightarrow q$

2. 부정식 삼단논법 $\sim q \wedge (p \to q) \Rightarrow \sim p$

3. 배리법 $(p \to q) \Longleftrightarrow (p \wedge \sim q \to c)$

4. 논리합 삼단논법 $(p \vee q) \wedge \sim p \Rightarrow q$

5. 정리 7(c) $c \Rightarrow p$

6. $(p \to q) \Longleftrightarrow (p \to p \wedge q)$

7. $(p \to q) \Longleftrightarrow (p \vee q \to q)$

8. $(p \to q) \Longleftrightarrow \sim p \vee q$

9. $(p \to r) \wedge (q \to r) \Longleftrightarrow (p \vee q \to r)$

10. $(p \to q) \wedge (p \to r) \Longleftrightarrow (p \to q \wedge r)$

11. $(p \to q) \wedge (p \to \sim q) \Longleftrightarrow \sim p$

12. $(p \to q) \vee (p \to r) \Longleftrightarrow (p \to q \vee r)$

13. $(p \to r) \vee (q \to r) \Longleftrightarrow (p \wedge q \to r)$

14.* 조건문

$$p \to q \qquad \cdots\cdots ①$$

에 대하여 다음 조건문

$q \to p$ ······② $\sim p \to \sim q$ ······③ $\sim q \to \sim p$ ······④

를 차례로 ①의 **역**(converse), **이**(reverse)이라 한다. 한편 ④는 ②의 대우(contraposition : 29쪽 각주 ** 참조)임을 알고 있다. ①이 참일 때 ②-④ 중 참인 것이 있으면 지적하고 거짓인 것에 대해서는 반례를 들어라.

15.* 임의의 실수 a, b에 대한 조건문

$$a = b \to a^2 = b^2 \qquad \cdots\cdots ⑤$$

의 역, 이, 대우는 차례로 다음과 같다.

역 : $a^2 = b^2 \to a = b$ 이** : $a \neq b \to a^2 \neq b^2$ 대우 : $a^2 \neq b^2 \to a \neq b$

(a) ⑤ 및 그것의 역, 이, 대우 중 참인 것은 증명을 하고 거짓인 것에 대해선 반례를 들어라.

(b) (a)에서의 네 명제 중 동치인 것이 있으면 그 쌍을 지적하여라.

---

* 이 문제는 옮긴이가 추가한 것이다. ** $a \neq b$, $a^2 \neq b^2$의 부정은 각각 $a = b$, $a^2 = b^2$이다.

1.6 Quantification Rules

In any general discussion, we have in mind a particular universal set or domain of discourse, that is, a collection of objects whose properties are under consideration. For example, in the statement

"All humans are mortal,"

the universe is the collection of all humans. With this understanding of the universe, the statement "All humans are mortal" may be alternatively expressed as:

For all x in the universe, x is mortal.

The phrase "For all x in the universe" is called a universal quantifier, and is symbolized as $\forall \boldsymbol{x}$. The sentence "x is mortal" says something about x; we symbolize this as $p(x)$. Using these new symbols we can now write the general statement "All humans are mortal" as $\forall \boldsymbol{x}\, \boldsymbol{p}(\boldsymbol{x})$.

Next, consider the statement "Some humans are mortal." Here the universe (or domain of discourse) is still the same as for the previous statement. With this universe in mind, we can restate the statement "Some humans are mortal" successively as:

There exists at least one individual who is mortal.

There exists at least one x such that x is mortal.

and as There exists at least one x such that $p(x)$.

The phrase "There exists at least one x such that" is called an existential quantifier and is symbolized as $\exists \boldsymbol{x}$. Using this new symbol we can now rewrite the statement "Some humans are mortal" as $\exists \boldsymbol{x}\, \boldsymbol{p}(\boldsymbol{x})$.

In general, suppose we have a domain of discourse U and a general statement $p(x)$, called a propositional function, whose "variable" x ranges over U. Then $\forall x\, p(x)$ asserts that for all x in U, the statement $p(x)$ about x is true, and $\exists x\, p(x)$ means that there exists at least one x in U such that $p(x)$ is true.

1.6 한정규칙 *

대체로 토론을 진행할 때 미리 염두에 둔 성질을 지닌 것들의 모임을 **전체집합**이라고 한다. 이를테면 하나의 명제

"모든 사람은 죽기 마련이다." ……①

의 전체집합은 인류이다. 이와 같이 우리는 전체집합을 감안할 때 명제 ①을 다음과 같이 나타낼 수 있다.

전체집합의 모든 x에 대하여 x는 죽기 마련이다.

여기서 하나의 표현 "전체집합의 모든 x에 대하여"를 **전칭기호**라 하고 이것을 $\forall \boldsymbol{x}$로 나타낸다. 이 경우 "x는 죽기 마련이다."라는 주장은 다름아닌 x에 관한 조건을 제시하고 있다. 여기서 그러한 조건을 기호 $p(x)$로 나타냄으로써 명제 ①을 $\forall \boldsymbol{x}\, \boldsymbol{p}(\boldsymbol{x})$와 같이 나타낸다. 한편

"어느 사람은 죽기 마련이다." ……②

와 같은 명제의 전체집합도 인류이다. 이런 전체집합을 염두에 둘 때 ②를 다음 중 어느 한 가지를 주장하는 명제로 받아 들인다.

죽기 마련인 사람이 적어도 한 명 존재한다.
적어도 하나의 x가 존재함으로써 x는 죽는다.
적어도 하나의 x가 존재함으로써 $p(x)$.

이 경우 "적어도 하나의 x가 존재함으로써"를 $\exists \boldsymbol{x}$로 나타내고 이것을 **존재기호**라고 한다. 그러할 때 명제 ②를 $\exists \boldsymbol{x} \boldsymbol{p}(\boldsymbol{x})$와 같이 나타낼 수 있다.

대체로 $p(x)$**를 전체집합 U에서의 **명제함수**(3쪽 각주* 참조)로 보면 그것의 참, 거짓은 물론 $x \in U$ 하나 하나에 따라 정해진다.

앞으로는 $\forall x\, p(x)$, $\exists x\, p(x)$를 각각 모든 x에 대한, 적어도 하나의 $x \in U$에 대한(하나의) 명제로 다루기로 하고 전자를 **전칭명제**, 후자를 **특칭명제**로 부르도록 한다.

---

* 이병무, 《논리와 집합의 세상》, 경문사, 2003년, 제7장 술어논리(89-95쪽) 참조.

** 명제함수 $p(x)$가 참일 때의 $x \in U$의 집합을 $p(x)$의 **진리집합**(truth set)이라 하고 이것을 $\{\boldsymbol{x} \mid \boldsymbol{p}(\boldsymbol{x})\}$로 나타낸다. 이를테면 $U=\{1, 2, 3, 4\}$, $p(x)$: $1 \leq |x| < 4$에 대하여 $\{x \mid p(x)\} = \{1, 2, 3\}$. 진리집합이 전체집합 U, 공집합 ϕ와 각각 같은 명제함수를 들어라.

In elementary mathematics, quantifiers are often suppressed for the sake of simplicity. For example,

$$(x+1)(x-1)=x^2-1$$

in high school algebra books should be understood to say that "for every real number x, $(x+1)(x-1)=x^2-1$." In mathematics, "for every" and "for all" mean the same and both are symbolized by $\forall$; and "for some" means the same as "there exists" and is symbolized by $\exists$. In less formal expressions, we often put the quantifier after the statement. For example, the statement

"$f(x)=0$ for all x"

is just the same as "$\forall x\, f(x)=0$."

In logic and in mathematics, the negation of the statement "$p(x)$ is true for every x (in U)," $\sim\forall x\, p(x)$, is considered to be the same as the assertion "there exists at least one x (in U) for which $p(x)$ is false," $\exists x \sim p(x)$. Similarly, $\sim\exists x\, p(x)$ is considered the same as "there is no x (in U) such that $p(x)$ is true" : or in other words, "$p(x)$ is false for all x (in U)," or $\forall x \sim p(x)$. We summarize in the following.

Rule of Quantifier Negation (Q. N.). Let $p(x)$ be a propositional predicate, that is, a statement about an unspecified object x in a given universal set. Then

$$\sim\forall x\, p(x) \equiv \exists x \sim p(x) \qquad \text{and} \qquad \sim\exists x\, p(x) \equiv \forall x \sim p(x)$$

We have used "$\equiv$" to denote that two quantified statements on both sides of $\equiv$ are considered the same in logic ; this usage is consistent with the usage of $\equiv$ for logical equivalences, as will be seen in the next paragraph.

To better understand the quantified statements $\forall x\, p(x)$ and $\exists x\, p(x)$, let us inspect the case in which the universal set consists of finitely many elements denoted by $a_1, a_2, a_3, \cdots, a_n$. Then, since $\forall x(p(x))$ asserts that $p(x)$ is true for every $a_1, a_2, a_3, \cdots, a_n$, the statement $\forall x\, p(x)$ is true if and only if the conjunction of

전칭기호 $\forall x$, 존재기호 $\exists x$를 모두 **한정기호**라고 한다. 사실 우리는 초·중등수학에서는 한정기호 없이 다루어왔다. 이를테면

$$(x+1)(x-1)=x^2-1 \qquad \cdots\cdots ③$$

은 모든 실수 x에 대하여 성립하는 등식(사실은 항등식!)으로 받아들인다. 여기서 "모든 x에 대하여", "각 x에 대하여"는 같은 뜻으로 기호 $\forall x$로 나타낸다. 한편 "적당한 x에 대하여", "x가 존재함으로써"는 모두 기호 $\exists x$로 나타낸다. 다만 형식적인 표현을 피하고 싶을 때 한정기호를 명제 뒤에 적기도 한다. 예를 들어

"$f(x)=0$, 모든 x에 대하여"는 바로 "$\forall x f(x)=0$"

을 뜻한다. 한편 명제 "모든 $x\in U$에 대하여 $p(x)$."의 부정, 즉 $\sim\forall x\, p(x)$은

"적어도 하나의 $x\in U$가 존재함으로써 $\sim p(x)$."

곧 $\exists x \sim p(x)$를 나타내는 명제를 뜻한다.

마찬가지로 $\sim\exists x\, p(x)$는

"적당한 $x\in U$에 대하여 $p(x)$는 아니다."

또는 "모든 $x\in U$에 대하여 $p(x)$는 아니다."

로서 $\forall x \sim p(x)$와 동치인 것으로 생각한다.

여기서 지금까지의 부정들을 다음과 같이 요약할 수 있다.

한정기호의 부정규칙 전체집합 U의 임의의 원소 x에 관한 명제함수 $p(x)$*에 대하여 다음이 성립한다.

$$\sim\forall x\, p(x)\equiv\exists x\sim p(x), \qquad \sim\exists x\, p(x)\equiv\forall x\sim p(x)$$

말할 것도 없이 하나의 명제함수 $p(x)$의 진리집합은 유한집합 아니면 무한집합이다. 특히 유한집합 $U=\{a_1, a_2, a_3, \cdots, a_n\}$일 경우

$$p(a_1),\ p(a_2),\ p(a_3),\ \cdots,\ p(a_n) \qquad \cdots\cdots ④$$

이 모두 참이면 그리고 그때에만 $\forall x\, p(x)$는 참이다. 따라서 우리는 다음과 같이 주장할 수 있다.

* 명제함수의 기호로는 $p(x)$와에 $q(x)$, $r(x)$ 등도 인용된다.

$$p(a_1),\ p(a_2),\ p(a_3),\ \cdots,\ p(a_n)$$

is true. Thus, $\forall x\ p(x)$ amounts to

$$p(a_1)\wedge p(a_2)\wedge\cdots\wedge p(a_n)$$

Similarly, $\exists x\ p(x)$ means

$$p(a_1)\vee p(a_2)\vee\cdots\vee p(a_n)$$

Thus the Rule of Quantifier Negation may be viewed as a generalization of De Morgan's Laws (Theorem 3).

Example 9 Which of the following is equivalent to the negation of the statement "All snakes are poisonous"?

(a) All snakes are not poisonous.

(b) Some snakes are poisonous.

(c) Some snakes are not poisonous.

[Solution] The universal set U is the collection of all snakes. Let $p(x)$ be the propositional predicate which asserts that x is poisonous. (where the variable x ranges over U). The statement "All snakes are poisonous" is then translated into $\forall x\ p(x)$. According to Q.N., $\sim\forall x\ p(x)$ is equivalent to $\exists x \sim p(x)$, which represents "Some snakes are not poisonous."

Exercise 1.6

1. Translate the statement from elementary algebra

" The equation $x^2-3x+2=0$ has solutions "

into logical language using a quantifier. What is the universal set here?

앞의 ④의 논리곱이 참이면 그리고 그때에만 $\forall x\, p(x)$는 참이다. 따라서 $\forall x\, p(x)$는

$$p(a_1)\wedge p(a_2)\wedge\cdots\wedge p(a_n)$$

를 뜻한다.

마찬가지로 $\exists x\, p(x)$는 논리합 $p(a_1)\vee p(a_2)\vee\cdots\vee p(a_n)$을 뜻한다.

이상으로 미루어 보아 한정기호의 부정규칙은 드 모르간의 법칙[정리 3(29쪽)]을 일반화한 것임을 알 수 있다.

예제 9 다음 중 어느 것이 명제 "모든 뱀은 독을 지닌다."의 부정과 동치인가를 지적하여라.

(a) 모든 뱀은 독을 지니지 않는다.

(b) 독을 지닌 뱀이 존재한다.

(c) 독을 지니지 않은 뱀이 존재한다.

[풀이] 모든 뱀의 모임을 전체집합 U라고 놓고 (U의) 임의의 원소 x에 대한 명제함수 "x는 독을 지닌다."를 $p(x)$라 할 때, 주어진 명제의 부정은 $\sim\forall x\, p(x)$로 나타낸다. 한편 명제 (c)는 $\exists x\sim p(x)$로 옮겨 적을 수 있다. 그런데 한정기호의 부정규칙(첫 번째)에 따라 $\sim\forall x\, p(x)\equiv\exists x\sim p(x)$이므로 구하는 명제는 (c)이다.

연습문제 1.6

1. (a) 한정기호를 써서 명제

"이차방정식 $x^2-3x+2=0$의 해가 존재한다."

를 논리적 언어로 옮겨라. 또 여기서의 전체집합*을 말하여라.

(b)** $p(x)$: $0<x^2-1$의 진리집합을 말하여라. 다만 $U=\mathbb{R}$.

---

* 전체집합으로서 집합 $\mathbb{R}$을 염두에 두고 있다. 지금 $\mathbb{R}$ 대신 집합 $\mathbb{N}$을 전체집합이라 할 때 해가 존재하는 경우, 존재하지 않는 경우의 이차방정식의 보기를 하나씩 더 들어라.

** 옮긴이가 추가함.

2. Find the equivalent statement of the negation for each of the following by using Q. N.

(a) All snakes are reptiles. (b) Some horses are gentle.

(c) Some mathematicians are not sociable.

(d) All female students are either attractive or smart.

(e) No baby is not cute.

3. Find the universal set for each of the five statements in Problem 2.

4. Derive $\sim \exists x\, p(x) \equiv \forall x \sim p(x)$ from $\sim \forall x\, p(x) \equiv \exists x \sim p(x)$

5. Derive $\sim \forall x\, p(x) \equiv \exists x \sim p(x)$ from $\sim \exists x\, p(x) \equiv \forall x \sim p(x)$

6. Prove that $\sim \forall x \sim q(x) \equiv \exists x\, q(x)$ and $\sim \exists x \sim q(x) \equiv \forall x\, q(x)$.
[Hint : Use Q.N.]

1.7 Proof of Validity

One of the most important tasks of a logician is the testing of arguments. an argument is assertion that a statement, called the conclusion, follows from other statements, called the hypotheses or premises. An argument is considered to be valid if the conjunction of the hypotheses implies the conclusion. For examples, the following is an argument in which the first four statements are hypotheses, and the last statement is the conclusion.

If he studies medicine, then he prepares to earn a good living.
If he studies the arts, then he prepares to live a good life.
If he prepares to earn a good living or he prepares to live a good life, then his college tuition is not wasted.
His college tuition is wasted.
Therefore, he studies neither medicine nor the arts.

2. 다음 각각의 부정과 동치인 명제를 한정기호의 부정규칙을 써서 구하여라.
(a) 모든 뱀은 파충류이다. (b) 말 중에는 백마도 있다.
(c) 한국에는 30세 미만의 수학자도 있다.
(d) 모든 여자의 머리는 검거나 갈색이다.
(e) 세상에 울지 않는 아기는 없다.

3. 문제 2의 각 명제에 대한 전체집합을 말하여라.

4. $\sim\forall x\, p(x) \equiv \exists x \sim p(x)$로부터 다음을 유도하여라.

$$\sim\exists x\, p(x) \equiv \forall x \sim p(x)$$

5. $\sim\exists x\, p(x) \equiv \forall x \sim p(x)$로부터 다음을 유도하여라.

$$\sim\forall x\, p(x) \equiv \exists x \sim p(x)$$

6. 다음을 각각 증명하여라(힌트 : 한정기호의 부정규칙을 이용하여라).

$$\sim\forall x \sim q(x) \equiv \exists x\, q(x), \qquad \sim\exists x \sim q(x) \equiv \forall x\, q(x)$$

1.7 타당성 밝힘

논리학자의 중요한 과제 중의 하나는 **논증**을 점검하는 일이다. 논증이란 **가정** 또는 **전제**로 불리는 명제로부터 **결론**이라고 불리는 명제를 논리적으로 유도하는 과정을 뜻한다. 여기서 가정들의 논리곱이 결론을 함의하면 그 논증은 **타당**한 것으로 여긴다. 이를테면 다음은 처음 네 명제 H_1—H_4가 모두 가정이고 마지막 명제가 결론인 논증을 나타낸다.

H_1. 명수가 의학공부를 한다면 부유*한 생활을 하게 될 것이다.
H_2. 명수가 예술공부를 한다면 멋진* 생애를 보낼 것이다.
H_3. 명수가 부유한 생활 또는 멋진 생애를 보내려면 대학등록금을 낭비*해서는 안 된다.
H_4. 명수는 대학등록금을 낭비하고 있다.
따라서 명수는 의학과 예술을 공부하지 않는다.

위의 논증에 대한 가정과 결론을 각각 다음과 같은 문자로 나타내보자.

M : 명수는 의학공부를 한다. E : 명수는 부유한 생활을 할 것이다.
A : 명수는 예술공부를 한다. L : 명수는 멋진 생애를 보낼 것이다.
W : 명수는 대학등록금을 낭비한다.

* 모호하기 짝이 없는 용어 "부유", "멋진", "낭비"를 모두 독자 나름대로 바꾸어 보아라.

It may be symbolized as:

$$\begin{array}{ll} H_1. & M \to E \\ H_2. & A \to L \\ H_3. & (E \vee L) \to \sim W \\ H_4. & W \\ \hline C. & \therefore^* \sim M \wedge \sim A \end{array}$$

To establish the validity of this argument by means of a truth table would require a table with $32(=2^5)$ rows. But we can prove this argument valid by deducing the conclusion from the hypotheses in a few steps using the rules of inference.

From the hypotheses H_3 and H_4, $(E \vee L) \to \sim W$ and W, we infer $\sim(E \vee L)$, or equivalently $\sim E \wedge \sim L$, by Modus Tollens and De Morgan's Law. From $\sim E \wedge \sim L$ we validly infer $\sim E$ (and also $\sim L$) by the Law of Simplification. From H_1, $M \to E$, and $\sim E$ we validly obtain $\sim M$.

Similarly, from H_2, $A \to L$, and $\sim L$, we validly infer $\sim A$. Finally, the conjunction of $\sim M$ and $\sim A$ gives the conclusion $\sim M \wedge \sim A$. In this proof, the rules of inferences Modus Tollens (M.T.), De Morgan's Law (De. M.), and Laws of Simplification (Simp.) are used.

A formal and more concise way of expressing this proof of validity is to list the hypotheses and the statements deduced from them on one side, with the justification of each step written beside it. In each step the "justification" indicates the preceding statements from which, and the rules of inference by which, the statement given in that step was obtained. For easy reference, it is convenient to number the hypotheses and the statements deduced from them and to put the conclusion to the right of the last premise, separated from it by a slash / which indicates that all the statements above it are hypotheses. The formal proof of validity for the above argument may thus be written as

이러한 문자들과 결합자를 써서 주어진 논증을 다음과 같은 형식으로 진행할 수 있다.

H_1. $M \rightarrow E$

H_2. $A \rightarrow L$

H_3. $(E \vee L) \rightarrow \sim W$

H_4. W

C. 그러므로* $\sim M \wedge \sim A$

한편 이 논증의 **타당성**을 진리표로 밝히려면 무려 $32 = (2^5)$행을 지닌 표를 작성해야 한다. 그러나 추론의 규칙을 이용하면 불과 몇 단계만으로 가정으로부터 결론을 이끌어냄으로써 논증의 타당성을 밝힐 수 있다. 이제부터 그 과정을 소개하여 보자.

우선 부정식 삼단논법과 드 모르간의 법칙에 따라 가정 H_3, H_4, 즉 $(E \wedge L) \rightarrow \sim W$, W로부터 $\sim(E \vee L)$ 혹은 이것과 동치인 $\sim E \wedge \sim L$을 이끌어낸다. 또 단순화 법칙에 따라 $\sim E \wedge \sim L$로부터 $\sim E$, $\sim L$을 각각 이끌어내는 것은 타당하다. 한편 가정 H_1인 $M \rightarrow E$, $\sim E$로부터 $\sim M$을 각각 얻는 것도 타당하다.

마찬가지로 가정 H_2 $A \rightarrow L$, $\sim L$로부터 $\sim A$를 이끌어내다보면 마침내 $\sim M$, $\sim A$의 논리곱에 따른 결론 $\sim M \wedge \sim A$를 내릴 수 있다.

여기서 인용된 추론의 규칙으로는 다음에서 볼 수 있듯이 이중부정, 부정식 삼단논법, 드 모르간의 법칙, 단순화 법칙, 논리곱 등을 들 수 있다.

한 가지 논증의 타당성 여부를 간편한 형식으로 나타내려면 한 쪽에 가정과 그로부터 유도된 명제들을 나열하고 각 단계마다의 **정당성**을 적도록 한다. 이 경우 각 단계에서의 정당성이란 바로 그 단계에 주어진 명제에 앞서 있는 명제이거나 추론의 규칙을 가르킨다. 특히 쉬운 추론의 경우에는 가정과 그로부터 유도된 명제에 일일이 번호를 매기고 결론만은 마지막 단계의 오른쪽에 적어 두면 편리하다. 이때 마지막 단계와 결론 사이에 빗금을 쳐서 그 위의 명제는 모두 가정을 나타내도록 한다.

다음은 위의 논증에 대한 타당성을 형식적 과정으로 보여주고 있다.

---

* 원서에는 기호 "∴"가 실려있으나 최근에는 이를 기피하는 경향을 감안하여 그 대신 "그러므로" 아니면 "따라서"를 쓰기로 한다.

1. $M \to E$	Hyp.
2. $A \to L$	Hyp.
3. $(E \vee L) \to \sim W$	Hyp.
4. $W / \therefore\ \sim M \wedge \sim A$	Hyp. /Concl.
5. $\sim(E \wedge L)$	3, 4 M.T.
6. $\sim E \wedge \sim L$	5, De. M.
7. $\sim E$	6, Simp.
8. $\sim L$	6, Simp.
9. $\sim M$	1, 7, M.T.
10. $\sim A$	2, 8, M.T.
11. $\sim M \wedge \sim A$	9, 10, Conj.

A formal proof of validity for a given argument is a sequence of statements each of which is either a premise of the argument or follows from preceding statements by a known valid argument, ending with the conclusion of the argument.

Example 10 Construct a formal proof of validity for the following argument, using the suggested symbols :

Either Winston is elected president of the board or both Halbert and Luke are elected vice presidents of the board. If either Winston is elected president or Halbert is elected vice president of the board, then David will file a protest. Therefore, either Winston is elected president of the board or David files a protest. (W, H, L, D)

[Proof]

1. $W \vee (H \wedge L)$	
2. $W \vee H \to D / \therefore\ W \vee D$	
3. $(W \vee \mathrm{H}) \wedge (W \vee L)$	1, Dist.
4. $W \vee H$	3, Simp.
5. D	2, 4, M.P.
6. $D \vee W$	5, Add.
7. $W \vee D$	6, Com.

1. $M \to E$	가정
2. $A \to L$	가정
3. $(E \vee L) \to \sim W$	가정
4. W/그러므로 $\sim M \wedge \sim A$	가정/결론
5. $\sim(\sim W)$	4, 이중부정
6. $\sim(E \vee L)$	3·5 및 부정식 삼단논법
7. $\sim E \wedge \sim L$	6에 대한 드 모르간의 법칙
8. $\sim E$	7에 대한 단순화 법칙
9. $\sim L$	7에 대한 단순화 법칙
10. $\sim M$	1·8 및 부정식 삼단논법
11. $\sim A$	2·9 및 부정식 삼단논법
12.* $\sim M \wedge \sim A$	10·11 및 논리곱

논증의 **형식적 밝힘**이란 논증에서의 결론을 이끌어 낼 수 있는 일련의 명제를 가르키며 그 각각은 전제이거나 아니면 앞선 명제로부터 이미 알려진 타당한 논증을 거쳐 유도된 명제를 말한다.

예제 10 다음 소괄호 안의 문자를 써서 아래 논증에 대한 타당성을 형식적으로 밝혀보아라.

Winston이 위원회 회장으로 선출되거나 Halbert와 Luke는 부회장으로 선출된다. Winston이 회장으로 선출되거나 Halbert가 부회장으로 선출되면 David는 항의서를 제출할 것이다. 그러므로 Winston이 회장으로 선출되거나 David가 항의서를 제출할 것이다(W, H, L, D).

[증명]

1. $W \vee (H \wedge L)$	
2. $W \vee H \to D$/그러므로 $W \vee D$	
3. $(W \vee H) \wedge (W \vee L)$	1, 분배법칙
4. $W \vee H$	3에 대한 단순화 법칙
5. D	2·4 및 긍정식 삼단논법
6. $D \vee W$	5에 대한 합의법칙
7. $W \vee D$	6에 대한 교환법칙

* 사실 1-12까지의 모든 과정은 한 마디로 다음과 같은 함의를 증명한 것에 불과하다.

$$(M \to E) \wedge (A \to C) \wedge [(E \vee C) \to \sim W] \wedge W \Rightarrow (\sim M \wedge \sim A)$$

There is another method of proof called indirect proof, or the method of proof by reductio ad absurdum. An indirect proof of validity for a given argument is done by including, as an additional premise, the negation of its conclusion, and then deriving a contradictions ; as soon as a contradiction is obtained, the proof is complete.

Example 11 Give an indirect proof of validity for the following argument :

$$p \vee q \rightarrow r$$
$$s \rightarrow p \wedge u$$
$$q \vee s / \therefore\ r$$

[Proof]

1. $p \vee q \rightarrow r$	
2. $s \rightarrow p \wedge u$	
3. $q \vee s / \therefore\ r$	
4. $\sim r$	I.P. (Indirect Proof)
5. $\sim(p \vee q)$	1, 4, M.T.
6. $\sim p \wedge \sim q$	5, De M.
7. $\sim p$	6, Simp.
8. $\sim q$	6, simp.
9. s	3, 8, D.S.
10. $p \wedge u$	2, 9 M.P.
11. p	10, Simp.
12. $p \wedge \sim p$	7, 11, conj.

The statement $p \wedge \sim p$ in step 12 is a contradiction ; therefore the indirect proof of validity is complete.

In contrast to an "indirect proof" the formal proof of validity introduced earlier may be called a "direct proof." In a mathematical proof, a direct proof or an indirect proof may be used. The choice of the method of proof for a given mathematical argument depends on taste and convenience.

논증에 대한 타당성의 형식적밝힘으로써 또다른 방법이 있다. 곧 **간접증명** 또는 배리법 등이 그것이다.

그런데 간접증명법은 논증의 결론의 부정을 하나의 새로운 전제로 세워 주어진 가정에 덧붙임으로써 모순을 이끌어내는 방법을 말한다. 이 경우 일단 모순이 유도되자마자 소기의 증명은 끝이 난다.

예제 11 간접증명법으로 다음 논증의 타당성을 밝혀라.

$$p \vee q \to r$$
$$s \to p \wedge u$$
$$q \vee s / \text{그러므로} \quad r$$

[증명]

1. $p \vee q \to r$	
2. $s \to p \wedge u$	
3. $q \vee s /$ 그러므로 r	
4. $\sim r$	간접증명(결론 부정)
5. $\sim(p \vee q)$	1·4 및 부정식 삼단논법
6. $\sim p \wedge \sim q$	5에 대한 드 모르간의 법칙
7. $\sim p$	6에 대한 단순화 법칙
8. $\sim q$	6에 대한 단순화 법칙
9. s	3·8 및 논리합의 삼단논법
10. $p \wedge u$	2·9 및 긍정식 삼단논법
11. p	10에 대한 단순화법칙
12. $p \wedge \sim p$	7·11 및 논리곱

위 명제 $p \wedge \sim p$(단계 12)는 모순이므로 우리가 밝히려했던 간접증명은 끝을 맺는다. 한편 간접증명에 반하여 앞에서 배운 논증의 타당성의 밝힘은 **직접증명**이라고 부른다.

대체로 수학이나 논리에서 다룰 증명에서는 직접증명 또는 간접증명을 활용할 수 있지만 그 선택은 주로 편리함에 달려 있다.*

* 예를 들어 조건문 $p \to q$의 증명이 까다로울 때 그것의 대우 또는 배리법(29쪽 및 35쪽)을 이용하는 쪽이 훨씬 쉬울 때가 있다. 이 시점에서 그런 경우를 독자는 찾아낼 수 있을까?

Exercise 1.7

For each of the following arguments give both a direct proof and an indirect proof of validity, and compare their lengths.

1. $A \vee (B \wedge C)$
$B \to D$
$C \to E$
$D \wedge E \to A \vee C$
$\sim A / \therefore\ C$

2. $B \vee (C \to E)$
$B \to D$
$\sim D \to (E \to A)$
$\sim D / \therefore\ C \to A$

3. $(A \vee B) \to (A \to D \wedge E)$
$A \wedge C / \therefore\ E \vee F$

4. $A \vee B$
$\sim B \vee C / \therefore\ A \vee C$

5. $B \vee C \to B \wedge A$
$\sim B / \therefore\ \sim C$

6. $A \wedge B \to C$
$(A \to C) \to D$
$\sim B \vee E / \therefore\ B \to D \wedge E$

In the proofs of the following arguments, use the suggested symbols.

7. If the population increases rapidly and production remains constant, then prices rise. If prices rise, then the government will control prices. If I am rich, then I do not care about increase in prices. It is not true that I am not rich. Either the government does not control prices or I do care about increase in prices. Therefore, it is not true that the population increases rapidly and production remains constant. (P : The population increases rapidly. C : Production remains constant. R : Prices rise. G : The government will control prices. H : I am rich. I : I care about increase in prices.)

8. If Winston or Halbert wins, then Luke and Susan cry. Susan does not cry. Therefore, Halbert did not win. (W : Winston wins. H : Halbert wins. L : Luke cries. S : Susan cries.)

연습문제 1.7*

다음 각 논증에 대한 직접증명과 간접증명을 하고 두 단계의 길이를 비교하여라.

1. $A \vee (B \wedge C)$
$B \rightarrow D$
$C \rightarrow E$
$D \wedge E \rightarrow A \vee C$
$\sim A$/ 그러므로 C

2. $B \vee (C \rightarrow E)$
$B \rightarrow D$
$\sim D \rightarrow (E \rightarrow A)$
$\sim D$/ 그러므로 $C \rightarrow A$

3. $(A \vee B) \rightarrow (A \rightarrow D \wedge E)$
$A \wedge C$/ 그러므로 $E \vee F$

4. $A \vee B$
$\sim B \vee C$/ 그러므로 $A \vee C$

5. $B \vee C \rightarrow B \wedge A$
$\sim D$/ 그러므로 $\sim C$

6. $A \wedge B \rightarrow C$
$(A \rightarrow C) \rightarrow D$
$\sim B \vee E$/ 그러므로 $B \rightarrow D \wedge E$

다음 소괄호 안의 기호를 이용하여 논증 7·8·9를 밝혀보아라.

7. 인구가 급속히 증가하고 생신량이 머물러 있으면 물가는 오른다. 물가가 오르면 정부가 통제할 것이다. 내 재산이 10억 원 이상이면 물가상승을 걱정하지 않을 것이다. 내 재산이 10억 원 미만이라 함은 참이 아니다. 정부가 물가상승을 통제하지 않거나, 나는 물가상승을 걱정한다. 그러므로 인구는 급속히 증가하고 생산량은 그대로 머물러 있다는 것은 참이 아니다(P: 인구가 급속히 증가한다. C: 생산량은 머물러 있다. R: 물가가 오른다. G: 정부가 물가를 통제한다. H: 내 재산은 10억 원 이상이다. I: 나는 물가상승을 걱정한다).

8. 오늘 탁구시합에서 Winston 또는 Halbert가 이긴다면 Luke와 Susan은 소리칠 것이다. Susan은 소리치지 않는다. 그러므로 Winston은 이기지 못한다(W: Winston이 이긴다. H: Halbert가 이긴다. L: Luke가 소리친다. S: Susan이 소리친다).

---

* 수학에서 다루는 증명방법 중 수학적 귀납법(61쪽), 초한귀납법의 원리(321쪽)가 인용되는 경우도 있다. 아무튼 논증의 밝힘에 있어서 많은 문제를 스스로 풀어 볼수록 그 길이 쉽게 엿보인다. 그러기에 일찍이 "기하학에는 왕도(王道)가 없다."라고 하지 않던가.

9. If I enroll in this course and study hard, then I make good grades. If I make good grades, then I am happy. I am not happy. Therefore, either I did not enroll in this course or I did not study hard. (E : I enroll in this course. S : I study hard. G : I make good grades. H : I am happy.)

1.8 Mathematical Induction

Another method of proof that is very useful in proving the validity of a mathematical statement $p(n)$ involving the natural number n is the following principle of mathematical induction.

Mathematical Induction If $p(n)$ is a statement involving the natural number n such that

(1) $p(1)$ is true, and

(2) $p(k) \Rightarrow p(k+1)$ for any arbitary natural number k, then $p(n)$ is true for every natural number n.

The preceding principle is a consequence of one of Peano's Axioms for the natural numbers, which are included in the Appendix for reference.

In order to apply the principle of mathematical induction to prove a theorem, the theorem must be capable of being broken down into cases, one case for each natural number. Then, we must verify both conditions (1) and (2). The verification of (1), that usually is easy, assures us that the theorem is true for at least the case $n=1$. To verify the condition (2), we must prove an auxiliary theorem whose premise is "$p(k)$ is true" and whose conclusion is "$p(k+1)$ is true." The premise "$p(k)$ is true" is called the induction hypothesis.

Example 12 Prove by mathematical induction that

$$1+2+3+\cdots+n=\frac{n(n+1)}{2} \qquad \text{for every natural number } n.$$

9. 내가 이 과목을 신청하여 매일 2시간씩 공부하면 A학점을 받을 것이다. 내가 A학점을 받으면 장학금을 받을 것이다. 나는 장학금을 받지 못하고 있다. 그러므로 나는 이 과목을 신청하지 않았거나 매일 2시간씩 공부하지 않았다(E : 나는 이 과목을 신청한다. S : 나는 이 과목을 매일 2시간씩 공부한다. G : 나는 A학점을 받는다. H : 나는 장학금을 받는다).

1.8 수학적 귀납법

논증의 타당성을 밝히는 또다른 방법으로는 자연수 n에 관한 명제(즉 명제함수 : 3쪽 * 참조) $p(n)$의 타당성을 증명하는 데 쓰이는 수학적 귀납법의 원리란 것이 있다.

수학적 귀납법 자연수 n에 관한 명제함수 $p(n)$*에 대하여

(1) $n=1$일 때 $p(1)$이 참이다.

(2) 임의의 자연수 k에 대하여 $p(k) \Longrightarrow p(k+1)$

이들 두 가정하에 모든 자연수 n에 대한 $p(n)$은 참이다.

이 원리는 자연수에 관한 페아노의 공리들 중 다섯 번째인 것으로서 그 공리들은 이 책의 부록에 실려 있다(367쪽 참조).

지금 수학적 귀납법의 원리를 적용하여 어느 정리의 증명을 시도할 때 우선 그 정리는 두 전제 (1), (2)를 모두 만족하는지 확인할 필요가 있다. 그러려면 (1)의 확인을 통하여 주어진 정리는 $n=1$의 경우 참임을 판단할 수 있어야 한다. 한편 (2)를 밝히려면 하나의 보조정리로써 "$n=k$에 대하여 $p(k)$는 참이다."라는 가정하에 "$p(k+1)$는 참이다."를 유도할 수 있음을 보여야 한다. 여기서 가정 "$p(k)$는 참이다."를 귀납적 가정이라고 한다.

예제 12 임의의 자연수 n에 대한 다음 등식이 성립한다. 이것을 수학적 귀납법으로 증명하여라.

$$1+2+3+\cdots+n=\frac{n(n+1)}{2} \qquad \cdots\cdots ①^{**}$$

* 새삼 언급이 없는한 $p(n)$에 대한 전체집합은 자연수의 집합 N과 같다.

** 특히 $n=19$인 경우 ①의 좌변을 직접 구함으로써 등식 ①이 성립하는지 확인하여라. 사실 ①은 모든 자연수 n에 대해서도 성립한다는데에 그 뜻이 있다.

[Proof] Here, $p(n)$ represents the statement

$$\text{“ } 1+2+3+\cdots+n=\frac{n(n+1)}{2} \text{ ”}$$

In particular, $p(1)$ represents " $1=(1\cdot2)/2$ " which obviously is a true statement. Therefore, condition (1) for mathematical induction is satisfied.

To prove that condition (2) is satisfied, we assume that $p(k)$, which is " $1+2+3+\cdots+k=k(k+1)/2$, " is true. Then, add $k+1$ to both sides of this equality. We have

$$\begin{aligned} 1+2+3+\cdots+k+(k+1) &= \frac{k(k+1)}{2}+(k+1) \\ &= \frac{k(k+1)}{2}+\frac{2(k+1)}{2} \\ &= \frac{(k+2)(k+1)}{2} \\ &= \frac{(k+1)(k+2)}{2} \end{aligned}$$

which shows that $p(k+1)$ is true. We have now shown that the conditions (1) and (2) of mathematical induction are satisfied. Therefore, by the principle of mathematical induction, $1+2+3+\cdots+n=n(n+1)/2$ is true for every natural number n.

The idea of mathematical induction may be used in making definitions involving natural numbers. For example, the definition of powers of an unknown number x may be defined by:

$$x^1=x, \quad \cdots \quad x^{n+1}=x^n\cdot x, \qquad \text{for any natural number } n$$

The two equations above indicate that

$$x^1=x, \quad x^2=x\cdot x, \quad x^3=x^2\cdot x, \cdots$$

and so forth. As another application we give the following inductive definition of the symbol $\boldsymbol{C(n, r)}$.

[증명] 임의의 자연수 n에 관한 등식 ①을 $p(n)$으로 놓는다.

(1) $n=1$일 때 $p(1)$, 즉 ①의 좌변은 1, 우변은 $\dfrac{1(1+1)}{2}=1$

그러므로 $1=\dfrac{1(1+1)}{2}$. 따라서 $p(1)$은 참이다.

(2) $n=k$에 대하여 $p(k)$, 즉 $1+2+3+\cdots+k=\dfrac{k(k+1)}{2}$이 참이라고 가정한다. 지금 이 등식의 양변에 $k+1$을 더하면

$$\begin{aligned}1+2+3+\cdots+k+(k+1)&=\frac{k(k+1)}{2}+(k+1)\\&=\frac{k(k+1)}{2}+\frac{2(k+1)}{2}\\&=\frac{(k+1)\{(k+1)+1\}}{2}\end{aligned}$$

곧 $$1+2+3+\cdots+k+(k+1)=\frac{(k+1)(k+2)}{2}$$

그러므로 $p(k+1)$은 참이다. 따라서 수학적 귀납법에서의 전제 (1), (2)는 모두 성립한다.

이상으로 수학적 귀납법에 따라 임의의 자연수 n에 관한 등식 ①이 성립한다.

수학적 귀납법은 자연수에 관련된 정의를 내리는 데에도 이용된다. 다음은 우리에게 익숙한 문자 x에 관한 **거듭제곱**의 정의이다.

$$x^1=x, \qquad \text{임의의 자연수 } n\text{에 대하여} \qquad x^{n+1}=x^n\cdot x$$

위의 두 등식은 다름아닌

$$x^1=x, \quad x^2=x\cdot x, \quad x^3=x^2\cdot x, \ \cdots \qquad\cdots\cdots ②$$

와 같은 사실을 요약해 놓은 것이다.

또다른 응용으로써 기호 $\boldsymbol{C(n, r)}$에 대한 귀납적 정의가 있다.

Definition 7. Let n be a natural number and r an integer. The symbol $C(n, r)$ is defined by

$$C(0, 0)=1, \quad C(0, r)=0 \text{ for each } r \neq 0, \text{ and}$$
$$C(n+1, r)=C(n, r)+C(n, r-1)$$

Theorem 8. If n and r are integers such that $0 \leq r \leq n$, then

$$C(n, r)=\frac{n!}{r!(n-r)!}$$

where $n!$ denotes the product $n \cdot (n-1) \cdots 3 \cdot 2 \cdot 1$ of the first n consecutive natural numbers if $n>0$ and $0!=1$ by convention.

[Proof] Exercise.

Theorem 9. The Binomial Theorem If x and y are two variables and n is a natural number, then

$$(x+y)^n=C(n, 0)x^n+C(n, 1)x^{n-1}y+\cdots+C(n, r)x^{n-r}y^r$$
$$+\cdots+C(n, n)y^n$$

[Proof] We shall prove the validity of this theorem by mathematical induction. First, the theorem is clearly true for $n=1$. To complete the proof, we assume the validity of the theorem for $n=k$; that is, we assume that

$$(x+y)^k=C(k, 0)+x^k+C(k, 1)x^{k-1}y+\cdots+C(k, r)x^{k-r}y^r$$
$$+\cdots+C(k, k)y^k$$

Then, by multiplying both sides of the above equality by $(x+y)$, we have

정의 7 자연수 n과 정수 $r(0 \leqq r)$에 관한 기호 $C(n, r)$을 다음과 같이 정의한다.

$$C(n+1, r) = C(n, r) + C(n, r-1)$$

특히 $$C(0, 0) = 1, \quad C(0, r) = 0$$

정리 8 정수 n, $r(0 \leqq r \leqq n)$에 대하여 다음 등식이 성립한다.

$$C(n, r) = \frac{n!}{r!(n-r)!} \qquad \cdots\cdots ③$$

여기서 $n>0$일 때의 $n!$은 1부터 n까지의 잇따른 자연수의 곱, 즉

$$1 \cdot 2 \cdot 3 \cdots (n-1) \cdot n$$

을 나타낸다. 특히 $n=0$이면 $0!=1$인 것으로 약속한다.

[증명] 독자의 연습문제로 돌린다.

정리 9 이항정리 임의의 자연수 n과 두 문자 x, y에 대하여 다음 등식이 성립한다.

$$(x+y)^n = C(n, 0)x^n + C(n, 1)x^{n-1}y + \cdots + C(n, r)x^{n-r}y^r + \cdots + C(n, n)y^n \qquad \cdots\cdots ④^*$$

[증명] 이 정리를 수학적 귀납법으로 증명하자.

우선 $n=1$일 때 ④는 분명히 성립한다(과연?).

이제 $n=k$에 대하여 다음 등식이 성립한다고 가정하자.

$$(x+y)^k = C(k, 0) + x^k + C(k, 1)x^{k-1}y + \cdots + C(k, r)x^{k-r}y^r + \cdots + C(k, k)y^k$$

이 등식의 양변에 각각 $x+y$를 곱하면

* 등식 ④의 좌변을 정의역, 공역 모두 실수의 집합인 변수 x, y에 관한 함수 $f(x, y)$로 볼 수 있다. 변수 x에 관한 함수 $f(x)$에 대해서는 157-159쪽 참조.

$$
\begin{aligned}
(x+y)^{k+1} &= (x+y)\{x^k + C(k,1)x^{k-1}y + \cdots + C(k,r)x^{k-r}y^r \\
&\qquad + \cdots + y^k\} \\
&= x^{k+1} + \{C(k,0) + C(k,1)\}x^k y \\
&\qquad + \cdots \{C(k,r-1) + C(k,r)\}x^{(k+1)-r}y^r + \cdots + y^{k+1} \\
&= C(k+1,0)x^{k+1} + C(k+1,1)x^k y + \cdots \\
&\qquad + C(k+1,r)x^{(k+1)-r}y^r + \cdots + C(k+1,k+1)y^{k+1}
\end{aligned}
$$

which shows that the theorem is valid for $n=k+1$ if it is valid for $n=k$. Thus, by mathematical induction, the binomial theorem is true for all natural numbers n.

The numbers $C(n,r)$ in the binomial theorem are called binomial coefficients.

Exercise 1.8

1. Prove Theorem 8 by mathematical induction.

2. Show that $C(n,0)=1=C(n,n)$ for all natural numbers n.

Prove, by mathematical induction, that the following equalities 3 through 7 hold for all natural numbers n.

3. $1\cdot 2+2\cdot 3+\cdots+r\cdot(r+1)+\cdots+n(n+1)=\frac{1}{3}n(n+1)(n+2)$

4. $1^2+2^2+3^2+\cdots+n^2=\frac{1}{6}n(n+1)(2n+1)$

5. $1^3+2^3+3^3+\cdots+n^3=\frac{1}{4}n^2(n+1)^2$

6. $1+3+5+\cdots+(2n-1)=n^2$

7. $\frac{1}{1\cdot 2}+\frac{1}{2\cdot 3}+\frac{1}{3\cdot 4}+\cdots+\frac{1}{n(n+1)}=\frac{n}{n+1}$

8. Prove the following Generalized De Morgan's Laws.
(a) $\sim(p_1\wedge p_2\wedge\cdots\wedge p_n)\Longleftrightarrow\sim p_1\vee\sim p_2\vee\cdots\vee\sim p_n$
(b) $\sim(p_1\vee p_2\vee\cdots\vee p_n)\Longleftrightarrow\sim p_1\wedge\sim p_2\wedge\cdots\wedge\sim p_n$

9. Prove the following Generalized Distributive Laws.
(a) $p\wedge(q_1\vee q_2\vee\cdots\vee q_n)\Longleftrightarrow(p\wedge q_1)\vee(p\wedge q_2)\vee\cdots\vee(p\wedge q_n)$
(b) $p\vee(q_1\wedge q_2\wedge\cdots\wedge q_n)\Longleftrightarrow(p\vee q_1)\wedge(p\vee q_2)\wedge\cdots\wedge(p\vee q_n)$

$$\begin{aligned}(x+y)^{k+1} &= (x+y)\{x^k + C(k,1)x^{k-1}y + \cdots + C(k,r)x^{k-r}y^r + \cdots + y^k\} \\ &= x^{k+1} + \{C(k,0)+C(k,1)\}x^k y + \cdots \\ &\qquad + \{C(k,r-1)+C(k,r)\}x^{(k+1)-r}y^r + \cdots + y^{k+1} \\ &= C(k+1,0)x^{k+1} + C(k+1,1)x^k y + \cdots \\ &\qquad + C(k+1,r)x^{(k+1)-r}y^r + \cdots + C(k+1,k+1)y^{k+1}\end{aligned}$$

즉 $$\begin{aligned}(x+y)^{k+1} &= C(k+1,0)x^{k+1} + C(k+1,1)x^{(k+1)-1}y + \cdots \\ &\qquad + C(k+1,r)x^{(k+1)-r}y^r + \cdots + C(k+1,k+1)y^{k+1}\end{aligned}$$

그러므로 $n=k+1$에 대하여도 ④는 성립한다. 따라서 수학적 귀납법에 따라 임의의 자연수 n에 대한 등식 ④, 곧 이항정리를 얻는다.

여기서 $C(n, r)$을 **이항계수**라고 한다.

연습문제 1.8

1. 정리 8을 수학적 귀납법으로 증명하여라.

2. 각 자연수 n에 대하여 $C(n, 0)=C(n, n)=1$이 성립함을 보여라.

임의의 자연수 n에 대하여 다음 3−7이 성립함을 증명하여라.

3. $1\cdot2+2\cdot3+\cdots+n(n+1)=\dfrac{1}{3}n(n+1)(n+2)$

4. $1^2+2^2+3^2+\cdots+n^2=\dfrac{1}{6}n(n+1)(2n+1)$

5. $1^3+2^3+3^3+\cdots+n^3=\dfrac{1}{4}n^2(n+1)^2$

6. $1+3+5+\cdots+(2n-1)=n^2$

7. $\dfrac{1}{1\cdot2}+\dfrac{1}{2\cdot3}+\dfrac{1}{3\cdot4}+\cdots+\dfrac{1}{n(n+1)}=\dfrac{n}{n+1}$

8. 임의의 자연수 $n(2\leqq n)$에 대한 다음 명제는 드 모르간의 법칙을 확장한 것에 지나지 않는다(29쪽 정리 3). 이를 증명하여라.

(a) $\sim(p_1\wedge p_2\wedge\cdots\wedge p_n)\Longleftrightarrow\sim p_1\vee\sim p_2\vee\cdots\vee\sim p_n$

(b) $\sim(p_1\vee p_2\vee\cdots\vee p_n)\Longleftrightarrow\sim p_1\wedge\sim p_2\wedge\cdots\wedge\sim p_n$

9. 다음은 분배법칙[31쪽 정리(b)]을 확장한 것이다. 이를 증명하여라.

(a) $p\wedge(q_1\vee q_2\vee\cdots\vee q_n)\Longleftrightarrow(p\wedge q_1)\vee(p\wedge q_2)\vee\cdots\vee(p\wedge q_n)$

(b) $p\vee(q_1\wedge q_2\wedge\cdots\wedge q_n)\Longleftrightarrow(p\vee q_1)\wedge(p\vee q_2)\wedge\cdots\wedge(p\vee q_n)$

Chapter 2
The Concept of Sets

제 2 장
집합의 개념

칸토어(Georg Cantor : 1845~1918)
집합론의 창시자*

---

* 초보자에게는 부담이 될 수 있으므로 다음 책의 내용을 참고하기 바란다.
이장우, 《현대집합론의 원리》, 경문사, 1998, § 2.7 소박한 집합론의 배경과 러셀의 역리(69~75쪽).
한편 영문에 능한 독자에게 다음을 권한다.
Charles C. Pinter, *Set Theory*, Kyung Moon, 2002, Chapter 01. The background of the set theory(pp. 1~4).

2

The Concept of Sets

In this chapter we introduce the concept of sets, subsets, and set operations (union, intersection, and complementation) together with the fundamental rules governing these operations. These are developed in parallel with Chapter 1 on logic. Indexed families of sets are discussed. The chapter ends with the Russell Paradox and a historical remark.

2.1 Sets and Subsets

"What is a set?" is a very difficult question to answer.[1] In this elementary book, we shall not go into any complicated axiomatic approach to Set Theory, but shall content ourself to accept the following : a set is any collection into a whole of definite, distinguishable objects, called elements, of our intuition or thought. This intuitive definition of a set was first given by Georg Cantor (1845-1918), who originated the theory of sets in 1895. Examples :

(a) The set of all chairs in this classroom.*

(b) The set of all students in this university.*

(c) The set of letters a, b, c and d.

(d) The set of rules in our dormitories.

(e) The set of all rational numbers whose square is equal to 2.

[1] Students will realize the difficulty when they come to Section 7 and 8.

집합의 개념

2

여기서는 제1장에서 다루었던 논리와 병행하여 집합, 부분집합, 집합의 연산(합집합, 교집합, 여집합)의 뜻과 그것들 연산에 관한 기본성질을 소개한다. 한편 첨수집합족을 논한 다음 러셀의 역리와 역사적 단평으로 끝을 맺는다.

2.1 집합과 부분집합

대체로 "무엇이 집합인가?"란 물음에 답하기는 쉽지 않다.[1] 집합론의 기초를 다룰 이 책에서 공리적 방법의 이론전개보다는 다음과 같은 칸토어*(1845-1918)가 착상한 직관적 정의를 받아들이는 것으로 만족한다.

"우리의 직관 또는 사고의 대상으로서 서로 뚜렷이 구분되는 원소의 모임을 집합이라고 한다". 다음의 모임은 모두 집합을 나타낸다.

(a) 서강대학교 R관 5호 강의실 의자의 모임**

(b) 지구상의 국가**

(c) 문자 a, b, c, d

(d) 철수의 기숙사의 규칙

(e) 제곱하면 2와 같은 유리수

---

[1] 그 어려움을 독자는 2.7절(121쪽)과 2.8절(123쪽)을 공부할 때 깨달을 것이다.

* 일본의 수학자로서 1970년 노벨상에 버금가는 필즈상을 탄 바 있는 히로나가 헤이스케(広中平祐 1931~)는 그의 저서《학문의 발견(学問の発見)》(Kosei 출판사, 2001)에서 다음과 같이 적고 있다. 집합론의 창시자로서 유명한 독일의 수학자 칸토어는 "수학의 본질이 자유성(*freedom*)에 있다"라고 하였다. 이 저자는 근래 서울대학교 개원교수를 지낸 바 있다.

** 원서 (a)의 "이 교실", (b)의 "이 대학"은 모두 모호한 표현이므로 각각 위와 같이 바꾸어 보았거니와 앞으로도 이런 수정이 가끔 있을 것이다.

(f) The set of all natural numbers.

(g) The set of all real numbers between 0 and 1.

A set which contains only finitely many elements is called a finite set* ; an infinite set is one which is not a finite set. Examples (a) to (e) above are all finite sets, and Examples (f) and (g) are infinite sets.

Sets are frequently designated by enclosing symbols representing their elements in braces when it is possible to do so. Thus, the set in Example (c) is $\{a, b, c, d\}$ and the set in Example (f) may be denoted by $\{1, 2, 3, \cdots\}$. The set described in Example (e) has no elements; such a set is called the empty set, which will be denoted by the symbol ϕ.

We shall also use capital letters to denote sets, and lower-case letters to denote elements. If a is an element of the set A, we write $a \in A$ (read : "a is an element of A" or "a belongs to A," whereas $b \notin A$ means that b is not an element of A.

Definition 1. Two sets A and B are said to be equal or identical, in symbols : $\boldsymbol{A = B}$, provided that they contain the same elements. That is, $A = B$ means that

$$\forall_x (x \in A \Leftrightarrow x \in B)$$

The order of appearance of the elements of a set is of no consequence. Thus, the set $\{a, b, c\}$ is the same as $\{b, c, a\}$ or $\{c, b, a\}$, etc. Furthermore, since elements in a set are distinct, $\{a, a, b\}$, for example, is not a proper notation of a set and should be replaced by $\{a, b\}$. If a is an element of a set, a and $\{a\}$ are to be considered different, that is, $a \neq \{a\}$. For, $\{a\}$ denotes the *set* consisting of the single element a alone, whereas a is just the *element* in the set $\{a\}$.

(f) 자연수 (g) 0과 1 사이의 실수

집합의 종류를 구분짓는 기준은 여럿 있지만 우선 다음부터 알아보자.

원소의 개수를 자연수로 나타낼 수 있는 집합을 **유한집합***이라 하고, 유한집합이 아닌 집합을 **무한집합**이라고 한다.

위의 보기 중 (a)−(e)는 모두 유한집합이고, (f), (g)는 무한집합이다(과연?). 흔히 집합은 원소를 나타내는 기호를 중괄호로 묶어 나타낸다. 이를테면 위 보기 (c), (f)를 각각 $\{a, b, c, d\}$, $\{1, 2, 3, \cdots\}$와 같이 나타낸다. 특히 (e)와 같이 원소가 하나도 존재하지 않는 것 조차 **공집합****이라 하고 이것을 ϕ로 나타내며 유한집합으로 여긴다(여기서 다룰 집합은 새삼 언급이 없는 한 공집합 ϕ이 아닌 것으로 생각한다).

앞으로 집합은 대문자로, 원소는 소문자로 나타내며 a가 집합 A의 원소이면 $a \in A$(이것을 "a는 A의 원소이다." 또는 "a는 A에 속한다."라고 읽는다)와 같이 나타내고 b가 A의 원소가 아니면 $b \notin A$로 나타낸다.

정의 1 두 집합 A, B에 대하여 A의 원소와 B의 원소가 같을 때 A와 B는 **같다**고 하고 이것을 $\boldsymbol{A=B}$와 같이 나타낸다. 즉 $A=B$는

$$\forall_x (x \in A \Longleftrightarrow x \in B)$$

임을 뜻한다. 한편 A와 B가 같지 않을 때는 $\boldsymbol{A \neq B}$로 나타낸다.

집합의 원소를 표시할 때 그 순서는 개의하지 않는다. 다시 말해서 집합 $\{a, b, c\}$는 $\{b, c, a\}$ 또는 $\{c, b, a\}$와 같다. 그러나 집합의 원소는 서로 구분되어야 한다. 이를테면, $\{a, a, b\}$에서 원소 a의 중복을 피하여 이 집합을 $\{a, b\}$로 나타내도록 한다. 한편 집합 $\{a\}$는 그 원소 a와 구분지어야 한다. 즉 $\{a\}$는 a만의 원소의 집합인데 반하여 a는 바로 그 집합의 원소이다***.

* "유한이 많은 원소"에서와 같이 "유한"이란 용어를 빌려 유한집합의 정의를 내린 원서의 표현은 적절하지 못하다. 더 엄밀한 정의는 4.1절 정의 1(205쪽) 참조.

** 97쪽 정리 5(b), (c) 참조.

*** 원서의 $a \neq \{a\}$와 같은 표현은 피하도록 하자. 왜냐하면 부등호 $\neq$는 원소와 원소, 집합과 집합 사이의 관계를 나타낼 때 의미가 있지만 원소와 집합 사이에서는 그렇지 못하기 때문이다.

Definition 2. Let A and B be sets. If every element of A is an element of B, then A is called a subset of B, in symbols: $A \subseteq B$ or $B \supseteq A$. If A is a subset of B, then B is called a superset of A. Thus, logically speaking,

$$A \subseteq B \equiv \forall_x (x \in A \Rightarrow x \in B)$$

Obviously, every set is a subset (and a superset) of itself. When $A \subseteq B$ and $A \neq B$, we write $A \subset B$, or $B \supset A$, which reads: A is a proper subset of B, or B is a proper superset of A. In other words, A is a proper subset of B provided that every element of A is an element of B, and there exists an element of B which is not an element of A. If A is not a subset of B, we write $A \nsubseteq B$.

Theorem 1. The empty set ϕ is a subset of every set.

[Proof] Let A be any set. We are to prove that the conditional statement

$$x \in \phi \rightarrow x \in A$$

is true for every x. Since the empty set ϕ has no elements, the statement "$x \in \phi$" is false, whereas "$x \in A$" may be true or may be false. In either case, the conditional statement $x \in \phi \rightarrow x \in A$ is true according to the truth table for the conditional (case 3 and 4 of 5, Chapter 1). Thus, $\phi \subseteq A$ for any set A.

Theorem 2. If $A \subseteq B$ and $B \subseteq C$, then $A \subseteq C$.

[Proof] We are to show that $x \in A \Rightarrow x \in C$:

$$\begin{aligned} x \in A &\Rightarrow x \in B, \quad \text{because } A \subseteq B \\ &\Rightarrow x \in C, \quad \text{because } B \subseteq C \end{aligned}$$

Hence, by the Transitive Law(Theorem 4(c) of Chapter 1), we have $x \in A \Rightarrow x \in C$. Thus, we have proved that $A \subseteq C$.

정의 2 임의의 집합 A, B에 대하여 A의 모든 원소가 B의 원소일 때 A는 B의 **부분집합**, B는 A의 **초집합** 또는 **포함집합**이라 하고 이것을 $\boldsymbol{A \subseteq B}$* 또는 $\boldsymbol{B \supseteq A}$와 같이 나타낸다. 이 사실을 다음과 같이 풀어 적을 수 있다.

$$A \subseteq B \equiv \forall_x (x \in A \Rightarrow x \in B)$$

임의의 집합은 분명히 그 자신의 부분집합(이고 초집합)이다. 두 집합 A, B에 대하여 $A \subseteq B$이고 $A \neq B$일 때 A를 B의 **진부분집합**, B는 A의 **진포함집합**이라 하고 이것을 $A \subset B$ 또는 $B \supset A$와 같이 나타낸다. 다시 말하면 A의 모든 원소는 B의 원소이지만 B의 원소 중에는 A의 원소가 아닌 것이 존재할 때 A는 B의 진부분집합이다. 한편 A가 B의 부분집합이 아닐 때 이것을 $A \not\subseteq B$로 나타낸다.

정리 1** 공집합 ϕ는 임의의 집합의 부분집합이다.

[증명] 임의의 집합을 A라 놓은 다음 전칭명제가 참인 것을 밝히면 된다.

$$\forall_x (x \in \phi \rightarrow x \in A)$$

공집합 ϕ에는 원소가 존재하지 않으므로 임의의 x에 대하여 명제함수 "$x \in \phi$"***는 거짓인 반면에 명제 "$x \in A$"***는 참 아니면 거짓이다. 그 어느 경우에도 조건문 "$x \in \phi \rightarrow x \in A$"는 참이다[제 1장 표 5: 경우 3, 4(19쪽) 참조]. 그러므로 임의의 집합 A에 대하여 $\phi \subseteq A$.

정리 2 임의의 집합 A, B, C에 대하여 $A \subseteq B \wedge B \subseteq C$이면 $A \subseteq C$

[증명] $\forall_x (x \in A \Rightarrow x \in C)$인 것을 밝히면 충분하다. 가정으로부터 $A \subseteq B$이므로 $\forall_x (x \in A \Rightarrow x \in B)$. 마찬가지로 $\forall_x (x \in B \Rightarrow x \in C)$.
그러므로 $\Rightarrow$에 관한 추이율[제 1장 정리 4(c)]에 따라

$$\forall_x (x \in A \Rightarrow x \in C) \quad \text{즉} \quad A \subseteq C.$$

---

* $A \subseteq B$는 $(A \subset B) \vee (A = B)$를 간단히 나타낸 것이다. 오늘날 기호 $A \subseteq B$ 대신 $A \subset B$를 흔히 사용하지만 혼돈의 여지가 있어 이 책에서는 그대로 $A \subseteq B$를 고수하기로 한다.

** 이 정리를 배리법으로 증명하여 보아라. 매우 간단할 것이다.

*** 여기서의 명제함수의 전체집합은 A 자신이다.

Exercise 2.1

1. Show that the set of letters needed to spell "cataract" and the set of letters needed to spell " tract " are equal.

2. Decide, among the following sets, which are subsets of which.
   (a) $A=\{$all real numbers satisfying $x^2-8x+12=0\}$
   (b) $B=\{2, 4, 6\}$ (c) $C=\{2, 4, 6, 8, \cdots\}$ (d) $D=\{6\}$

3. List all the subsets of the set $\{-1, 0, 1\}$.

4. Prove that $(A\subseteq B)\wedge(B\subseteq A)\Leftrightarrow A=B$.
   [*Remark*: Frequently in mathematics the best way to show that $A=B$ is to show that $A\subseteq B$ and $B\subseteq A$.]

5. Prove that $A\subseteq\phi\Rightarrow A=\phi$.

6. Prove that
   (a) $(A\subset B)\wedge(B\subseteq C)\Rightarrow A\subset C$ (b) $(A\subseteq B)\wedge(B\subset C)\Rightarrow A\subset C$

7. Give an example of a set whose elements are themselves sets.

8. In each of the following, determine whether the statement is true or false. If it is true, prove it. If it is false, disprove it by an example (such an example which disproves a statement is called a counterexample).
   (a) If $x\in A$ and $A\in B$, then $x\in B$.
   (b) If $A\subseteq B$ and $B\in C$, then $A\in C$.
   (c) If $A\nsubseteq B$ and $B\subseteq C$, then $A\nsubseteq C$.
   (d) If $A\nsubseteq B$ and $B\nsubseteq C$, then $A\nsubseteq C$.
   (e) If $x\in A$ and $A\nsubseteq B$, then $x\notin B$.
   (f) If $A\subseteq B$ and $x\notin B$, then $x\notin A$.

9. Given a set with n elements, prove that there are exactly $C(n, r)$ subsets with r elements.

연습문제 2.1

1. 두 단어 cataract, tract를 각각 적는 데 필요한 글자의 집합은 같다. 이것을 확인하여라.

2. 다음 집합 (a)−(d) 중 어느 것이 어느 것의 부분집합인가를 지적하여라.

(a) $A=\{$이차방정식 $x^2-8x+12=0$의 해$\}$ (b) $B=\{2, 4, 6\}$

(c) $C=\{2, 4, 6, 8, \cdots\}$ (d) $D=\{6\}$

3. 집합 $\{-1, 0, 1\}$의 부분집합을 모두 열거하여라.

4.* 임의의 집합 A, B에 대하여 $(A\subseteq B)\wedge(B\subseteq A)\Leftrightarrow A=B$임을 증명하여라(주의: $A=B$인 것을 밝히려면 정의 1를 만족하는가를 확인해야 한다).

5. 집합 A에 대하여 $A\subseteq\phi\Rightarrow A=\phi$임을 증명하여라.

6. 집합 A, B, C에 대하여 다음을 증명하여라.

(a) $(A\subset B)\wedge(B\subseteq C)\Rightarrow A\subset C$ (b) $(A\subseteq B)\wedge(B\subset C)\Rightarrow A\subset C$

7. 집합의 원소로서 집합인 것도 있다. 그런 것의 보기 둘을 들어라.

8. 다음 명제가 참인지 거짓인지를 가려라. 그것이 참이면 증명을 하고 거짓이면 그에 따른 보기를 들어 반증하여라(명제를 반증하는 보기를 그 명제에 대한 **반례****라고 한다).

(a) $x\in A\wedge A\in B\Rightarrow x\in B$ (b) $A\subseteq B\wedge B\in C\Rightarrow A\in C$

(c) $A\not\subseteq B\wedge B\subseteq C\Rightarrow A\not\subseteq C$ (d) $A\not\subseteq B\wedge B\not\subseteq C\Rightarrow A\not\subseteq C$

(e) $x\in A\wedge A\not\subseteq B\Rightarrow x\notin B$ (f) $A\subseteq B\wedge x\notin B\Rightarrow x\notin A$

9.*** 원소가 n개인 집합의 부분집합으로서 r개의 원소인 것의 수는 $C(n, r)$와 같다. 이를 증명하여라.

---

* 책에 따라서는 문제 4의 주장을 상등관계 "$A=B$"에 대한 정의로 삼는다. 그럴 경우 정의 1(75쪽)은 정리 아니면 문제의 구실을 한다.

** 반례는 많이 존재하더라도 그 중 하나만 들면 충분하다.

*** 수학적 귀납법(61쪽)을 인용하여라.

2.2 Specification of Sets

One way of making a new set out of a given set is to specify those elements of the given set that satisfy a particular property. For example, let A be the set of all students in this university. The statement "x is female" is true for some element x of A and false for others. We shall use the notation

$$\{x \in A \mid x \text{ is female}\}$$

to specify the set of all female students in this university. Similarly,

$$\{x \in A \mid x \text{ is not female}\}$$

specifies the set of all male students in this university.

As a rule, to every set A and to every statement $p(x)$ about $x \in A$, there exists a set $\{x \in A \mid p(x)\}$ whose elements are precisely those elements x of A for which the statement $p(x)$ is true. In an axiomatic approach to set theory, this rule is usually postulated as an axiom, called the Axiom of Specification. The symbol $\{x \in A \mid p(x)\}$ reads: the set of all x in A such that $p(x)$ is true. The notation of the form $\{x \in A \mid p(x)\}$ which describes a set is called the set builder notation.

Example 1 Let $\mathbb{R}$ denote the set of all real numbers. Then

(a) $\{x \in \mathbb{R} \mid x = x + 1\}$ is equal to the empty set.
(b) $\{x \in \mathbb{R} \mid 2x^2 - 5x - 3 = 0\}$ is equal to the set $\{-1/2, 3\}$.
(c) $\{x \in \mathbb{R} \mid x^2 + 1 = 0\}$ is equal to the empty set.

Because of their frequent appearance throughout the remainder of this book and in other topics of mathematics, the following special symbols will be reserved for the sets described:

2.2 집합의 규정

주어진 집합으로부터 새로운 집합을 구성할 수 있는 방법으로는 그 집합의 원소로서 같은 성질(조건)을 지닌 것을 한데 모으는 일이다.

이를테면 H대학교의 2010학년도 1학기 학생들의 집합을 A라고 할 때 명제함수 "x는 여학생이다."는 A의 일부분을 이루는 원소 x에 대하여 참이지만 그 나머지 부분의 원소(남학생)에 대해서는 거짓이다.

이러한 여학생, 남학생의 집합을 각각 다음과 같이 나타낸다.

$$\{x \in A \mid x\text{는 여학생이다}\}, \quad \{x \in A \mid x\text{는 여학생이 아니다}\}$$

대체로 우리는 집합 A의 임의의 원소와 $\forall x \in A$에 관한 명제함수 $p(x)$에 대하여 A의 원소로서 $p(x)$가 참인 것으로만 이루어진 집합

$$\{x \in A \mid p(x)\text{는 참이다}\} \qquad \cdots\cdots ①$$

이 존재하는 것(①이 공집합 ϕ과 같을 수도 있지만)으로 받아들인다. 사실 집합론을 공리적으로 다루려면 이런 전제의 **분출공리**를 내세워야 한다. 여기서 ①과 같은 꼴의 표기를 집합의 **조건제시법**[$p(x)$를 집합 A의 하나의 조건으로 봄]이라고 한다.

예제 1 실수의 집합 $\mathbb{R}$에서의 다음 등식이 성립한다. 이를 밝혀라.

(a) $\{x \in \mathbb{R} \mid \text{일차방정식 } x = x+1\text{의 해}\} = \phi$

(b) $\{x \in \mathbb{R} \mid \text{이차방정식 } 2x^2 - 5x - 3 = 0\text{의 해}\} = \left\{-\frac{1}{2}, 3\right\}$

(c) $\{x \in \mathbb{R} \mid \text{이차방정식 } x^2 + 1 = 0\text{의 해}\} = \phi$

[풀이] (a) $x = x+1$의 해가 존재한다면 $0=1$(왜?) 이것은 모순이므로 해는 존재하지 않는다. 따라서 (a)가 성립한다.

(b) $2x^2 - 5x - 3 = 0$을 직접 풀어 해를 구한다.

(c) $\forall_{x \in \mathbb{R}}$일 때 $0 \leq x^2$ 한편 $0 < 1$. 그러므로 $0 < x^2 + 1$.
이처럼 $x^2 + 1 = 0$의 해는 존재하지 않으므로 (c)가 성립한다.

다음은 우리에게 낯익은 수의 집합을 특별한 기호로 나타낸 것으로서 앞으로 곳곳에서 인용되고 있다.

$\mathbb{R}=\{x \mid x \text{ is a real number}\}$	$\mathbb{Q}=\{x \mid x \text{ is a rational number}\}$
$\mathbb{Z}=\{x \mid x \text{ is an integer}\}$	$\mathbb{I}=\{x\in\mathbb{R} \mid 0\leqq x\leqq 1\}$
$\mathbb{N}=\{n \mid n \text{ is a natural number}\}$	$\mathbb{R}_+=\{x\in\mathbb{R} \mid x>0\}$

It should be noticed that $\mathbb{N}\subset\mathbb{Z}\subset\mathbb{Q}\subset\mathbb{R}$ and $\mathbb{N}\subset\mathbb{R}_+\subset\mathbb{R}$.

It is quite possible that elements of a set may themselves be sets. For example, the set of all subsets of a given set A has sets as its elements. This set is called the power set[2] of A and is denoted by $\mathscr{P}(A)$.

Example 2 $\mathscr{P}(\{a\})=\{\phi, \{a\}\}$, $\mathscr{P}(\phi)=\{\phi\}$, and
$\mathscr{P}(\{a, b\})=\{\phi, \{a\}, \{b\}, \{a, b\}\}$.

The name "power set" is motivated by the following theorem.

Theorem 3. If a set A consists of n elements, then its power set $\mathscr{P}(A)$ contains exactly 2^n elements.

[Proof] The theorem is clearly true for $A=\phi$. For a nonempty set A, we let $A=\{a_1, a_2, a_3, \cdots, a_n\}$. Given an element a_k of A, each subset of A has two possibilities: it either contains a_k or it does not. Therefore, the problem of finding the number of subsets of A may be considered as the problem of filling a list of n blank spaces □□□⋯□ at random with the numbers 0 and 1, one number in each space. Each filling of the n blanks determines a subset X of A in the following manner: $a_k\in X$ if and only if 1 appears in the k-th space. Since there are exactly 2^n different such fillings, there are exactly 2^n subsets of A.

---

[2] In an axiomatic set theory, the existence of the power set is not taken for granted. Since the existence of a power set does not follow from the axiom of specification, a new axiom is needed; this axiom is usually called the Axiom of Power Sets and may be stated:

For each set there exists a set of sets that consists of all the subsets of the given set.

$$\mathbb{R}=\{x \mid x\text{는 실수이다}\}, \qquad \mathbb{Q}=\{x \mid x\text{는 유리수이다}\},$$
$$\mathbb{Z}=\{x \mid x\text{는 정수이다}\}, \qquad \mathbb{N}=\{x \mid x\text{는 자연수이다}\},$$
$$\mathbb{I}=\{x\in\mathbb{R} \mid 0\leqq x\leqq 1\},^{*} \qquad \mathbb{R}_{+}=\{x\in\mathbb{R} \mid 0<x\}$$

여기서 $\mathbb{N}\subset\mathbb{Z}\subset\mathbb{Q}\subset\mathbb{R}$**이고 $\mathbb{N}\subset\mathbb{R}_{+}\subset\mathbb{R}$**인 것에 유의하자.

집합 중에는 그 원소가 또한 집합인 것(연습문제 2.1 문제 7)이 있다. 특히 주어진 집합 A의 모든 부분집합이 원소인 집합이 존재한다. 그런 집합을 A의 **멱집합**[2]이라 하고 이것을 $\mathscr{P}(A)$와 같이 나타낸다.

예제 2 $\mathscr{P}(\{a\})=\{\phi, \{a\}\}$, $\mathscr{P}(\phi)=\{\phi\}$, $\mathscr{P}(\{a, b\})=\{\phi, \{a\}, \{b\}, \{a, b\}\}$

멱집합이란 이름은 다음 정리로부터 유래된 것이다.

정리 3 원소가 n개인 집합 A의 멱집합 $\mathscr{P}(A)$의 원소의 수는 2^n와 같다.

[증명] $A=\phi$인 경우 분명히 정리 3은 참이다(과연?).

$A\neq\phi$인 경우 $A=\{a_1, a_2, \cdots, a_n\}$으로 놓는다면 A의 하나의 원소 a_k에 대하여 A의 각 부분집합으로서는 두 가지의 가능성, 즉 a_k가 속하는 경우와 속하지 않는 경우가 있다. 그러므로 A의 부분집합의 수를 구하는 문제는 n개의 빈 칸 □□□…□의 각각에 0, 1 중에서 무작정 하나씩을 택하여 채우는 일로 바꿔 생각할 수 있다. 다시 말하면 이러한 n개의 빈 칸을 채우는 각 경우마다 A의 부분집합 X가 하나씩 정해진다. 즉 $a_k\in X$이면 그리고 그때에만 k번째의 빈 칸에 1이 나타난다. 여기서 빈 칸을 채울 수 있는 경우는 2^n가지가 있으므로 A의 부분집합의 수는 2^n와 같다.

---

* 책에 따라서는 부등호 ≦, ≧를 각각 ≤, ≥로 나타내기도 한다.

** 첫째 식은 $\mathbb{N}\subset\mathbb{Z}$, $\mathbb{Z}\subset\mathbb{Q}$, $\mathbb{Q}\subset\mathbb{R}$을 한꺼번에 나타낸 것으로 피하도록 하자. 왜냐하면 기호 "⊂"는 두 집합 사이의 포함관계를 나타낼 뿐이기 때문이다.

[2] 공리적 집합론에서는 멱집합의 존재를 당연시 하지 않는다. 사실 멱집합의 존재성이 분출공리(81쪽 중간구절)로부터는 유도되지 않으므로 다음과 같이 **멱집합의 공리**를 내세워야 한다.

"임의의 집합 A에 대하여 A의 부분집합인 것을 원소로 하는 집합이 존재한다."

It may be interesting to know the following alternative proof of Theorem 3.

[Alternative Proof] First the empty set ϕ belongs to $\mathscr{P}(A)$. Next, each element $x \in A$ forms a subset $\{x\}$ belonging to $\mathscr{P}(A)$. Observe that the number of these singleton subsets is $C(n, 1)$. Continuing, there are exactly $C(n, 2)$ subsets of A containing exactly two elements of A[3]. Finally, there is exactly $C(n, n)=1$ subset of A containing n elements of A, namely the set A itself. Counting the empty set, the total number of subsets of A is equal to

$$C(n,0)+C(n,1)+\cdots+C(n,n)$$

Then using the binomial expansion for $(1+1)^n$, we have

$$(1+1)^n = C(n,0)+C(n,1)+\cdots+C(n,n)$$

Thus, the number of elements in $\mathscr{P}(A)$ is $(1+1)^n(=2^n)$.

Exercise 2.2

1. Display within braces the elements of each of the following sets.

$A=\{x\in\mathrm{N} \mid x<5\}$, $\quad B=\{x\in\mathrm{Z} \mid x^2\leqq 25\}$,

$C=\{x\in\mathrm{Q} \mid 10x^2+3x-1=0\}$, $\quad D=\{x\in\mathrm{R} \mid x^3+1=0\}$,

$E=\{x\in\mathrm{R}_+ \mid 4x^2-4x-1=0\}$.

2. Denote each of the following sets by the set builder notation.

$A=\{1, 2, 3\}$, $\quad B=\left\{-1, -\frac{2}{3}, -\frac{1}{3}, 0\right\}$,

$C=\{1, 3, 5, 7, 9, \cdots\}$, $\quad D=\{1-\sqrt{3}, 1+\sqrt{3}\}$.

3. What are the elements of the power set of the set $\{x, \{y, z\}\}$? How many elements does this power set contain?

[3] See Problem 9, Exercise 2.1.

여기서 정리 3을 다음과 같이 별도로 증명하여 보자.

[별증]* 공집합 ϕ는 임의의 집합의 부분집합이므로 $\phi \in \mathscr{P}(A)$. 한편 각 원소 $x \in A$에 대하여 $\mathscr{P}(A)$에 속하는 A의 부분집합 $\{x\}$가 하나씩 대응한다. 이와같이 A의 부분집합 중 원소가 하나인 부분집합은 $C(n,1)$개, A의 원소가 2개인 것은 $C(n,2)$[3]개 각각 존재한다. 마침내 A의 원소가 n개인 것, 즉 A 자신은 $C(n,n)=1$개 존재하므로 공집합 ϕ마저 생각하면 A의 부분집합의 수는 다음과 같다.

$$C(n,0)+C(n,1)+\cdots+C(n,n)$$

여기서 $(1+1)^n$에 대한 이항정리로는

$$(1+1)^n=C(n,0)+C(x,1)+\cdots+C(n,n)$$

이상으로 $\mathscr{P}(A)$의 원소의 수는 $(1+1)^n$ 곧 2^n과 같다.

연습문제 2.2

1. 다음 집합이 원소 낱낱을 중괄호로 묶어 나타내어라(이러한 표시법을 **원소나열법**이라고 한다).

$A=\{x \in \mathbb{N} \mid x<5\}$, $B=\{x \in \mathbb{Z} \mid x^2 \leqq 25\}$,
$C=\{x \in \mathbb{Q} \mid 10x^2+3x-1=0\}$, $D=\{x \in \mathbb{R} \mid x^3+1=0\}$,
$E=\{x \in \mathbb{R}_+ \mid 4x^2-4x-1=0\}$.

2. 다음 집합을 조건제시법(81쪽 중간 구절 참조)으로 나타내어라.

$A=\{1, 2, 3\}$, $B=\left\{-1, -\frac{2}{3}, -\frac{1}{3}, 0\right\}$,
$C=\{1, 3, 5, 7, 9, \cdots\}$, $D=\{1-\sqrt{3}, 1+\sqrt{3}\}$.

3. 다음 집합의 멱집합을 원소나열법으로 나타내고 그것의 원소의 수를 말하여라.

(a)** $A=\{a, b, c\}$ (b) $B=\{x, \{y, z\}\}$

* 83쪽의 증명보다는 이 별증이 정리 3에 대한 수학적 증명이라 볼 수 있다.
[3] 연습문제 2.1 문제 9 참조.
** 이 문제는 옮긴이가 추가한 것이다.

4. Let B be a subset of A, and let $\mathscr{P}(A:B)=\{X\in\mathscr{P}(A)\,|\,B\subseteq X\}$.

(a) Let $B=\{a, b\}$ and $A=\{a, b, c, d\}$. List all the members of the set $\mathscr{P}(A:B)$; how many are there?

(b) Show that $\mathscr{P}(A:\phi)=\mathscr{P}(A)$.

5. Let A be a set with n elements and B a subset with m elements, $n\geqq m$.

(a) Find the number of elements in the set $\mathscr{P}(A:B)$.

(b) Deduce Theorem 3 from (a) by setting $B=\phi$.

2.3 Unions and Intersections

In arithmetic we can add, multiply, or subtract any two numbers. In set theory, there are three operations — union, intersection, and complementation — analogous respectively to the addition, multiplication, and subtraction of numbers.

Definition 3. The union of any two sets A and B, denoted by $\boldsymbol{A\cup B}$, is the set of all elements x such that x belongs to at least one of the two sets A and B. That is, $x\in A\cup B$ if and only if $x\in A\vee x\in B$.

Definition 4. The intersection of any two sets A and B, denoted by $\boldsymbol{A\cap B}$, is the set of all elements x which belong to both A and B. In symbols,

$$A\cap B=\{x\,|\,(x\in A)\wedge(x\in B)\},\quad \text{or}\quad \{x\in A\,|\,x\in B\}.$$

If $A\cap B=\phi$, then A and B are said to be disjoint.

For example, if $A=\{1,2,3,4\}$ and $B=\{3,4,5\}$, then $A\cup B=\{1,2,3,4,5\}$ and $A\cap B=\{3,4\}$; if Im denotes the set of imaginary numbers, then the sets Im and $\mathbb{R}$ are disjoint.

4. 집합 A의 부분집합 B에 대하여 $\mathscr{P}(A:B)=\{X\in\mathscr{P}(A)\mid B\subseteq X\}$로 놓았을 때 다음 물음 (a)에 답하고 (b)는 밝혀 보아라.
(a) $A=\{a, b, c, d, e\}$, $B=\{a, b\}$에 대한 $\mathscr{P}(A:B)$를 원소나열법으로 나타내고 그 원소의 수를 말하여라.
(b) $\mathscr{P}(A:\phi)=\mathscr{P}(A)$

5. 원소가 n개인 집합 A의 부분집합 B의 원소는 m개($n\geqq m$)라고 한다.
(a) 집합 $\mathscr{P}(A:B)$의 원소의 수를 구하여라.
(b) (a)에서 $B=\phi$이라 놓고 정리 3의 결론을 이끌어 내어라.

2.3 합집합과 교집합

우리는 집합에서 합집합, 교집합, 여집합을 배울 수 있거니와 이들 연산은 마치 실수의 덧셈, 곱셈, 뺄셈과 유사함을 납득할 것이다.

정의 3 임의의 두 집합 A, B에 대하여 적어도 A, B 각각에 속하는 원소들의 집합을 A와 B의 **합집합**이라 하고 이것을 $\boldsymbol{A\cup B}$로 나타낸다. 다시 말하면

$$x\in A\vee x\in B \Longleftrightarrow x\in A\cup B \qquad \cdots\cdots ①^*$$

정의 4 임의의 두 집합 A, B에 대하여 A, B 각각에 동시에 속하는 원소의 집합을 A와 B의 **교집합**이라 하고 이것을 $\boldsymbol{A\cap B}$와 같이 나타낸다. 즉

$$A\cap B=\{x\mid x\in A\wedge x\in B\} \quad \text{또는} \quad A\cap B=\{x\in A\mid x\in B\}.$$

특히 $A\cap B=\phi$일 때 A와 B는 **서로 소**라고 한다.

이를테면 $A=\{1, 2, 3, 4\}$, $B=\{3, 4, 5\}$에 대하여

$$A\cup B=\{1, 2, 3, 4, 5\}, \qquad A\cap B=\{3, 4\}.$$

한편 허수의 집합을 Im라고 할 때 $\mathrm{Im}\cap\mathbb{R}=\phi$, 즉 Im과 $\mathbb{R}$은 서로 소이다.

---

* 수학에서 정의는 흔히 ①에서와 같이 동치 $p\Longleftrightarrow q$의 형식을 띤다(27쪽 및 28쪽).

Example 3 In the following, the sets I, N, Z, $\cdots$ are as defined in the last section.

(a) $\mathrm{I}\cap\mathrm{Z}=\{0,1\}$ and $\mathrm{N}\cap\mathrm{I}=\{1\}$. (b) $\mathrm{Z}\cup\mathrm{Q}=\mathrm{Q}$ and $\mathrm{Z}\cap\mathrm{Q}=\mathrm{Z}$.
(c) $\mathrm{I}\cup\mathrm{I}=\mathrm{I}$ and $\mathrm{I}\cap\mathrm{I}=\mathrm{I}$.

Theorem 4. Let X be a set and let A, B, and C be subsets of X. Then we have

(a) The unities: $A\cup\phi=A \quad A\cap X=A$

(b) The idempotency laws: $A\cup A=A \quad A\cap A=A$

(c) The commutative laws: $A\cup B=B\cup A \quad A\cap B=B\cap A$

(d) The associative laws: $A\cup(B\cup C)=(A\cup B)\cup C$
$A\cap(B\cap C)=(A\cap B)\cap C$

(e) The distributive laws: $A\cap(B\cup C)=(A\cap B)\cup(A\cap C)$
$A\cup(B\cap C)=(A\cup B)\cap(A\cup C)$

[Proof] We leave the proofs of parts (a), (b), and (c) to the reader as an exercise.

(d) According to Definition 3, an element

$$x\in A\cup(B\cup C)\Leftrightarrow x\in A\vee x\in B\cup C$$

and $$x\in B\cup C\Leftrightarrow x\in B\vee x\in C$$

so $$x\in A\cup(B\cup C)\Leftrightarrow x\in A\vee(x\in B\vee x\in C)$$

By the Associative Law (for the disjunction),

$$(x\in A)\vee(x\in B\vee x\in C)$$

is equivalent to $$(x\in A\vee x\in B)\vee(x\in C).$$

The last statement, by Definition 3, is equivalent to

$(x\in A\cup B)\vee x\in C$, and hence to $x\in(A\cup B)\cup C$.

예제 3 집합 I, N, Z, Q에 대하여 다음 등식을 얻는다.

(a) $\mathrm{I}\cap\mathrm{Z}=\{0, 1\}$, $\mathrm{N}\cap\mathrm{I}=\{1\}$ (b) $\mathrm{Z}\cup\mathrm{Q}=\mathrm{Q}$, $\mathrm{Z}\cap\mathrm{Q}=\mathrm{Z}$

(c) $\mathrm{I}\cup\mathrm{I}=\mathrm{I}$, $\mathrm{I}\cap\mathrm{I}=\mathrm{I}$ (d)* $\mathrm{N}\cap\{m\in\mathrm{Z}\mid m<0\}=\phi$

정리 4** 집합 X의 부분집합 A, B, C에 대하여 다음이 성립한다.

(a) **항등법칙** $A\cup\phi=A$, $A\cap X=A$

(b) **멱등법칙** $A\cup A=A$, $A\cap A=A$

(c) **교환법칙** $A\cup B=B\cup A$, $A\cap B=B\cap A$

(d) **결합법칙** $A\cup(B\cup C)=(A\cup B)\cup C$,
$A\cap(B\cap C)=(A\cap B)\cap C$

(e) **분배법칙** $A\cap(B\cup C)=(A\cap B)\cup(A\cap C)$,
$A\cup(B\cap C)=(A\cup B)\cap(A\cup C)$

[증명] (a), (b), (c)와 (d), (e)의 두 번째 등식의 증명은 모두 독자의 연습문제로 돌린다.

(d) 정의 3에 따라 $x\in A\cup(B\cup C)\Longleftrightarrow x\in A\vee x\in B\cup C$

여기서 $x\in B\cup C\Rightarrow x\in B\vee x\in C$

그러므로 $x\in A\cup(B\cup C)\Longleftrightarrow x\in A\vee(x\in B\vee x\in C)$

논리합에 대한 결합법칙[31쪽 정리 4(a) 두 번째]에 따라

$$x\in A\vee(x\in B\vee x\in C)\Longleftrightarrow(x\in A\vee x\in B)\vee x\in C$$

한편 정의 3에 따라

$$(x\in A\vee x\in B)\vee x\in C\Longleftrightarrow(x\in A\cup B)\vee x\in C$$

$$(x\in A\cup B)\vee x\in C\Longleftrightarrow x\in(A\cup B)\cup C$$

따라서 $\Longleftrightarrow$에 관한 추이율(단 한 번의 적용)에 의하여

$$x\in A\vee(x\in B\vee x\in C)\Longleftrightarrow x\in(A\cup B)\cup C$$

* 이것은 옮긴이가 추가한 것이다.

** '2.5 벤 다이어그램(103–105쪽)'을 공부한 뒤 정리 4의 각 법칙[다만 (a)의 첫 번째는 제외]을 그것으로 재확인하여라. 정리 4의 (c), (d) 및 (e)의 두 번째 법칙은 $\cup$ 대신 덧셈 $+$, $\cap$ 대신 곱셈 $\times$로 각각 생각할 때 실수의 덧셈, 곱셈에 대한 법칙들과 유사하다.

Thus, we have

$$x \in A \cup (B \cup C) \Longleftrightarrow x \in (A \cup B) \cup C$$

By Definition 1,

$$A \cup (B \cup C) = (A \cup B) \cup C$$

The above proof may be condensed into a neat display of essential logical steps, with the justification of each step written on the right for easy reference :

$x \in A \cup (B \cup C)$	$\Longleftrightarrow x \in A \vee (x \in B \cup C)$	Def. $\cup$
	$\Longleftrightarrow x \in A \vee (x \in B \vee x \in C)$	Def. $\cup$
	$\Longleftrightarrow (x \in A \vee x \in B) \vee x \in C$	Assoc. for $\vee$
	$\Longleftrightarrow x \in A \cup B \vee x \in C$	Def. $\cup$
	$\Longleftrightarrow x \in (A \cup B) \cup C$	Def. $\cup$

Hence, by Definition 1, we have proved that

$$A \cup (B \cup C) = (A \cup B) \cup C$$

The students should try to appreciate this kind of orderly precise proof by logic.

We leave the proof of $A \cap (B \cap C) = (A \cap B) \cap C$ to the reader as an exercise.

(e) Again, only the first half of (e) is proved ; the other half is left to the reader as an exercise.

$x \in A \cap (B \cup C)$	$\Longleftrightarrow x \in A \wedge x \in B \cup C$	Def. $\cap$
	$\Longleftrightarrow x \in A \wedge (x \in B \vee x \in C)$	Def. $\cup$
	$\Longleftrightarrow (x \in A \wedge x \in B) \vee (x \in A \wedge x \in C)$	Dist. Law of logic (Ch. 1)
	$\Longleftrightarrow x \in A \cap B \vee x \in A \cap C$	Def. $\cap$
	$\Longleftrightarrow x \in (A \cap B) \cup (A \cap C)$	Def. $\cup$

Hence, by Definition 1,

$$A \cap (B \cup C) = (A \cap B) \cup (A \cap B)$$

그러므로

$$x \in A \cup (B \cup C) \Leftrightarrow x \in (A \cup B) \cup C$$

여기서 정의 1에 따라

$$A \cup (B \cup C) = (A \cup B) \cup C$$

정리 4의 증명(55쪽)에서와 같이 논리적 각 단계에 따른 정당성을 오른쪽에 적어 놓으면 편리하다. 이를테면 다음을 보자.

$x \in A \cup (B \cup C) \Leftrightarrow x \in A \vee (x \in B \cup C)$	$\cup$의 정의
$\Leftrightarrow x \in A \vee (x \in B \vee x \in C)$	$\cup$의 정의
$\Leftrightarrow (x \in A \vee x \in B) \vee x \in C$	$\vee$에 대한 결합법칙
$\Leftrightarrow x \in A \cup B \vee x \in C$	$\cup$의 정의
$\Leftrightarrow x \in (A \cup B) \cup C$	$\cup$의 정의
그러므로 $x \in A \cup (B \cup C) \Leftrightarrow x \in (A \cup B) \cup C$	$\Leftrightarrow$에 관한 추이율

여기서 정의 1에 의하여 $A \cup (B \cup C) = (A \cup B) \cup C$

(e) 첫 번째 등식만을 증명하자.

$x \in A \cap (B \cup C)$	
$\Leftrightarrow x \in A \wedge x \in B \cup C$	$\cap$의 정의
$\Leftrightarrow x \in A \wedge (x \in B \vee x \in C)$	$\cup$의 정의
$\Leftrightarrow (x \in A \wedge x \in B) \vee (x \in A \wedge x \in C)$	논리에 관한 분배법칙
$\Leftrightarrow x \in A \cap B \vee x \in A \cap C$	$\cap$의 정의
$\Leftrightarrow x \in (A \cap B) \cup (A \cap C)$	$\cup$의 정의

따라서 $\Leftrightarrow$에 관한 추이율(몇 번의 반복?)에 따라

$$x \in A \cap (B \cup C) \Leftrightarrow x \in (A \cap B) \cup (A \cap C)$$

그러므로 정의 1에 의하여 $A \cap (B \cup C) = (A \cap B) \cup (A \cap C)$

Exercise 2.3

Prove the following problems 1 through 11.

1. $A \subseteq B \Leftrightarrow A \cup B = B$.

2. $A \subseteq B \Leftrightarrow A \cap B = A$.

3. The parts (a), (b), and (c) of Theorem 4.

4. The second half of Theorem 4(d).

5. The second half of Theorem 4(e).

6. (a) $A \subseteq C$ and $B \subseteq C$ implies $(A \cup B) \subseteq C$

(b) $A \subseteq B$ and $A \subseteq C$ implies $A \subseteq (B \cap C)$. [Hint: Use Theorem 5 in Chapter 1, if you wish.]

7. $(A \cap B) \cup C = A \cap (B \cup C) \Leftrightarrow C \subseteq A$.

8. If $A \subseteq B$, then $\mathscr{P}(A) \subseteq \mathscr{P}(B)$.

9. $A \cup B = A \cap B \Leftrightarrow A = B$.

10. If $A \subseteq B$, then $A \cup C \subseteq B \cup C$ and $A \cap C \subseteq B \cap C$ for any set C.

11. If $A \subseteq C$ and $B \subseteq D$, then $A \cup B \subseteq C \cup D$.

연습문제 2.3

집합 A, B, C, D에 대한 다음 1−11를 증명하여라.

1. $A \subseteq B \Leftrightarrow A \cup B = B$

2. $A \subseteq B \Leftrightarrow A \cap B = A$

3. 정리 4의 (a), (b), (c)

4. 정리 4(d)의 두 번째 것

5. 정리 4(e)의 두 번째 것

6. (a) $A \subseteq C \wedge B \subseteq C \Leftrightarrow (A \cup B) \subseteq C$를 함의한다.
(b) $A \subseteq B \wedge A \subseteq C \Leftrightarrow A \subseteq (B \cap C)$[필요하면 제 1 장 정리 5(34쪽) 참조].

7. $(A \cap B) \cup C = A \cap (B \cup C) \Leftrightarrow C \subseteq A$

8. $A \subseteq B \Rightarrow \mathscr{P}(A) \subseteq \mathscr{P}(B)$

9. $A \cup B = A \cap B \Leftrightarrow A = B$

10. $A \subseteq B$이면 임의의 집합 C에 대하여
(a) $A \cup C \subseteq B \cup C$
(b) $A \cap C \subseteq B \cap C$

11. $A \subseteq C \wedge B \subseteq D \Rightarrow A \cup B \subseteq C \cup D$

12.* 문제 10 (a), (b) 각각의 역이 참이면 그 사실을 증명하고 거짓이면 그에 따른 반례 하나씩을 들어라.

13.* 집합 $A = \{a, b, c\}$의 멱집합 $\mathscr{P}(A)$의 원소 e로 하여금 다음을 성립토록 하는 것은 몇 개 존재하는가?
(a) $A \cap e = e \cap A = A$
(b) $A \cup e = e \cup A = A$

14.* 하나의 동치의 $p \Longleftrightarrow q$에 대하여

q를 p이기 위한 **필요조건**(necessary condition),
p를 q이기 위한 **충분조건**(sufficient condition)이라고 한다.

(a) 이차방정식 $x^2 + x - 2 = 0$을 풀고 그에 따른 필요조건, 충분조건을 지적하여라.
(b) 참인 조건문 $p \to q$(즉 $p \Rightarrow q$)의 역 $q \to p$로 하여금 참, 거짓 각각의 경우에 대한 보기를 하나씩 들어라(역시 책의 것은 제외).

---

* 문제 12, 13, 14는 옮긴이가 추가한 것이다. 사실 14(b)에서의 $q \to p$가 참인 경우 q는 p이기 위한 필요충분조건이다(87쪽 각주 * 참조).

2.4 Complements

There is, in set theory, an operation known as complementation, which is similar to the operation of subtraction in arithmetic.

Definition 5. If A and B are sets, the relative complement of B in A is the set $\boldsymbol{A-B}$ defined by

$$A-B=\{x\in A \mid x\notin B\}$$

In this definition it is not assumed that $B\subseteq A$.

Example 4 Let $A=\{a,b,c,d\}$ and $B=\{c,d,e,f\}$. Find

$$A-B \quad \text{and} \quad A-(A\cap B)$$

[Solution] $A-B=\{a,b,c,d\}-\{c,d,e,f\}=\{a,b\}$ and
$A-(A\cap B)=\{a,b,c,d\}-\{c,d\}=\{a,b\}$

Although the universal set in the absolute sense, the set of all sets, does not exist (see the Russell Paradox in Section 7), it does no harm to assume temporarily that

all the sets to be mentioned in the rest of this book are subsets of a fixed set U

which may be regarded (temporarily) as a universal set in a restricted sense. In order to state the basic rules concerning complementation as simply as possible, we assume unless otherwise stated, that all complements are formed relative to this set U. We shall then write A^c for $U-A$.

2.4 여집합

집합론에는 수학에서의 뺄셈과 흡사한 연산으로서 주어진 집합의 여집합을 구하는 일이 허다하다.

정의 5 집합 A, B에 대하여 A에 속하지만 B에는 속하지 않는 원소의 집합을 A에 대한 B의 **차집합**이라 하고 이것을 $\boldsymbol{A-B}$*로 나타낸다. 다시 말하면

$$A-B=\{x\in A\mid x\notin B\}.$$

이 정의에서 집합 B로 하여금 집합 A의 부분집합이 아니더라도 무방하다.**

예제 4 집합 $A=\{a, b, c, d\}$, $B=\{c, d, e, f\}$에 대하여 다음을 구하여라.

$$A-B, \quad A-(A\cap B)$$

[풀이]

$$A-B=\{a, b, c, d\}-\{c, d, e, f\}=\{a, b\},$$
$$A-(A\cap B)=\{a, b, c, d\}-\{c, d\}=\{a, b\}$$

우리는 이미 논리의 관점(1.6절)에서 전체집합을 소개한 바 있으나 그 것의 뜻을 여기서 좀더 확실히 다져보자.

절대적인 뜻에서의 전체집합, 즉 모든 집합의 집합은 존재하지 않으나[2.7절 러셀의 역리(121쪽) 참조.] 이 절에서는 앞으로 언급되는 각 집합은 일시적으로 고정된 하나의 집합 U의 부분집합인 것으로 가정한다. 그럴 경우의 집합 U를 제한된 뜻에서의 **전체집합**으로 생각한다. 한편 여집합에 관한 별도의 언급이 없는 한 전체집합 U에 대한 A의 차집합 $U-A$를 A의 **여집합**이라 하고 이것을 A^c***로 나타낸다.

이를테면 $U=\mathbb{R}$에 대하여 $\mathbb{Q}^c=\mathbb{R}-\mathbb{Q}=\{x\mid x$는 무리수임$\}$,

한편 $U=\mathbb{I}$에 대하여 $\mathbb{Z}^c=\{x\in\mathbb{I}\mid x\le 0\}$.

---

* $A-B$를 A에 대한 B의 여집합이라고도 한다.

** 그와 같은 차집합 $A-B$의 보기를 하나 들어라. 특히 $A-B=\phi$인 것은 어느 경우인가?

*** 책에 따라서는 집합 A의 여집합을 A'로 나타내기도 한다.

Example 5 Show that $A-B=A\cap B^c$

[Solution]
$$
\begin{aligned}
x\in A\cap B^c &\equiv x\in A\wedge x\in U-B && \text{Def. } \cap\text{, Def. of } ^c\\
&\equiv x\in A\wedge(x\in U\wedge x\notin B) && \text{Def. 5}\\
&\equiv (x\in A\wedge x\in U)\wedge x\notin B && \text{Assoc.}\\
&\equiv x\in A\cap U\wedge x\notin B && \text{Def. } \wedge\\
&\equiv x\in A\wedge x\notin B && A\cap U=A\\
&\equiv x\in A-B && \text{Def. 5}
\end{aligned}
$$

Therefore, by Definition 1, $A\cap B^c=A-B$

> ***Theorem 5.*** Let A and B be sets. Then
> (a) $(A^c)^c=A$. (b) $\phi^c=U$ and $U^c=\phi$.
> (c) $A\cap A^c=\phi$ and $A\cup A^c=U$.
> (d) $A\subseteq B$ if and only if $B^c\subseteq A^c$.

[Proof] The proofs for parts (a), (b), and (c) use only the definitions and are left to the reader as exercises. We give a proof for part (d) :

$$
\begin{aligned}
A\subseteq B &\equiv (x\in A\Rightarrow x\in B) && \text{Def. } \subseteq\\
&\equiv (x\notin B\Rightarrow x\notin A)^{4)} && \text{Contrap.}\\
&\equiv (x\in B^c\Rightarrow x\in A^c) && \text{Def. of } ^c\\
&\equiv B^c\subseteq A^c && \text{Def. } \subseteq
\end{aligned}
$$

We have thus proved that $A\subseteq B\equiv B^c\subseteq A^c$

In the preceding proof, again symbols and laws of logic (from Chapter 1) are used, which enable us to display each step of the proof neatly and precisely with supporting reasons on the right-hand side. The reader is encouraged to make full use of Chapter 1 for proofs whenever possible.

[4)] Recall that the negation of $x\in B$, $\sim(x\in B)$, is denoted by $x\notin B$.

예제 5 $A-B=A\cap B^c$임을 보여라.

[풀이] $x\in A\cap B^c \equiv x\in A \wedge x\in U-B$ ∩의 정의, ^{c}의 정의

$\equiv x\in A \wedge (x\in U \wedge x\notin B)$ 정의 5

$\equiv (x\in A \wedge x\in U) \wedge x\notin B$ 결합법칙

$\equiv x\in A\cap U \wedge x\notin B$ ∧의 정의

$\equiv x\in A \wedge x\notin B$ $A\cap U=A$

$\equiv x\in A-B$ 정의 5

그러므로 $x\in A\cap B^c \equiv x\in A-B$ ≡에 관한 추이율

따라서 $x\in A-B \equiv x\in A\cap B^c$ ≡에 관한 대칭률

정의 1에 의하여 $A-B=A\cap B^c$

정리 5 임의의 집합 A, B에 대하여

(a) $(A^c)^c=A$ (b)* $\phi^c=U$, $U^c=\phi$

(c) $A\cap A^c=\phi$, $A\cup A^c=U$ (d) $A\subseteq B \Longleftrightarrow B^c\subseteq A^c$

[증명] (a), (b), (c)는 모두 정의만의 반복으로 증명되는 것으로 독자의 연습문제로 맡기고 여기서는 (d)만을 증명한다.

$A\subseteq B\equiv(x\in A \Rightarrow x\in B)$ ⊆의 정의

$\equiv(x\notin B \Rightarrow x\notin A)$[4] 대우법칙

$\equiv(x\in B^c \Rightarrow x\in A^c)$ ^{c}의 정의

$\equiv B^c\subseteq A^c$ ⊆의 정의

그러므로 $A\subseteq B\equiv B^c\subseteq A^c$ ≡에 관한 추이율

이 증명에서도 제1장의 기호와 논리법칙을 인용하고 있으며 추론과정에 대한 뒷받침을 그 오른쪽에 적어 각 단계마다의 근거로 삼고 있다. 이처럼 독자는 제1장에서 배운 내용을 활용함으로써 앞으로의 여러 증명에 충분히 대비해 나가야 한다.

* 정리 5의 등식 중 특히 (b)의 첫 번째 등식에 대해서는 쉽게 와닿지 않을지 모른다.

[4] $x\in B$의 부정 $\sim(x\in B)$를 $x\notin B$로도 나타낸다.

The most useful property of complements is the following De Morgan's Theorem. The reader should compare this theorem with De Morgan's Law in Chapter 1.

Theorem 6. De Morgan's Theorem For any two sets A and B,

(a) $(A\cup B)^c = A^c\cap B^c$. (b) $(A\cap B)^c = A^c\cup B^c$.

[Proof] (a)

$$
\begin{aligned}
x\in(A\cup B)^c &\equiv \sim(x\in A\cup B) && \text{Def. of } {}^c\\
&\equiv \sim(x\in A\cup x\in B) && \text{Def. } \cup\\
&\equiv \sim(x\in A)\wedge\sim(x\in B) && \text{De M. of logic}\\
&\equiv x\in A^c\wedge x\in B^c && \text{Def. of } {}^c\\
&\equiv x\in(A^c\cap B^c) && \text{Def. } \cap
\end{aligned}
$$

Therefore, by Definition 1, $(A\cup B)^c = A^c\cap B^c$

The proof for (b) is left to the reader.

Example 6 Let A, B, and C be any three sets. Decide whether the set $A\cap(B-C)$ is equal to the set $(A\cap B)-(A\cap C)$.

[Solution]

$$
\begin{aligned}
(A\cap B)-(A\cap C) &= (A\cap B)\cap(A\cap C)^c && \text{Example 5}\\
&= (A\cap B)\cap(A^c\cup C^c) && \text{De M. Th. (Th. 6)}\\
&= (A\cap B\cap A^c)\cup(A\cap B\cap C^c) && \text{Dist.}\\
&= (A\cap A^c\cap B)\cup(A\cap B\cap C^c) && \text{Com.}\\
&= \phi\cup[A\cap(B\cap C^c)] && \text{Th. 5(c)}: A\cap A^c=\phi\\
&= A\cap(B-C) && \text{Th. 4(a), Example 5}
\end{aligned}
$$

Hence, we have proved that

$$A\cap(B-C) = (A\cap B)-(A\cap C).$$

여집합의 성질 중 드 모르간의 정리가 있다. 독자는 이 중요한 정리를 제 1 장의 드 모르간의 법칙(29쪽)과 비교해보길 바란다.

정리 6 드 모르간의 정리 임의의 두 집합 A, B에 대하여

(a) $(A\cup B)^c = A^c\cap B^c$ (b) $(A\cap B)^c = A^c\cup B^c$

[증명] (a)

$x\in(A\cup B)^c \equiv (\sim x\in A\cup B)$	^{c}의 정의
$\equiv \sim(x\in A\cup x\in B)$	$\cup$의 정의
$\equiv \sim(x\in A)\wedge\sim(x\in B)$	논리에 관한 드 모르간의 법칙
$\equiv x\in A^c \wedge x\in B^c$	^{c}의 정의
$\equiv x\in(A^c\cap B^c)$	$\cap$의 정의

그러므로 $x\in(A\cup B)^c \equiv x\in(A^c\cap B^c)$ $\equiv$에 관한 추이율

따라서 정의 1에 의하여 $(A\cup B)^c = A^c\cap B^c$

(b)의 증명은 독자의 연습문제로 돌린다.

예제 6 임의의 집합 A, B, C에 대한 다음 등식

$$A\cap(B-C) = (A\cap B)-(A\cap C)$$

을 증명하여라.

[증명]

$(A\cap B)-(A\cap C) = (A\cap B)\cap(A\cap C)^c$	예제 5
$= (A\cap B)\cap(A^c\cup C^c)$	정리 6(b)
$= (A\cap B\cap A^c)\cup(A\cap B\cap C^c)$	분배법칙
$= (A\cap A^c\cap B)\cup(A\cap B\cap C^c)$	교환법칙
$= (\phi\cap B)\cup(A\cap B\cap C^c)$	정리 5(c)
$= \phi\cup[A\cap(B\cap C^c)]$	$\phi\cap B=\phi$
$= A\cap[(B\cap C^c)]$	정리 4(a)
$= A\cap(B-C)$	예제 5

그러므로 $(A\cap B)-(A\cap C) = A\cap(B-C)$ $=$에 관한 추이율

따라서 $A\cap(B-C) = (A\cap B)-(A\cap C)$ $=$에 관한 대칭률

Exercise 2.4

Prove the following problems 1 through 6.

1. Let A and B be sets. Then $A-B=A-(B\cap A)$.

2. The parts (a), (b), and (c) of Theorem 5.

3. Let A and B be sets. Then $B\subseteq A^c$ if and only if $A\cap B=\phi$.

4. Let A and B be sets. Then $(A-B)\cup B=A$ if and only if $B\subseteq A$.

5. Theorem 6(b).

6. Let A, B, and C be any three sets, then
(a) $(A-C)\cup(B-C)=(A\cup B)-C$
(b) $(A-C)\cap(B-C)=(A\cap B)-C$

7. Let A and B be two sets. Prove that A and $B-A$ are disjoint and that $A\cup B=A\cup(B-A)$(This shows how to represent the union $A\cup B$ as a disjoint union.).

8. Let A, B, and C be any three sets. Prove that
(a) $(A\cap B\cap C)^c=A^c\cup B^c\cup C^c$ (b) $(A\cup B\cup C)^c=A^c\cap B^c\cap C^c$
Generalize these results to statements involving n sets

$$A_1,\ A_2,\ A_3,\ \cdots,\ A_n$$

9. For any sets A and B, prove or disprove that
(a) $\mathscr{P}(A)\cap\mathscr{P}(B)=\mathscr{P}(A\cap B)$ (b) $\mathscr{P}(A)\cup\mathscr{P}(B)=\mathscr{P}(A\cup B)$.

10. Prove that if $A\subseteq C$, $B\subseteq C$, $A\cup B=C$, and $A\cap B=\phi$, then $A=C-B$.

11. Let A and B be any two sets. Prove that

$$(A-B)\cup(B-A)=(A\cup B)-(A\cap B).$$

연습문제 2.4

임의의 집합 A, B, C에 대한 다음 1~8을 증명하여라.

1. $A-B=A-(B\cap A)$

2. 정리 5의 (a), (b), (c)

3. $A\cap B=\phi \Longleftrightarrow B\subseteq A^c$

4. $B\subseteq A \Longleftrightarrow (A-B)\cup B=A$

5. 정리 6(b)

6. (a) $(A-C)\cup(B-C)=(A\cup B)-C$
(b) $(A-C)\cap(B-C)=(A\cap B)-C$

7. A와 $B-A$가 서로 소일 때 $A\cup B=A\cup(B-A)$(이처럼 합집합 $A\cup B$를 서로 소인 두 집합의 합집합으로 나타낼 수 있음을 보여준다).

8. (a) $(A\cap B\cap C)^c=A^c\cup B^c\cup C^c$ (b) $(A\cup B\cup C)^c=A^c\cap B^c\cap C^c$
이 결과를 $n(n\geqq 4)$개의 집합

$$A_1,\ A_2,\ A_3,\ \cdots,\ A_n$$

의 경우로 일반화하여라.*

9. 임의의 집합 A, B에 대하여 다음 등식이 성립하면 그를 증명하고, 그렇지 않으면 반례 하나를 들어라.
(a) $\mathscr{P}(A)\cap\mathscr{P}(B)=\mathscr{P}(A\cap B)$ (b) $\mathscr{P}(A)\cup\mathscr{P}(B)=\mathscr{P}(A\cup B)$

아래 10, 11을 각각 증명하여라.

10. 집합 A, B, C에 대하여

$$A\subseteq C,\ B\subseteq C,\ A\cup B=C,\ A\cap B=\phi\text{이면}\quad A=C-B$$

11. 임의의 집합 A, B에 대하여 $(A-B)\cup(B-A)=(A\cup B)-(A\cap B)$

---

* 수학적 귀납법을 인용하여라.

2.5 Venn Diagrams

To help in visualizing set operations and their results we introduce diagrams, called Venn diagrams, which illustrate them. We shall represent the imaginary relative universal set U by a rectangle and let subsets (of U) be circles drawn inside the rectangle. For example, in Figure 1, we represent two sets A and B as two shaded circles ; the doubly crosshatched part is the intersection $A \cap B$ and the total shaded area is the union $A \cup B$.

Figure 2 shows two sets A and B that are disjoint. The shaded area in Figure 3 represents the complement A^c of the set A.

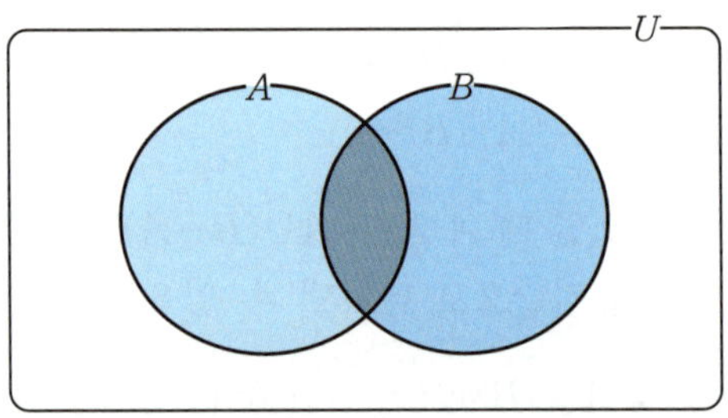

Figure 1

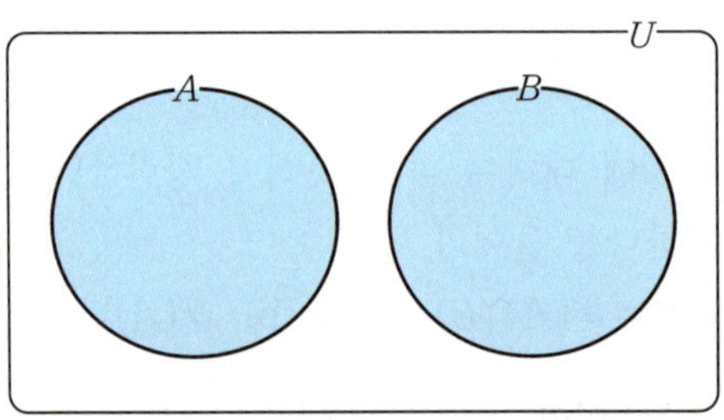

Figure 2

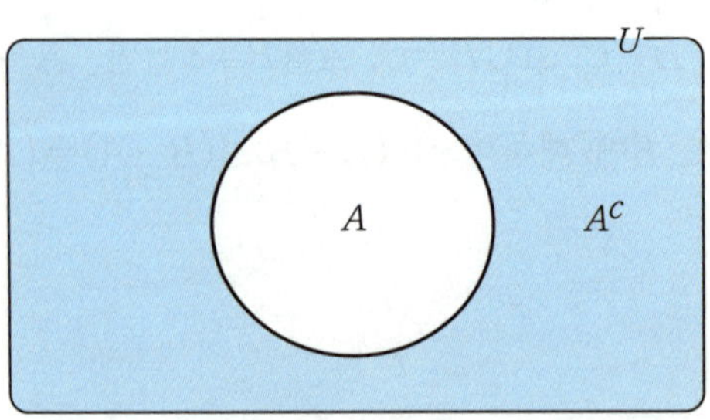

Figure 3

2.5 벤 다이어그램

집합의 연산과 그 결과를 쉽게 볼 수 있는 벤 다이어그램*을 소개하거니와 전체집합 U는 직사각형으로, 그 부분집합은 직사각형 내부의 원(꼭 원이 아니더라도 무방)으로 나타내고 있다. 이를테면 다음 그림 1에서 그늘진 두 원은 두 집합 A, B를, 짙게 그늘진 부분은 교집합 $A \cap B$를, 또 그늘진 전체는 합집합 $A \cup B$를 나타낸다. 한편 그림 2는 서로 소인 두 집합 A, B를, 그림 3의 그늘진 부분은 집합 A의 여집합 A^c를 각각 나타낸다.

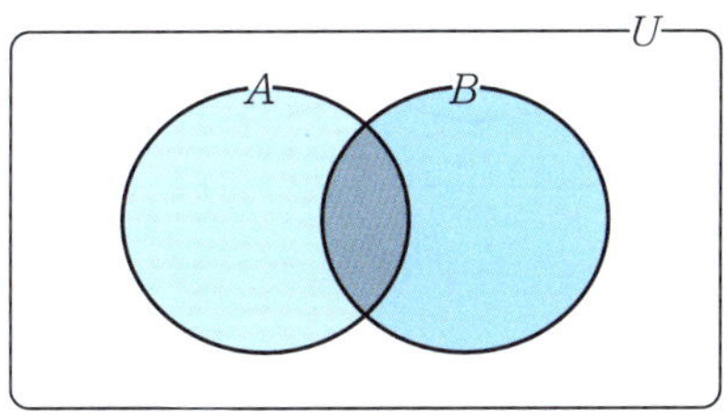

그림 1

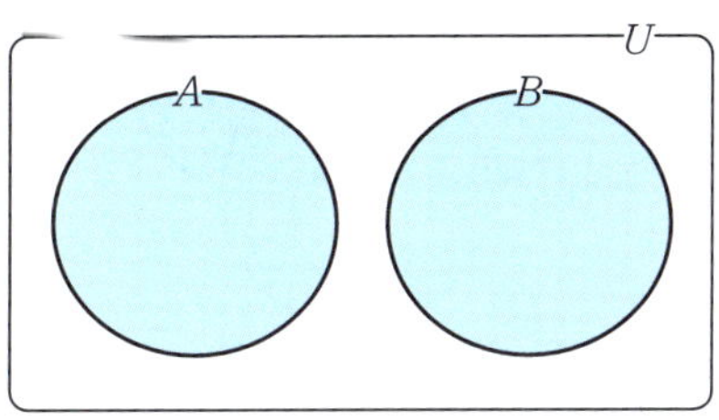

그림 2

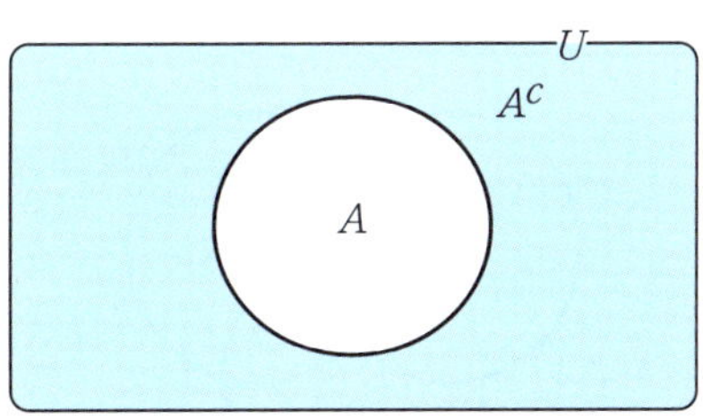

그림 3

* 벤 다이어그램은 어디까지나 우리의 시각적 이해를 도울 뿐 수학적 증명이 아님을 주의할 필요가 있다(107쪽 각주 * 참조).
처음 도형으로 집합을 나타낸 수학자는 스위스 오일러(Leonard Euler : 1707-1783)이며 영국의 드모르간(De Morgan : 1801-1871) 및 여기에 등장하는 벤(John Venn : 1831-1823)에 의하여 더욱 진전을 보기에 이르렀다.

The set $A-B$, the relative complement of B in A, is represented by the shaded part in Figure 4.

A typical Venn diagram of three sets A, B and C may be drawn as Figure 5. These three sets divide the universal set U into 8 parts as indicated in Figure 6.

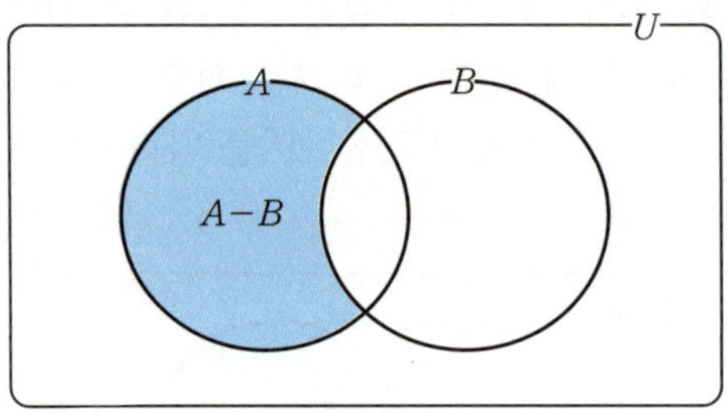

Figure 4

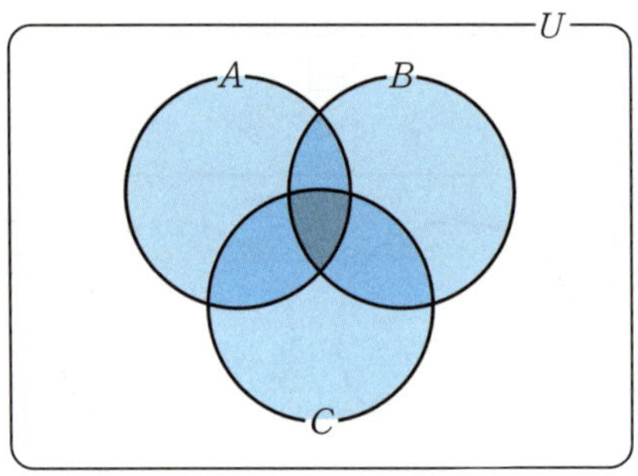

Figure 5

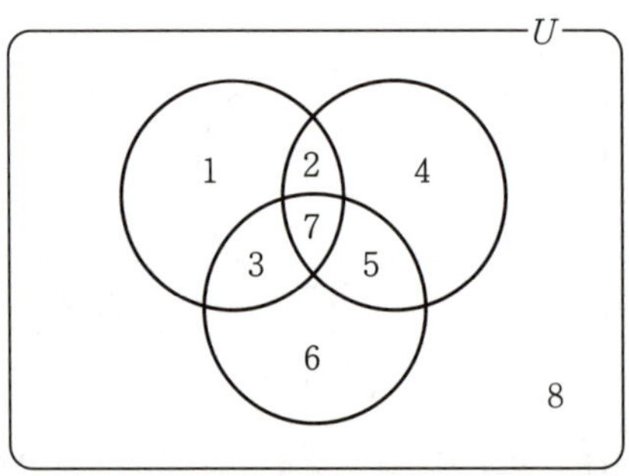

Figure 6

다음 그림 4의 그늘진 부분은 집합 A에 대한 집합 B의 차집합 $A-B$를 나타낸다. 특히 그림 5는 세 집합 A, B, C에 대한 벤 다이어그램으로서 이 세 집합은 그림 6에서와 같이 전체집합 U를 여덟 부분집합으로 분할하고 있음을 알 수 있다(147쪽 정의 5 및 보기 6 참조).*

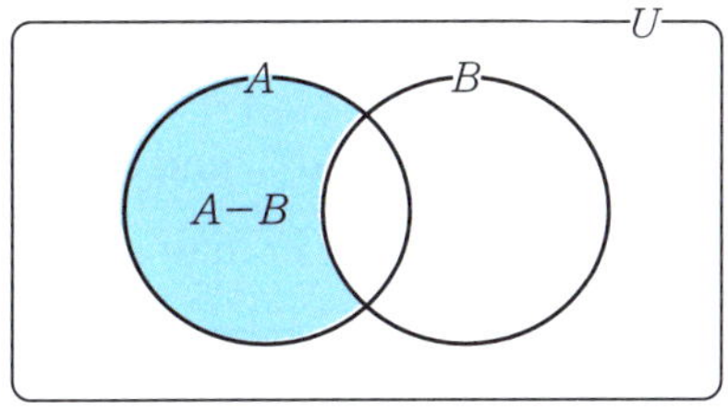

그림 4

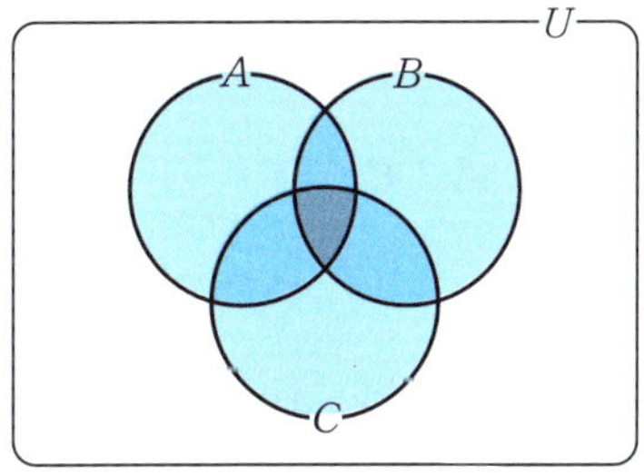

그림 5

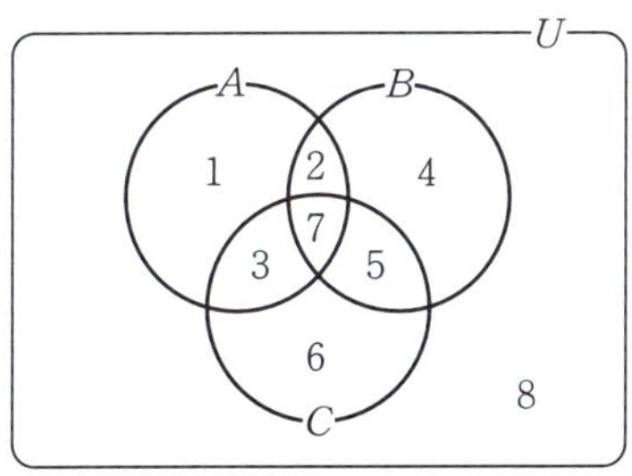

그림 6**

* 그림 4·5·6에서의 전체집합 U와 집합 A, B, C의 예를 들어 이들 각 그림의 뜻을 재확인하여라.

** 그림 6의 8개의 부분집합 1−8 중 어느 둘도 서로 소임을 주목하고 그림 4에 대해서도 같은 사실을 확인하여라. 한편 정리 3(83쪽)에 있어서 $n=3$으로 놓고 여기의 그림 6과 연관지어 보아라. 그 정리의 주장은 마침내 "3.3 분할과 동치관계"(147쪽)로 이어진다.

Using the preceding diagram, we can give a simple heuristic argument for the validity of, for example, the distributive law,

$$A\cap(B\cup C)=(A\cap B)\cup(A\cap C),$$

as follows: From Figure 6, $A\cap(B\cup C)$ consists of areas 2, 3, and 7. On the other hand,

$$(A\cap B)\cup(A\cap C)$$

is represented by the union of areas 2 and 7, and areas 3 and 7. Therefore, the above equality seems plausible. However, in mathematics a heuristic argument cannot be accepted as a proof.

Exercise 2.5

1. Draw a Venn diagram for $A\subset B$.

2. Draw Venn diagrams for $A\cap B^c$, $A^c\cap B$, and $A^c\cap B^c$.

3. Draw Venn diagrams for $A\cup B^c$, $A^c\cup B$, and $A^c\cup B^c$.

In Problems 4 through 10, draw Venn diagrams and give heuristic arguments that each of these statements is plausible.

4. $A\cap(B\cap C)=(A\cap B)\cap C$

5. $A\cup(B\cup C)=(A\cup B)\cup C$

6. $A\cup(B\cap C)=(A\cup B)\cap(A\cup C)$

7. $(A\cup B)^c=A^c\cap B^c$

8. $(A\cap B)^c=A^c\cup B^c$

9. $A\cap(B-A)=\phi$ and $A\cup(B-A)=A\cup B$

10. $(A\cup B)-(A\cap B)=(A-B)\cup(B-A)$

여기서 우리는 벤 다이어그램*을 이용하여, 분배법칙

$$A\cap(B\cup C)=(A\cap B)\cup(A\cap C) \qquad \cdots\cdots ①$$

이 성립함을 다음과 같이 보일 수 있다.

먼저 그림 6에서 $A\cap(B\cup C)$는 부분집합 2, 3, 7로 구성되어 있는 반면 $(A\cap B)\cup(A\cap C)$는 부분집합 2, 7, 부분집합 3, 7의 합집합으로 구성되어 있으므로 결국 등식 ①이 성립한다.

연습문제 2.5

다음 1~3의 벤 다이어그램을 그려라.

1. $A\subset B$ **2.** $A\cap B^c,\ A^c\cap B,\ A^c\cap B^c$

3. $A\cup B^c,\ A^c\cup B,\ A^c\cup B^c$

다음 4~10에서의 등식의 좌변과 우변에 대한 벤 다이어그램을 그려 각 등식이 성립하는지 확인하여라.

4. $A\cap(B\cap C)=(A\cap B)\cap C$ **5.** $A\cup(B\cup C)=(A\cup B)\cup C$

6. $A\cup(B\cap C)=(A\cup B)\cap(A\cup C)$ **7.** $(A\cup B)^c=A^c\cap B^c$

8. $(A\cap B)^c=A^c\cup B^c$ **9.** $A\cap(B-A)=\phi,\ A\cup(B-A)=A\cup B$

10. $(A\cup B)-(A\cap B)=(A-B)\cup(B-A)$

11.* $X=\{a, b, c\}$, $Y=\{a, b\}$, $Z=\{a, b, d\}$, $W=\{c, d\}$, $V=\{a\}$일 때 다음 관계를 하세 다이어그램으로 나타내어라(아래 그림 참고).

(a) $Y\subset X$ (b) $V\subset X$ (c) $V\subset Z$ (d) $W\subset Y$

* 사실 오일러(L. Euler)와 함께 고안된 벤 다이어그램만으론 많은 집합을 다루기에 곤란하다. 그런 약점을 보완할 수 있도록 하나의 집합을 점으로 나타내고 각 집합은 그것을 포함하는 모든 집합과 화살표로 연결진 이른바 **하세 다이어그램**(Hasse's diagram) 또는 **선형 다이어그램**(line diagram)이란 것이 있다.

오른쪽 그림은 집합 $A=\{a, b, c\}$와 그 모든 부분집합 사이의 포함관계를 나타낸 하세 다이어그램이다(301쪽 문제 3 참조).

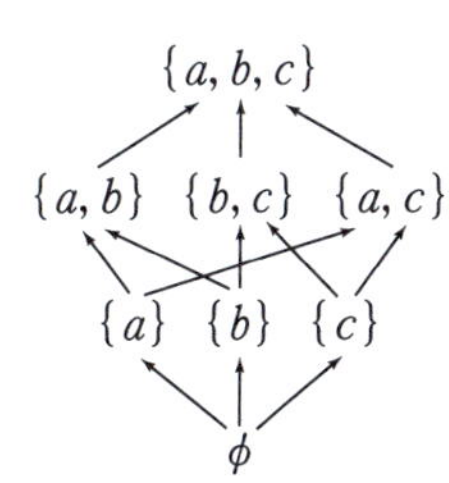

2.6 Indexed Families of Sets

Recall that a set is a collection of elements that are all distinct. Roughly speaking, a family is to be considered as a collection of not necessarily distinct objects called members. For example, $\{a, a, a\}$ is a family with three members $a, a,$ and a. But the same family $\{a, a, a\}$ considered as a set is just the singleton set $\{a\}$ with only one element, a.

Let Γ be a set and assume that with each element γ of Γ there is associated a set A_γ. The family of all such sets A_γ is called an indexed family of sets indexed by the set Γ and is denoted by $\{\boldsymbol{A_\gamma} \mid \boldsymbol{\gamma \in \Gamma}\}$. For example, the family of sets :

$$\{1,2\}, \{2,4\}, \{3,6\}, \cdots, \{n, 2n\}, \cdots$$

may be considered as an indexed family of sets indexed by the set $\mathbb{N}$ of natural numbers, where $A_n = \{n, 2n\}$ for each $n \in \mathbb{N}$. This family of sets may be denoted by $\{\{n, 2n\} \mid n \in \mathbb{N}\}$.

An arbitary family of sets might not appear to be indexed, but in most cases one can easily find a set Γ which can be used to index the given family of sets.

Example 7 Index the family $\mathscr{F}$ of sets ϕ, $\mathbb{N}$, $\mathbb{Z}$, $\mathbb{Q}$, $\mathbb{R}$ and $\mathbb{R}$.

[Solution] Since this family contains exactly six members (although two of them are the same), we choose $\Gamma = \{1, 2, 3, 4, 5, 6\}$ and let

$$A_1 = \phi,\ A_2 = \mathbb{N},\ A_3 = \mathbb{Z},\ A_4 = \mathbb{Q},\ A_5 = \mathbb{R}, \text{ and } A_6 = \mathbb{R}.$$

The family of sets is then indexed.

Virtually all symbols and notations used for sets will apply to families as well. For instance, $\phi \in \mathscr{F}$ and $\mathbb{R}_+ \notin \mathscr{F}$ will indicate respectively that ϕ is a member of the family $\mathscr{F}$ and $\mathbb{R}_+$ is not a member of $\mathscr{F}$. We may also write $\mathscr{F} = \{\phi, \mathbb{N}, \mathbb{Z}, \mathbb{Q}, \mathbb{R}, \mathbb{R}\}$.

Let us now extend the concepts of union $\cup$ and intersection $\cap$ defined in Definitions 3 and 4 to an arbitrary family of sets.

2.6 첨수집합족*

우리는 이 절과 다음 절에서 집합족, 즉 원소가 집합인 것의 모임을 다룬다.

이를테면 $\mathscr{P}=\{\{a\}, \{a, b\}, \{a, c\}, \{a, b, c\}\}$ ……①

는 하나의 집합족이다. 집합족에 대한 다음 정의부터 내려보자.

정의 하나의 집합 Γ의 각 $\gamma\in\Gamma$에 집합 A_γ가 대응할 때

$$\gamma,\ \Gamma,\ \{A_\gamma \mid \gamma\in\Gamma\}$$

를 차례로 **첨수, 첨수집합, 첨수집합족**이라고 한다.

지금 첨수집합으로 $\mathbb{N}$을 택하여 각 첨수 $n\in\mathbb{N}$에 집합 $A_n=\{n, 2n\}$을 대응시킴으로써 첨수집합족

$$\{A_n \mid n\in\mathbb{N}\}=\{A_1, A_2, A_3, \cdots, A_n, \cdots\} \mid n\in\mathbb{N}\} \quad \cdots\cdots②$$

을 얻는다. 이처럼 각 $n\in\mathbb{N}$에 대하여 $A_n=\{n, 2n\}$이라면 ②는 간단히 집합 $\{n, 2n \mid n\in\mathbb{N}\}$으로도 나타낼 수 있다.

예제 7 다음 족에 대한 첨수집합족을 구하여라.

$$\mathscr{F}=\{\phi, \mathbb{N}, \mathbb{Z}, \mathbb{Q}, \mathbb{R}, \mathbb{R}\}^{**} \quad \cdots\cdots③$$

[풀이] 첨수집합으로서 $\Gamma=\{1, 2, 3, 4, 5, 6\}$을 택하고

$$A_1=\phi,\ A_2=\mathbb{N},\ A_3=\mathbb{Z},\ A_4=\mathbb{Q},\ A_5=\mathbb{R},\ A_6=\mathbb{R}$$

와 같이 첨수를 매겨 다음과 같은 첨수집합족을 얻는다.

$$\mathscr{F}=\{A_1, A_2, A_3, A_4, A_5, A_6\} \quad \cdots\cdots④$$

실제로 족***을 다룰 때 용어와 기호는 집합의 경우와 같이 활용한다. 이를테면 "$\phi\in\mathscr{F}$, $\mathbb{R}_+\notin\mathscr{F}$"는 각각 집합에서와 같은 뜻을 지닌다.

이제 앞의 정의 3, 4에 따른 합집합 $\cup$, 교집합 $\cap$(87쪽)은 다음 정의 6에서 보듯이 각각 족으로 확장할 수 있다.

---

* 정의를 비롯하여 옮긴이는 본문을 간단 명료하게 수정해 놓았다.

** 집합족에는 같은 집합(③에서의 같은 집합 $\mathbb{R}$, $\mathbb{R}$)이 속할 수도 있다.

*** 집합족을 간단히 족이라고도 한다.

Definition 6. Let $\mathscr{F}$ be an arbitary family of sets. The union of the sets in $\mathscr{F}$, denoted by $\bigcup_{A\in\mathscr{F}} A$ or $\bigcup\mathscr{F}$, is the set of all elements that are in A for some $A\in\mathscr{F}$. That is,

$$\bigcup_{A\in\mathscr{F}} A=\{x\in U \mid x\in A \text{ for some } A\in\mathscr{F}\}$$

If the family $\mathscr{F}$ is indexed by the set Γ, the following alternate notation may be used:

$$\bigcup_{\gamma\in\Gamma} A_\gamma=\{x\in U \mid x\in A_\gamma \text{ for some } \gamma\in\Gamma\}$$

If the index set Γ is finite, $\Gamma=\{1, 2, 3, \cdots, n\}$ for some natural number n, more intuitive notations such as

$$\bigcup_{i=1}^{n} A_i \quad \text{or} \quad A_1\cup A_2\cup\cdots\cup A_n$$

are often used for $\bigcup_{\gamma\in\Gamma} A_\gamma$.

Example 8 Find the union of the family of sets

$$\{1\}, \{2, 3\}, \{3, 4, 5\}, \cdots, \{n, n+1, \cdots, 2n-1\}$$

[Solution] This family of sets may be considered as indexed by $\Gamma=\{1, 2, 3, \cdots, n\}$, where $A_i=\{i, i+1, \cdots, 2i-1\}$ for each $i\in\Gamma$. The problem reduces to that of finding $\bigcup_{i=1}^{n}\{i, i+1, \cdots, 2i-1\}$. Observe that each integer between 1 and $2n-1$ belongs to some A_i in the family, and no other element belongs to any of these A_i. Hence,

$$\bigcup_{i=1}^{n}\{i, i+1, \cdots, 2i-1\}=\{1, 2, 3, \cdots, 2n-1\}$$

정의 6 임의의 족 $\mathscr{F}$에 대하여 그것에 속하는 집합의 **합집합**을 $\bigcup_{A\in\mathscr{F}} A$ **또는** $\bigcup\mathscr{F}$로 나타낸다. 이것을 다음과 같은 등식으로 바꿔 놓을 수 있다.

$$\bigcup_{A\in\mathscr{F}} A=\{x\in U \mid \exists A\in\mathscr{F},\ x\in A\} \qquad \cdots\cdots ⑤^*$$

즉 ⑤는 족 $\mathscr{F}$에 속하는 모든 집합의 합집합을 뜻한다.

특히 $\mathscr{F}$가 첨수집합족일 때에는 ⑤를 다음과 같이 나타낸다.

$$\bigcup_{\gamma\in\Gamma} A_\gamma=\{x\in U \mid \exists\gamma\in\Gamma,\ x\in A_\gamma\} \qquad \cdots\cdots ⑥^*$$

여기서 Γ는 $\mathscr{F}$의 첨수집합이다. 지금 첨수집합 Γ가 유한집합

$$\Gamma=\{1, 2, 3, \cdots, n\} \qquad \cdots\cdots ⑦$$

인 경우 ⑥의 좌변의 기호 대신 다음과 같이 나타내기도 한다.

$$\bigcup_{i=1}^{n} A_i \quad \text{또는} \quad A_1\cup A_2\cup A_3\cup\cdots\cup A_n$$

보기 1 첨수집합족 ④(109쪽)의 합집합을 구하면 다음과 같다.

$$\begin{aligned}\bigcup_{i=1}^{n} A_i&=A_1\cup A_2\cup A_3\cup A_4\cup A_5\cup A_6\\&=\phi\cup\mathbb{N}\cup\mathbb{Z}\cup\mathbb{Q}\cup\mathbb{R}\cup\mathbb{R}=\mathbb{R}\end{aligned}$$

예제 8 다음과 같이 원소가 n개인 족 $\mathscr{F}$의 합집합을 구하여라.

$$\mathscr{F}=\{\{1\}, \{2, 3\}, \{3, 4, 5\}, \cdots, \{n, n+1, \cdots, 2n-1\}\}$$

[풀이] 족 $\mathscr{F}$는 첨수집합 ⑦에서의 각 $i\in\Gamma$에 $A_i=\{i, i+1, i+2, \cdots, 2i-1\}$가 대응하는 첨수집합족으로 볼 수 있다. 그러므로

$$\begin{aligned}\bigcup_{i=1}^{n}\{A_i \mid 1\le i\le 2n-1\}&=\bigcup_{i=1}^{n}\{i, i+1, i+2, \cdots, 2n-1\}\\&=\{1, 2, 3, \cdots, 2n-1\}\end{aligned}$$

따라서 구하는 합집합은 $\{1, 2, 3, \cdots, 2n-1\}$이다.

보기 2 원소가 n개인 집합 A와 그 멱집합 $\mathbf{P}(A)$(83쪽 정리 3)에 대하여 다음 등식이 성립한다(증명은 연습문제 2.6 문제 9*로 돌린다).

$$\bigcup\{\{a\} \mid a\in A\}=A, \quad \bigcup\{B \mid B\in\mathbf{P}(A)\}=A$$

* ⑤, ⑥에서의 A, A_γ는 각각 암암리에 받아들인 전체집합 U의 부분집합임을 상기할 필요가 있다.

Definition 7. Let **F** be an arbitrary family of sets. The intersection of sets in **F**, denoted by $\bigcap_{A \in \mathbf{F}} A$ or $\bigcap \mathscr{F}$, is the set of all elements that are in A for all $A \in \mathbf{F}$. That is,

$$\bigcap_{A \in \mathbf{F}} A = \{x \in U \mid x \in A \text{ for all } A \in \mathbf{F}\}$$

Here, the statement "$x \in A$ for all $A \in \mathbf{F}$" may be expressed alternately as "$A \in \mathbf{F} \rightarrow x \in A$." The latter expression has an advantage in proving theorems, as we shall see in the coming Theorem 7.

If the family **F** is indexed by the set Γ, the following alternate notation may be used:

$$\bigcap_{\gamma \in \Gamma} A_\gamma = \{x \in U \mid x \in A \text{ for all } \gamma \in \Gamma\}$$

If the index set Γ is finite, $\Gamma = \{1, 2, \cdots, n\}$ for some positive integer n, then as in the case of union we usually write

$$\bigcap_{i=1}^{n} A_i \quad \text{or} \quad A_1 \cap A_2 \cap \cdots \cap A_n$$

instead of $\bigcap_{\gamma \in \Gamma} A_\gamma$.

Let a and b be any two real numbers. By an open interval $(\boldsymbol{a}, \boldsymbol{b})$ we mean the subset $\{x \in \mathbb{R} \mid a < x < b\}$ of $\mathbb{R}$. It follows that if $a \geqq b$ then the interval $(a, b) = \phi$.

Example 9 Find the intersection of the family of open intervals

$$(0, 1), \quad (0, 1/2), \quad (0, 1/3), \quad \cdots$$

[Solution] We are to find the set $\bigcap_{n \in \mathrm{N}} (0, 1/n)$. Intuitively speaking, the given family is a sequence of "decreasing" intervals $(0, 1/n)$, where the interval $(0, 1/n)$ "approaches" the empty set ϕ as n becomes large. Therefore, we may conjecture that the intersection $\bigcap_{n \in \mathrm{N}} (0, 1/n)$ should be the empty set. We now prove that this conjecture is true. Suppose, on the contrary, that there exists some real number $a \in \bigcap_{n \in \mathrm{N}} (0, 1/n)$. Then we would have $0 < a < 1/n$ for all $n \in \mathrm{N}$. This contradicts the fact that for a fixed $a > 0$ there always exists a sufficiently large n in N such that $1/n < a$. The contradiction shows that $\bigcap_{n \in \mathrm{N}} (0, 1/n) = \phi$.

우리는 정의 6과 유사한 다음 정의를 내린다.

정의 7 임의의 족 $\mathscr{F}$에 대하여 그 것에 속하는 집합의 **교집합** $\bigcap_{A\in\mathscr{F}} A$ **또는** $\bigcap\mathscr{F}$ 를 다음과 같은 등식으로 바꿔 놓을 수 있다.

$$\bigcap_{A\in\mathscr{F}} A=\{x\in U \mid \forall A\in\mathscr{F},\ x\in A\} \quad \cdots\cdots ⑧$$

여기서 $\mathscr{F}$가 첨수집합족일 때에는 ⑧을 다음과 같이 나타내기도 한다.

$$\bigcap_{\gamma\in\Gamma} A_\gamma=\{x\in U \mid \forall\gamma\in\Gamma,\ x\in A_\gamma\}$$

이 때의 Γ는 물론 $\mathscr{F}$의 첨수집합이다.

여기서 첨수집합 Γ가 유한집합 ⑦인 경우에는 등식 ⑧의 좌변을 다음과 같이 나타내기도 한다.

$$\bigcap_{i=1}^{n} A_i \quad \text{또는} \quad A_1\cap A_2\cap\cdots\cap A_n$$

임의의 두 실수 a, b에 대한 집합 $\{x\in\mathbb{R} \mid a<x<b\}$를 **개구간**이라 하고 이것을 $\boldsymbol{(a, b)}$와 같이 나타낸다. 다만 $a\geqq b$인 경우 $(a, b)=\phi$으로 정한다.

예제 9 다음 개구간족의 교집합을 구하여라.

$$\{(0, 1),\ (0, 1/2),\ (0, 1/3),\ \cdots\}$$

[풀이] 교집합 $\bigcap_{n\in\mathrm{N}}(0, 1/n)$을 구하면 된다. 여기서 주어진 개구간의 열을 수직선 위에서 생각할 때 감소하는 구간 $(0, 1/n)$의 열로서 n이 커질수록 구간 $(0, 1/n)$은 공집합 ϕ에 점점 다가가는 것으로 볼 수 있어 구하는 교집합은 ϕ일 것으로 추측된다.

사실 우리는 이런 추측을 다음과 같이 배리법*으로 증명할 수 있다.

먼저 $\bigcap_{n\in\mathrm{N}}(0, 1/n)\neq\phi$라고 가정함으로써 구하려는 교집합에 속하는 실수 a가 존재한다고 가정하자. 즉 $a\in\bigcap_{n\in\mathrm{N}}(0, 1/n)$. 그러면 모든 $n\in\mathbb{N}$에 대하여 $0<a<1/n$. 이때 $1/a<n$, 즉 $1/n<a$를 만족할 만한 $n\in\mathbb{N}$이 존재한다. 이것은 $a<1/n$에 모순이다(왜?).

따라서 $$\bigcap_{n\in\mathrm{N}}(0, 1/n)=\phi$$

* 정리를 증명하는 데 집적증명보다 배리법이 편리할 때가 많다[35쪽 정리 6(c), 연습문제 1.4의 3 및 57쪽 예제 11 참조].

Theorem 7. Let $\{A_\gamma \mid \gamma \in \Gamma\}$ be the empty family of sets; that is, $\Gamma = \phi$. Then

(a) $\bigcup_{\gamma \in \phi} A_\gamma = \phi$ (b) $\bigcap_{\gamma \in \phi} A_\gamma = U$

[Proof] (a) To show $\bigcup_{\gamma \in \phi} A_\gamma = \phi$, we show equivalently that $x \notin \bigcup_{\gamma \in \phi} A_\gamma$ for all x (in U):

$$
\begin{aligned}
x \notin \bigcup_{\gamma \in \phi} A_\gamma &\equiv \sim (x \in \bigcup_{\gamma \in \phi} A_\gamma) && \text{Notation} \\
&\equiv \sim (x \in A_\gamma \text{ for some } \gamma \in \phi) && \text{Def. 6} \\
&\equiv x \notin A_\gamma \quad \text{for all} \quad \gamma \in \phi && \text{Q.N. (Ch. 1)} \\
&\equiv \gamma \in \phi \rightarrow x \notin A_\gamma
\end{aligned}
$$

The last statement is, by Theorem 7(c) of Chapter 1, true for all $x \in U$, because $\gamma \in \phi$ is a contradiction to $\gamma \notin \phi$. This completes the proof of part (a).

(b) We shall prove that $x \in \bigcap_{\gamma \in \phi} A_\gamma$ for all x in U. Observe that

$$
\begin{aligned}
x \in \bigcap_{\gamma \in \phi} A_\gamma &\equiv x \in A_\gamma \text{ for all } \gamma \in \phi && \text{Def. 7} \\
&\equiv \gamma \in \phi \rightarrow x \in A_\gamma
\end{aligned}
$$

The last assertion is, as we have explained in the proof of part (a), a true statement for all $x \in U$. The proof is complete.

Many theorems concerning operations of finitely many sets can be generalized to theorems concerning operations of an arbitary family of sets. For example, the following generalizes De Morgan's Theorem. The student should compare this theorem with Theorem 6.

Theorem 8. The Generalized De Morgan Theorem Let $\{A_\gamma \mid \gamma \in \Gamma\}$ be an arbitrary family of sets. Then

(a) $\left(\bigcup_{\gamma \in \Gamma} A_\gamma\right)^C = \bigcap_{\gamma \in \Gamma} A_\gamma^C$ (b) $\left(\bigcap_{\gamma \in \Gamma} A_\gamma\right)^C = \bigcup_{\gamma \in \Gamma} A_\gamma^C$

정리 7* 첨수집합족 $\{A_\gamma \mid \gamma \in \Gamma\}$에 대하여 $\Gamma = \phi$이면 다음이 성립한다.

(a) $\bigcup_{\gamma \in \phi} A_\gamma = \phi$ (b) $\bigcap_{\gamma \in \phi} A_\gamma = U$

[증명] (a) 모든 $x \in U$에 대하여 $x \notin \bigcup_{\gamma \in \phi} A_\gamma$임을 밝히면 충분하다.

$$x \notin \bigcup_{\gamma \in \phi} A_\gamma \equiv \sim\left(x \in \bigcup_{\gamma \in \phi} A_r\right) \qquad \text{기호 } \bigcup_{\gamma \in \phi} A_r$$

$$\equiv \sim(\text{적당한 } \gamma \in \phi\text{에 대하여 } x \in A_r) \qquad \text{정의 6}$$

$$\equiv (\text{모든 } \gamma \in \phi\text{에 대하여 } x \notin A_\gamma) \qquad \text{한정기호의 부정 규칙(제1장)}$$

$$\equiv (\gamma \in \phi \rightarrow x \notin A_r) \qquad \cdots\cdots ⑨^{**}$$

여기서 조건문 ⑨의 가정 $\gamma \in \phi$는 공집합의 정의에 따른 $\gamma \notin \phi$에 모순이다. 그러므로 제 1 장 정리 7(c)에 따라 모든 $x \in U$에서의 $(\gamma \in \phi \rightarrow x \notin A_\gamma)$는 참이다. 이로써 증명은 끝난다.

(b) 모든 $x \in U$에 대하여 $x \in \bigcap_{\gamma \in \phi} A_\gamma$임을 밝히기로 한다.

$$x \in \bigcap_{\gamma \in \phi} A_\gamma \equiv \text{모든 } \gamma \in \phi\text{에 대하여 } x \in A_\gamma \qquad \text{정의 7}$$

$$\equiv (\gamma \in \phi \rightarrow x \in A_\gamma)$$

그러므로 $$x \in \bigcap_{\gamma \in \phi} A_\gamma \equiv (\gamma \in \phi \rightarrow x \in A_\gamma)$$

여기서도 모든 $x \in U$에 대하여 $(\gamma \in \phi \rightarrow x \in A_\gamma)$는 참이다. 이로써 증명은 완료된다.

유한집합의 연산에 관한 정리 중 많은 것이 첨수집합족으로 확장될 수 있다. 이를테면 다음은 드 모르간의 정리(99쪽)를 일반화한 것이다.

정리 8 드 모르간의 정리 임의의 첨수집합족 $\{A_\gamma \mid \gamma \in \Gamma\}$에 대하여 다음이 성립한다.

(a) $\left(\bigcup_{\gamma \in \Gamma} A_\gamma\right)^C = \bigcap_{\gamma \in \Gamma} A_\gamma^C$ (b) $\left(\bigcap_{\gamma \in \Gamma} A_\gamma\right)^C = \bigcup_{\gamma \in \Gamma} A_r^C$

* (a), (b)의 증명은 모두 형식상 이해되지만 첨수집합 자체로 하여금 ϕ라는 가정에선 별 뜻을 찾지 못한다.

** 이를테면 "모든 $A \in \mathbf{F}$에 대하여 $x \in A$"를 "$A \in \mathbf{F} \rightarrow x \in A$"로 바꿔 놓으면 증명하는 데 편리하다.

[Proof] We shall only prove part (a), and leave part (b) to the student.

$$\begin{aligned} x\in\Big(\bigcup_{\gamma\in\Gamma} A_\gamma\Big)^C &\equiv \sim\Big(x\in\bigcup_{\gamma\in\Gamma} A_\gamma\Big) && \text{Def. of }^C \\ &\equiv \sim(\exists_{\gamma\in\Gamma}\, x\in A_\gamma) && \text{Def. 6} \\ &\equiv \forall_{\gamma\in\Gamma}\, x\notin A_\gamma && \text{Q.N. (Ch. 1)} \\ &\equiv \forall_{\gamma\in\Gamma}\, x\in A_\gamma^C && \text{Def. of }^C \\ &\equiv x\in\bigcap_{\gamma\in\Gamma} A_\gamma^C && \text{Def. 7} \end{aligned}$$

Therefore, by Definition 1, $\Big(\bigcup_{\gamma\in\Gamma} A_\gamma\Big)^C \equiv \bigcap_{\gamma\in\Gamma} A_\gamma^C$

The following theorem is a generalization of Theorem 4(e).

Theorem 9. Generalized distributive Laws Let A be a set and let $\mathscr{F}=\{B_\gamma \mid \gamma\in\Gamma\}$ be an arbitrary family of sets. Then

(a) $A\cap(\bigcup_{\gamma\in\Gamma} B_\gamma)=\bigcup_{\gamma\in\Gamma}(A\cap B_\gamma)$ (b) $A\cup(\bigcap_{\gamma\in\Gamma} B_\gamma)=\bigcap_{\gamma\in\Gamma}(A\cup B_\gamma)$

[Proof] (a) An element x is in the set $A\cap(\bigcup_{\gamma\in\Gamma} B_\gamma)$ if and only if

$$x\in A \quad\text{and}\quad x\in\bigcup_{\gamma\in\Gamma} B_\gamma,$$

which, according to Definition 6, is equivalent to

$$x\in A \quad\text{and}\quad x\in B_\gamma \quad\text{for some}\quad \gamma\in\Gamma$$

The last assertion may be expressed, by Definition 4, as

$$x\in A\cap B_\gamma \quad\text{for some}\quad \gamma\in\Gamma$$

which, by Definition 6, is precisely $x\in\bigcup_{\gamma\in\Gamma}(A\cap B_\gamma)$. Thus, by Definition 1,

$$A\cap(\bigcup_{\gamma\in\Gamma} B_\gamma)=\bigcup_{\gamma\in\Gamma}(A\cap B_\gamma)$$

The proof for part (b) is left as an exercise.

[증명] (a)만 증명하고 (b)는 연습문제로 돌린다.

$$x \in \Big(\bigcup_{\gamma\in\Gamma} A_\gamma\Big)^C \equiv \sim\Big(x \in \bigcup_{\gamma\in\Gamma} A_\gamma\Big) \qquad {}^C\text{의 정의}$$

$$\equiv \sim\Big(\exists_{\gamma\in\Gamma}\, x\in A_\gamma\Big) \qquad \text{정의 6}$$

$$\equiv \forall_{\gamma\in\Gamma}\, x\notin A_\gamma \qquad \text{한정기호의 부정규칙(제1장)}$$

$$\equiv \forall_{\gamma\in\Gamma}\, x\in A_\gamma^C \qquad {}^C\text{의 정의}$$

$$\equiv x\in \bigcap_{\gamma\in\Gamma} A_\gamma^C \qquad \text{정의 7}$$

따라서 $x\in(\bigcup_{\gamma\in\Gamma} A_\gamma)^C \equiv x\in \bigcap_{\gamma\in\Gamma} A_\gamma^C$ ≡에 관한 추이율

그러므로 정의 1에 따라 $\Big(\bigcup_{\gamma\in\Gamma} A_\gamma\Big)^C \equiv \bigcap_{\gamma\in\Gamma} A_\gamma^C$

다음 정리는 정리 4(e)(89쪽)를 일반화한 것이다.

정리 9 **분배법칙*** 집합 A와 임의의 첨수집합족 $\mathscr{F}=\{B_\gamma \mid \gamma\in\Gamma\}$에 대하여 다음이 성립한다.

(a) $A\cap(\bigcup_{\gamma\in\Gamma} B_\gamma)=\bigcup_{\gamma\in\Gamma}(A\cap B_\gamma)$ (b) $A\cup(\bigcap_{\gamma\in\Gamma} A_\gamma)=\bigcap_{\gamma\in\Gamma}(A\cup B_\gamma)$

[증명] (a) $x\in A\cap(\bigcup_{\gamma\in\Gamma} B_\gamma) \Longleftrightarrow x\in A \wedge x\in \bigcup_{\gamma\in\Gamma} B_\gamma$.

이 주장은 정의 6에 따라 다음과 동치이다.

$$x\in A \wedge \exists\gamma\in\Gamma\; x\in B_\gamma$$

이 마지막 주장은 정의 4에 따라 다음과 같이 나타낼 수 있다.

$$\exists\gamma\in\Gamma\; x\in A\cap B_\gamma$$

이것은 또한 정의 6에 따라 바로 $x\in\bigcup_{\gamma\in\Gamma}(A\cap B_\gamma)$임을 뜻하므로 정의 1에 따라 다음 등식을 얻는다.

$$A\cap(\bigcup_{\gamma\in\Gamma} B_\gamma)=\bigcup_{\gamma\in\Gamma}(A\cap B_\gamma)$$

(b)의 증명은 독자의 연습문제로 돌린다.

* 분배법칙을 분배율이라고도 한다.

Exercise 2.6

1. Let $\Gamma=\{1,2,3,4\}$ and $A_1=\{a, b, c, d\}$, $A_2=\{b, c, d\}$, $A_3=\{a, b, c\}$, $A_4=\{a, b\}$. Find the following.

(a) $\bigcup_{i=1}^{4} A_i$ (b) $\bigcap_{i=1}^{4} A_i$

2. For any two real numbers a and b, by the closed interval $[a, b]$ we mean the set $\{x\in\mathbb{R} \mid a\leqq x\leqq b\}$. If $a>b$, $[a, b]=\phi$. Find the following sets.

(a) $\bigcap_{n\in\mathrm{N}}[0, 1/n]$ (b) $\bigcup_{n\in\mathrm{N}}[0, 1/n]$ (c) $\bigcap_{n=1}^{99}[0, 1/n]$

3. Prove Theorem 8(b): $(\bigcap_{\gamma\in\Gamma} A_\gamma)^C=\bigcup_{\gamma\in\Gamma} A_\gamma{}^C$

4. Prove Theorem 9(b): $A\cup(\bigcap_{\gamma\in\Gamma} B_\gamma)=\bigcap_{\gamma\in\Gamma}(A\cup B_\gamma)$

5. Expand

(a) $(A_1\cup A_2)\cap(B_1\cup B_2\cup B_3)$ into a union of intersections, and

(b) $(A_1\cap A_2)\cup(B_1\cap B_2\cap B_3)$ into an intersection of unions.

[Hint: Use Theorem 9 several times.]

6. Expand

(a) $(\bigcup_{i=1}^{m} A_i)\cap(\bigcup_{j=1}^{m} B_j)$ into a union of intersections, and

(b) $(\bigcap_{i=1}^{m} A_i)\cup(\bigcap_{j=1}^{m} B_j)$ into an intersection of unions. (See Problems 5)

7. Let $\{A_\gamma \mid \gamma\in\Gamma\}$ and $\{B_\delta \mid \delta\in\Delta\}$ be any two families of sets. Expand

(a) $(\bigcup_{\gamma\in\Gamma} A_\gamma)\cap(\bigcup_{\delta\in\Delta} B_\delta)$ into a union of intersections, and

(b) $(\bigcap_{\gamma\in\Gamma} A_\gamma)\cup(\bigcap_{\delta\in\Delta} B_\delta)$ into an intersection of unions. (See Problems 5 and 6.)

연습문제 2.6

1. $\Gamma=\{1,2,3,4\}$, $A_1=\{a,b,c,d\}$, $A_2=\{b,c,d\}$, $A_3=\{a,b,c\}$, $A_4=\{a,b\}$에 대하여 다음을 구하여라.

(a) $\bigcup_{i=1}^{4} A_i$ (b) $\bigcap_{i=1}^{4} A_i$

2. 임의의 두 실수 a, b에 대하여 폐구간 $[a,b]$는 집합 $\{x\in\mathbb{R}\mid a\leqq x\leqq b\}$을 뜻한다. 여기서 $a>b$이면 $[a,b]=\phi$으로 정한다. 다음을 구하여라.

(a) $\bigcap_{n\in\mathbb{N}}[0,1/n]$ (b) $\bigcup_{n\in\mathbb{N}}[0,1/n]$ (c) $\bigcap_{n=1}^{99}[0,1/n]$

3. 정리 8(b)(115쪽), 정리 9(b)(117쪽)를 각각 증명하여라.

4.* 공집합 ϕ에 대하여 $\cup\mathscr{P}(\phi)=\cup\{\phi\}=\phi$(83쪽 보기 2와 비교하여라)이다. 다음을 구하여라.

(a) $\cup\mathscr{P}(\mathscr{P}(\phi))$ (b) $\cup\mathscr{P}(\mathscr{P}(\mathscr{P}(\phi)))$

5. 다음 (a), (c)는 교집합의 합집합으로, (b), (d)는 합집합의 교집합으로 각각 전개하여라(힌트: 정리 9(a), (b)를 반복 이용하여라).

(a) $(A_1\cup A_2)\cap(B_1\cup B_2\cup B_3)$ (b) $(A_1\cap A_2)\cup(B_1\cap B_2\cap B_3)$

(c) $(\bigcup_{i=1}^{m} A_i)\cap(\bigcup_{j=1}^{m} B_j)$ (d) $(\bigcap_{j=1}^{m} A_i)\cup(\bigcap_{j=1}^{m} B_j)$

6. 첨수집합족 $\{A_\gamma\mid\gamma\in\Gamma\}$, $\{B_\delta\mid\delta\in\Delta\}$에 대하여 다음 (a)는 교집합의 합집합으로, (b)는 합집합의 교집합으로 각각 전개하여라(문제 5 참조).

(a) $(\bigcup_{\gamma\in\Gamma} A_\gamma)\cap(\bigcup_{\delta\in\Delta} B_\delta)$ (b) $(\bigcap_{\gamma\in\Gamma} A_\gamma)\cup(\bigcap_{\delta\in\Delta} B_\delta)$

7.* 첨수집합족 $\{S_1,S_2,S_3,\cdots,S_n\}$과 임의의 집합 S에 대하여 다음이 성립한다. 이를 증명하여라.

(a) $\bigcup_{i=1}^{n}(S-S_i)=S-(\bigcap_{i=1}^{n} S_i)$ (b) $\bigcap_{i=1}^{n}(S-S_i)=S-(\bigcup_{i=1}^{n} S_i)$

8.* 족 $\{A_1,A_2,A_3,\cdots,A_n\}$에 대하여

$B_1=A_1, B_2=A_2-A_1, B_3=A_3-(A_1\cup A_2),\cdots, B_n-(A_1\cup A_2\cup A_3\cup\cdots\cup A_{n-1})$

라고 놓으면 다음이 성립한다. 이를 증명하여라.

(a) $B_i\cap B_j=\phi(i\neq j)$ (b) $\bigcup_{i=1}^{n} A_i=\bigcup_{i=1}^{n} B_i$

9.* 보기 5(111쪽)를 증명하여라.

* 원문에서의 문제 3,4를 문제 3에 또 5,6은 문제 5에 묶어 놓는다. 한편 문제 4,7 및 9는 옮긴이가 모두 보충한 것이니 혼돈없기 바란다.

2.7 The Russell Paradox

Now many of us may think we understand what is meant by a set — at least intuitively. Most of us taking a Set Theory course for the first time would not notice what is wrong in considering "the set of all sets" or the so-called "universal set" in the absolute sense. In fact, for a period of time (at least from 1895 when Georg Cantor first created a theory of sets, until 1902 when the Russell Paradox appeared), the existence of such a universal set was taken for granted. It was the famous English philosopher Bertrand Russell (1872-1970)[5] who shook the mathematics community in 1902 by declaring that the admission of a set of all sets would lead to a contradiction. This is the famous Russell Paradox. We present this paradox as two seemingly contradictory lemmas from which a theorem follows.

Lemma 1. Suppose that there is a set $\mathscr{U}$ of all sets. Let R be the set $R=\{ S\in\mathscr{U} \mid S\notin S \}$.[6] Then $R\notin R$.

[Proof] Suppose, on the contrary, that $R\in R$. Then by the specification of the set R, we must have $R\notin R$, which contradicts the assumption that $R\in R$. The contradiction proves that $R\notin R$.

---

[5] Bertrand Russell was born on May 18, 1872, at Trelleck, Wales. Before he was four, both of his parents died. He had been a shy, silent boy until he entered Trinity College, Cambridge University, in 1890. After three years of mathematics he concluded that what he was being taught was full of errors. He sold his mathematics books and changed to philosophy. In his Principia Mathematica(1910-1913), a three-volume monumental work co-authored with Alfred North Whitehead (1861-1947), he attempted to recast set theory so as to avoid paradoxes. In 1918 he wrote, "I want to stand at the rim of the world and peer into the darkness beyond, and see a little more than others have seen... I want to bring back into the world of men some little bit of wisdom,." He certainly did, more than just "some little bit." In the same year, he was put in prison for an unfavorable comment about the American Army. In 1950 he received the Order of Merit from the King of England and the Nobel prize for literature. In his later years he led a number of demonstrations against nuclear warfare.

[6] According to the Axiom of Specification, R is a set which is often called "the Russell set."

2.7 러셀의 역리

지금까지의 공부로 봐서 집합의 뜻을 분명히 이해한다고 자부할 독자가 있을지 모른다. 그러나 처음으로 집합론 과목을 택하는 독자 대부분은 "모든 집합의 집합" 또는 절대적 뜻에서의 "전체집합"을 논할 때 무슨 잘못이 잠재되 있을지를 알아차리지 못하였다. 사실 같은 전체집합의 존재는 한 시기(적어도 칸토어가 처음으로 집합론을 창조하던 1895년부터 러셀의 역리가 나타난 1902년까지)에는 당연한 것으로 받아들여졌다. 그러나 영국의 유명한 철학자 러셀(1872-1970)[5]은 1902년 집합을 원소로 하는 모든 집합의 집합을 허용한다면 모순이 일어난다는 사실을 주장함으로써 수학계에 충격을 주었다. 이러한 그의 주장은 유명한 러셀의 역리로 알려져 있다. 우리는 다음 두 보조정리에서 모순을 일으킴으로써 정리 10을 유도하기로 한다.

보조정리 1 모든 집합의 집합, 말하자면 $\mathscr{U}$가 존재한다고 전제하고

$$R=\{S\in\mathscr{U} \mid S\notin S\}^{6)} \qquad \cdots\cdots ①$$

라고 놓으면 $R\notin R$

[증명] $\mathscr{U}$는 모든 집합의 족이므로 $R\in\mathscr{U}$. 지금 ①의 결론의 부정 $\sim(R\notin R)$ 즉 $R\in R$이라고 가정하고 보자. 그럴 경우 등식 ①의 좌변 R와 조건제시법에 따른 우변에서의 문자 S, $\mathscr{U}$를 각각 감안하면 $R\notin R$. 이것은 가정 $R\in R$에 모순이다. 따라서 $R\notin R$.

---

5) 1872년 3월 18일 웨일스에서 태어난 러셀은 4세 전에 부모를 잃었다. 그는 1890년 케임브리지 대학교의 트리니티 컬리지에 입학할 때까지 수줍음을 타는 조용한 소년이었다. 3년간의 수학공부 끝에 교육받아온 것은 틀린 것으로 차 있다고 결론을 지은 그는 수학책들을 팔아버리고 철학으로 공부를 바꾸었다. 화이트헤드(1861-1913)와 공동으로 집필한 불멸의 세 권의 저서인 《수학의 원리》(1910-1913)에서 집합론을 재구성함으로써 역리를 제거하고자 노력하였던 그는 1918년에 다음과 같이 적고 있다. "나는 (지적)세계의 테두리에 서서 저 너머의 무지를 통찰함으로써 남들보다 좀더 많은 것을 알기 원하며 인류에게 작은 지혜를 상기시키고자 한다." 그는 분명히 "작은" 보다 더 많은 것을 이루어냈다. 그해에 러셀은 미군에 관해 거슬리는 논평을 한 이유로 옥고를 치르기도 하였다. 1950년에는 영국 왕으로부터 메리트 훈장을, 또 노벨문학상을 받기도 하였던 러셀은 말년에 핵전쟁을 반대하는 여러 시위운동에 앞장 서기도 하였다.

6) 분출공리(81쪽 참조)에 따라 R은 흔히 러셀집합이라고 하는 하나의 집합이다.

Lemma 2. Suppose that there is a set $\mathscr{U}$ of all sets. Let R be the set $\{S\in\mathscr{U} | S\notin S\}$. Then $R\in R$.

[Proof] Suppose, on the contrary, that $R\notin R$. Then since $R\in\mathscr{U}$, we have $R\in R$ by the definition of R. This is a contradiction. Thus, $R\in R$.

Theorem 10. There does not exist a set of all sets.

[Proof] In view of Lemmas 1 and 2, the set of all sets cannot exist. For, if it were otherwise, it would lead to the contradiction "$R\notin R$ and $R\in R$."

Paul R. Halmos puts it this way: "Nothing contains everything."[7)]

2.8 A Historical Remark

The modern theory of sets is generally considered to have been created in 1895 by the famous mathematician Georg Cantor[8)] (1845-1918), who noticed the need for such a theory while studying trigonometric series.

Cantor wrote; "By a 'set' we shall understand any collection into a whole of definite distinct objects of our intuition or thought." This definition does not prohibit anyone from considering the "set" of all sets, as Bertrand Russell did. The real difficulty in Cantor's definition of a set is the word. "collection." What is a collection? Of course we can look it up in a dictionary and find something like these definitions:

"collection: a group of collected objects."
"group: an aggregate or collection."
"aggregate: a collection."

7) Paul R. Halmos, Naive Set Theory, D. Van Nostrand Company, Inc., New York, 1960, p. 6.

8) Georg Cantor was born in St. Petersburg, Russia, in 1845, moved to Germany in 1856, studied mathematics at the University of Berlin (1863-1869), and taught at the University of Halle (1869-1905). One of Cantor's interests was trigonometric series, which led him to look into the foundation of analysis. As a result, he created the revolutionary work on set theory and an arithmetic of transfinite numbers.

보조정리 2* 모든 집합의 족 말하자면 $\mathscr{U}$가 존재한다고 전제하고

$$R=\{S\in\mathscr{U}\,|\,S\notin S\} \qquad \cdots\cdots②$$

라고 놓으면 $R\in R$

[증명] 독자에게 맡긴다(보조정리 1의 증명 참고).

정리 10 집합을 원소로 하는 모든 집합의 족 $\mathscr{U}$는 존재하지 않는다.

[증명] 보조정리 1, 2의 결론에 따라 우리는 다음을 얻는다.

$$(R\notin R)\wedge(R\in R)=[\sim(R\in R)]\wedge(R\in R)=c$$

이 모순은 모든 집합의 족, 즉 $\mathscr{U}$가 존재한다는 전제로 야기된 것인 만큼 $\mathscr{U}$는 존재하지 않는다.

이런 사실을 놓고 미국의 수학자 할모스라는 다음과 같이 말하고 있다.

"어떤 것도 모든 걸 담을 수 없다."[7)]

2.8 역사적 단평

대체로 현대집합론은 유명한 수학자 게오르크 칸토어(1845-1918)[8)]가 1895년에 창조한 것으로 알려져 있다. 그는 삼각함수열을 연구하다가 그 같은 이론에 대한 필요성을 주목하게 됐다면서 다음과 같이 적고 있다.

"집합을 두고 우리는 직관 또는 사고의 대상으로서 분명히 구분되는 것의 모든 집단으로 이해하려고 한다."

그런데 이 정의로 말미암아 러셀이 이미 지적한 대로 누구에게서나 모든 집합의 집합에 대한 생각을 가로막는다. 칸토어가 내린 집합의 정의에 대하여 참으로 곤란한 점은 "모임"이라는 단어에 있다. 그렇다면 모임이란 대체로 무엇일까? 물론 사전을 뒤져 다음과 같은 정의들과 동의어를 찾을 수 있다.

"집단: 수집한 대상들의 모임", "모임: 집합체 또는 집단", "집합체: 집단".

---

7) 할모스의 저서 : 《소박한 집합론》, D. Van Nostrand Company, Inc., New York 1960, 6쪽 참조.

8) 1845년 러시아의 성 페테르스부르크에서 태어난 칸토어는 1856년에 독일로 옮겨 베를린 대학교에서 수학공부를 하였으며(1863-1869) 대학에서 강의를 맡았다(1869-1905). 칸토어의 흥미 중의 하나는 해석학의 기초가 되는 삼각함수열이였으며, 그는 처음으로 집합론과 초한수의 셈에 관해 크나큰 업적을 이루었다.

These will hardly be of any help. When a mathematician gives a definition, it is not intended to be a mere synonym such as "collection" for "set", or a circular definition as we would find in a dictionary. Cantor apparently was not aware that the term "set" was really undefinable.

To avoid any difficulty such as the Russell Paradox in set theory, we must accept the terms "set" and "element" as undefined terms, or primitives, and guide these primitives by a number of axioms, including the Axiom of Specification and the Axiom of Power Sets that have been introduced in Section 2. Other axioms such as "$A=B$ if and only if A and B contain the same elements" (Axiom of Extension), "ϕ is a set" (Axiom of the Empty Set), "If A and B are sets then so is $\{A, B\}$" (Axiom of Pairing), and "If $\mathscr{F}$ is a set of sets then $\mathscr{F}$ is a set" (Axiom of Unions) are often given in axiomatic treatments of set theory.

The Russell Paradox was not the only one to arise in set theory. Shortly after the Russell Paradox appeared, many paradoxes were constructed by several mathematicians and logicians. As a consequence of all these paradoxes, many mathematicians and logicians have contributed to several brands of "axiomatic set theory" each designed to avoid these paradoxes and, at the same time, to preserve the main body of Cantor's set theory. However, at the time of this writing, no one has yet come up with a completely satisfactory axiomatic system for the set theory.

Despite the aforementioned difficulties, Cantor's set theory has today penetrated into every branch of modern mathematics, and it has proved to be of particular importance in the foundations of modern analysis and in topology. In fact, even the very simplest full-fledged axiomatic systems of set theory are entirely adequate for doing virtually all of classical mathematics (e.g., the theory of real and complex numbers, algebra, topology, etc.).

그러나 이런 정의들로는 부족하다. 왜냐하면 "집합"에 대한 "집단"처럼 단순한 동의어이거나 또는 사전에서와 같은 순환에 빠지는 정의이어서도 안 되기 때문이다. 추측하건데 칸토어는 "집합"이란 용어의 정의를 내릴 수 없는 사실을 알아차리지 못한것 같다.

집합론에 있어서 러셀의 역리와 같은 난점을 해소하려면 "집합"과 "원소"의 용어들은 무정의 용어로서 받아들여야 하고, 이들 무정의 용어는 제2장에서 소개한 것처럼 분출의 공리, 멱집합의 공리를 위시하여 몇개의 공리들로 하여금 집합에 관한 규제를 하도록 해야 한다. 여기서 공리적으로 집합론을 다룰 경우 필요로 하는 공리를 몇 개 더 들 수 있다.

외연성 공리: 집합 A, B의 원소가 같다. $A=B$
공집합의 공리: ϕ는 하나의 집합이다.
쌍의 공리: 집합 A, B에 대하여 $\{A, B\}$도 집합이다.
합집합의 공리: 집합이 원소인 것의 모임을 $\mathscr{F}$라면 그 $\mathscr{F}$는 집합이다.

집합론에서 대두하는 역리 중에는 러셀의 역리 이외의 것도 있다. 러셀의 역리가 나타난지 얼마 안가서 또 다른 역리들을 몇 명의 수학자와 논리학자들이 제시하였다. 이러한 모든 역리의 결과로써 많은 수학자와 논리학자는 공리적 집합론을 다양하게 구성하기에 이르렀지만 그 어느 집합론이건 역리를 극복하되 칸토어 집합론의 주요 골격만은 유지할 수 있도록 구성되었다. 그렇다고 하더라도 완전한 집합론을 구성한 수학자는 이 책을 집필하고 있는 현시점까지 아무도 없다*.

위에서 언급한 단점에도 불구하고 오늘날 현대 수학의 각 분야에 파급되어 있는 칸토어의 집합론은 각별히 현대해석학의 기초와 함께 위상수학에서의 중요성을 입증하여 왔다. 사실 제 구실을 다 하고 있는 집합론의 공리적 체계 중 가장 간단한 것일지라도 이는 실제로 모든 전통수학, 이를테면 실수와 복소수의 이론, 대수학, 위상수학 등을 다루어 나가는 데 있어서 전적으로 적당하다고 말할 수 있다.

---

* Charles C. Pinter, *Set theory*, Addison-Wesley Publishing Company, 1971, 1~20쪽 및 Chapter 1 Classes and Sets 중 2,3절 및 7절 참조.

Chapter 3
Relations and Functions

제 3 장
관계와 함수

3

Relations and Functions

The chapter begins with a discussion of ordered pairs and the Cartesian product of two sets. A relation is then defined as a set of ordered pairs. The intimate connection between a partition and an equivalence relation on a set is closely examined. The concept of a function is introduced as a special kind of relation. As a preparation for those readers who wish to pursue more modern mathematics, important properties of functions are studied. An abundance of examples is provided.

3.1 Cartesian Product of Two Sets

Given any two objects a and b, we may form a new object (a, b), called the ordered pair of a, b.[1] The adjective "ordered" here emphasizes that the order in which the objects a and b appear in the brackets is essential. Thus, (a, b) and (b, a) are two distinct ordered pairs. It should be noted that an ordered pair (a, b) is not the same as the set $\{a, b\}$. There is a satisfactory, but somewhat complicated, way of defining the ordered pair (a, b) as the set $\{\{a\}, \{a, b\}\}$, from which the property

$$(a, b)=(c, d) \Longleftrightarrow a=c \text{ and } b=d$$

follows(See Problem 11, Exercise 3.1.)

[1] Unfortunately the notation (a, b) for an ordered pair is the same as the notation for an open interval when a and b are real numbers. However, the careful reader should always be able to make the distinction from the context.

3

관계와 함수

이 장에서는 두 집합의 카테시안곱을 위시하여 하나의 *관계*를 순서쌍의 집합으로서 정의하고 집합 위에서의 *분할*과 *동치관계* 사이의 깊은 관련을 살펴본다.

한편 두 대상 사이의 특별한 관계로서 *함수*의 개념을 도입하고 현대수학을 더욱 공부하려는 학생들을 위하여 함수에 관한 중요한 성질도 다룬다. 특히 이 장에는 보기들도 여럿 실려 있다.

3.1 두 집합의 데카르트곱*

임의의 두 대상 a, b에 대하여 우리는 이들의 **순서쌍**[1)]이라고 하는 (a, b)를 구성할 수 있다. 여기서 형용사 "순서진"은 소괄호 속에서의 a, b의 순서의 중요함을 강조하고 있다. 한편 순서쌍 (a, b)는 집합 $\{a, b\}$와 다르다는 것을 주의해야 한다. 사실 순서쌍을 집합 $\{\{a\}, \{a, b\}\}$로서 정의하면 바람직할지 모르지만 다소 복잡하다. 어떻든 이 정의로부터 우리는 다음과 같은 성질을 얻는다[연습문제 3.1 문제 11(137쪽) 참조].

$$(a, b) = (c, d) \Longleftrightarrow a = c \wedge b = d$$

* 데카르트(Descartes René, 1596~1650)를 두고 근세철학의 아버지로 부르는 이유는 그가 중세 이후의 신학적 세계관 대신 일단 "모든 것을 의심해 본 뒤" 합리적으로 지식을 구성할 것을 주장하였기 때문이다. 이런 사상에 힘입어 근대수리문리학의 기초가 닦아졌다. 데카르트는 대수학을 활용하여 기하학의 문제를 다룸으로써 해석기하학(analitic geometry)을 개척한 수학자이기도 하다.

1) 공교롭게 순서쌍의 기호 (a, b)는 a, b가 실수일 때 개구간을 나타내는 기호와 같다. 그러나 주의 깊은 독자라면 앞뒤 문맥으로 미루어 이들을 쉽게 구분할 수 있을 것이다.

Two ordered pairs (a, b) and (c, d) are said to be equal $(=)$ if and only if $a=c$ and $b=d$. For example, $(x, y)=(7, 8)$ if and only if $x=7$ and $y=8$.

In analytic geometry, the Cartesian plane may be considered as the set of all ordered pairs of real numbers. We state this concept formally as follows :

Definition 1. Let A and B be any nonempty two sets. The set of all ordered pairs (x, y), with $x\in A$ and $y\in B$, is called the Cartesian product of A and B, and is denoted by $\boldsymbol{A\times B}$. In symbols

$$A\times B=\{(x, y)\mid x\in A\wedge y\in B\}$$

For the ordered pair (a, b), a is called the first coordinate and b is the second coordinate.

Example 1 Let $A=\{a, b, c\}$ and $B=\{1,2\}$. Find the Cartesian products $A\times B$ and $B\times A$.

[Solution] By Definition 1 above, we have

$$A\times B=\{(a, 1), (a, 2), (b, 1), (b, 2), (c, 1), (c, 2)\}$$

and $$B\times A=\{(1, a), (1, b), (1, c), (2, a), (1, b), (2, c)\}$$

We notice that $A\times B\neq B\times A$. We may picture the Cartesian product $A\times B$ as the set of dots in the following figure.

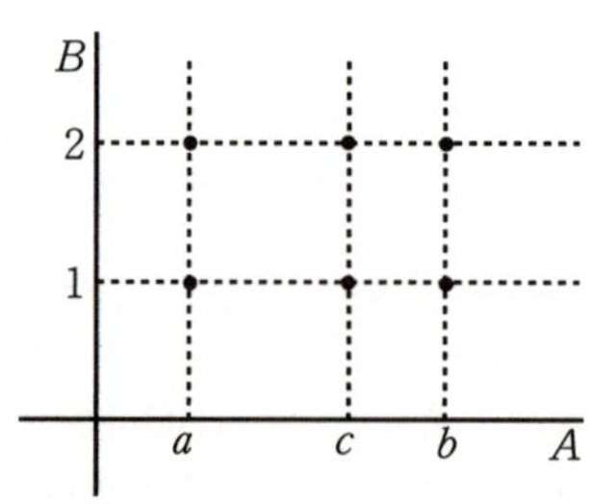

Figure 7

이를테면 $(x, y)=(7, 8) \Longleftrightarrow x=7 \wedge y=8.$

정의 1* 임의의 두 집합 A, B에 대하여 $x \in A$, $y \in B$의 순서쌍 (x, y)의 집합을 A, B의 **데카르트곱****이라 하고 이것을 $\boldsymbol{A \times B}$로 나타낸다. 다시 말하면

$$A \times B=\{(x, y) \mid x \in A \wedge y \in B\}$$

순서쌍 (a, b)에서의 a, b를 각각 **첫째 좌표**, **둘째 좌표*****라고 한다.

예제 1 $A=\{a, b, c\}$, $B=\{1,2\}$에 대하여 $A \times B$와 $B \times A$를 각각 구하여라.

[풀이] 정의 1에 따라 $A \times B=\{(a, 1), (a, 2), (b, 1), (b, 2), (c, 1), (c, 2)\}$,

$B \times A=\{(1, a), (1, b), (1, c), (2, a), (2, b), (2, c)\}$

데카르트곱 $A \times B$를 다음 그림과 같은 점의 집합으로 나타낼 수 있다. 여기서 $A \times B \neq B \times A$인 것에 주목하자.

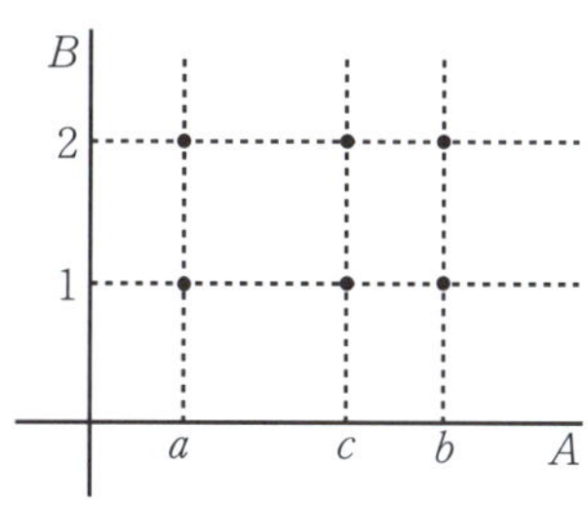

그림 7

위의 정의 1에 있어서 특히 $A=B=\mathbb{R}$인 경우

$$A \times B=\mathbb{R} \times \mathbb{R}=\{(x, y) \mid x \in \mathbb{R} \wedge y \in \mathbb{R}\}$$

이 데카르트곱은 실수의 모든 순서쌍의 집합으로서 해석기하학에서 말하는 **좌표평면**(또는 **데카르트평면**)을 뜻한다.

---

* 정의 1에서 특별한 언급이 없는 한 $A \neq \phi$, $B \neq \phi$인 것을 가정하고 있다.

** **카르테시안곱** 또는 **직적**이라고도 한다.

*** **첫째 성분**, **둘째 성분**이라고도 한다.

Example 2 Let A be any set. Find $A\times\phi$ and $\phi\times A$.

[Solution] Since $A\times\phi$ is the set of all ordered pairs (a,b) such that $a\in A$ and $b\in\phi$, and since the empty set ϕ contains no elements, there is no b in ϕ; therefore $A\times\phi=\phi$. Similarly, $\phi\times A=\phi$.

Theorem 1. Let A, B, and C be any three sets. Then

(a) $A\times(B\cap C)=(A\times B)\cap(A\times C)$.

(b) $A\times(B\cup C)=(A\times B)\cup(A\times C)$.

[Proof] (a) $(a, x)\in A\times(B\cap C)$

$\Longleftrightarrow a\in A\wedge x\in B\cap C$	Def. 1
$\Longleftrightarrow a\in A\wedge(x\in B\wedge x\in C)$	Def. $\cap$
$\Longleftrightarrow a\in A\wedge a\in A\wedge(x\in B\wedge x\in C)$	Idemp., Assoc. (Ch. 1)
$\Longleftrightarrow(a\in A\wedge x\in B)\wedge(a\in A\wedge x\in C)$	Com., Assoc. (Ch. 1)
$\Longleftrightarrow(a, x)\in(A\times B)\wedge(a, x)\in(A\times C)$	Def. 1
$\Longleftrightarrow(a, x)\in(A\times B)\cap(A\times C)$	Def. $\cap$

Hence, by Definition 1 of Chapter 2, we have proved that

$$A\times(B\cap C)=(A\times B)\cap(A\times C)$$

Informally, this equality may be stated:

The Cartesian product distributes over intersection.

We leave the proof of part (b) to the reader as an exercise.

Theorem 2. Let A, B, and C be sets. Then

$$A\times(B-C)=(A\times B)-(A\times C)$$

That is, the Cartesian product distributes over complementation.

예제 2* 임의의 집합 A에 대하여 $A\times\phi$, $\phi\times A$를 각각 구하여라.

[풀이] $A\times\phi$는 $a\in A$, $b\in\phi$인 순서쌍 (a, b)의 집합과 같다. 그러나 공집합 ϕ에는 원소가 존재하지 않으므로 b가 ϕ의 원소일 수 없다. 따라서 $A\times\phi=\phi$. 마찬가지로 $\phi\times A=\phi$.

정리 1 임의의 집합 A, B, C에 대하여 다음이 성립한다.
(a) $A\times(B\cap C)=(A\times B)\cap(A\times C)$.
(b) $A\times(B\cup C)=(A\times B)\cup(A\times C)$.

[증명] (a) $(a, x)\in A\times(B\cap C)$

$\Leftrightarrow a\in A\wedge x\in B\cap C$	정의 1
$\Leftrightarrow a\in A\wedge(x\in B\wedge x\in C)$	$\cap$의 정의
$\Leftrightarrow a\in A\wedge a\in A\wedge(x\in B\wedge x\in C)$	멱등법칙, 결합법칙(제1장)
$\Leftrightarrow(a\in A\wedge x\in B)\wedge(a\in A\wedge x\in C)$	교환법칙, 결합법칙(제1장)
$\Leftrightarrow(a, x)\in(A\times B)\wedge(a, x)\in(A\times C)$	정의 1
$\Leftrightarrow(a, x)\in(A\times B)\cap(A\times C)$	$\cap$의 정의

따라서 $(a, x)\in A\times(B\cap C)\Leftrightarrow(a, x)\in(A\times B)\cap(A\times C)$
그러므로 정의 1(제2장)에 의하여

$$A\times(B\cap C)=(A\times B)\cap(A\times C)$$

이 등식으로 보아 데카르트곱은 교집합에 관한 분배법칙이 성립함을 알 수 있다.

(b)의 증명은 독자의 연습문제로 돌린다.

정리 2 임의의 집합 A, B, C에 대하여 다음이 성립한다.

$$A\times(B-C)=(A\times B)-(A\times C)$$

즉 데카르트곱은 차집합에 관해서도 분배법칙이 성립한다.

* 예제 2는 형식상의 문제일 뿐 바람직한 것이 못된다. $A\times\phi\neq\phi$로 가정하고 배리법을 인용하여 보아라.

[Proof] $(a, x) \in A \times (B - C)$

$\Leftrightarrow a \in A \wedge x \in (B - C)$	Def. 1
$\Leftrightarrow a \in A \wedge (x \in B \wedge x \notin C)$	Def. 5(Ch. 2)
$\Leftrightarrow a \in A \wedge a \in A \wedge x \in B \wedge x \notin C$	Idemp., Assoc.(Ch. 1)
$\Leftrightarrow (a \in A \wedge x \in B) \wedge (a \in A \wedge x \notin C)$	Com., Assoc.(Ch. 1)
$\Leftrightarrow (a,\ x) \in (A \times B) \wedge (a,\ x) \notin (A \times C)$	Def. 1
$\Leftrightarrow (a,\ x) \in (A \times B - A \times C)$	Def. 5(Ch. 2)

Thus, we have proved that

$$A \times (B - C) = (A \times B) - (A \times C)$$

Exercise 3.1

1. Describe each of the following sets geometrically by sketching a graph on the Cartesian plane.
   (a) $\{(x, y) \in \mathbb{R} \times \mathbb{R} \mid x = y\}$ (b) $\{(x, y) \in \mathbb{R} \times \mathbb{R} \mid x > y\}$
   (c) $\{(x, y) \in \mathbb{R} \times \mathbb{R} \mid |x + y| \leqq 1\}$
2. Under what conditions on the sets A and B will it be true that $A \times B = B \times A$?
3. Prove Theorem 1(b) : $A \times (B \cup C) = (A \times B) \cup (A \times C)$.
4. Prove that $A \times B = \phi \Leftrightarrow A = \phi \vee B = \phi$
5. Prove that, if A, B and C are sets and $A \subseteq B$, then $A \times C \subseteq B \times C$.
6. If the set A has m elements and if the set B has n elements, how many elements (ordered pairs) does $A \times B$ have?
7. The Cartesian product $A \times A$ has nine elements among which $(-1, 0)$ and $(0, 1)$ are found. Find the remaining elements and the set A.
8. Prove or disprove (by giving a counterexample) each of the following statements.
   (a) $A \times B \subseteq C \times D$ if and only if $A \subseteq C$ and $B \subseteq D$.
   (b) The power set $\mathscr{P}(A \times B)$ of $A \times B$ is equal to the Cartesian product $\mathscr{P}(A) \times \mathscr{P}(B)$ of the power sets $\mathscr{P}(A)$ and $\mathscr{P}(B)$.
   (c) $A \times B \cup C \times D = (A \cup C) \times (B \cap D)$.
9. Prove that, if A, B, C, and D are any four sets, then
   $$(A \times C) \cap (B \times D) = (A \cap B) \times (C \times D)$$

[증명] $(a, x) \in A \times (B-C)$

$\Longleftrightarrow a \in A \wedge x \in (B-C)$	정의 1
$\Longleftrightarrow a \in A \wedge (x \in B \wedge x \notin C)$	정의 5(제2장)
$\Longleftrightarrow a \in A \wedge a \in A \wedge x \in B \wedge x \notin C$	멱등법칙, 결합법칙(제1장)
$\Longleftrightarrow (a \in A \wedge x \in B) \wedge (a \in A \wedge x \notin C)$	교환법칙, 결합법칙(제1장)
$\Longleftrightarrow (a, x) \in (A \times B) \wedge (a, x) \notin A \times C$	정의 1
$\Longleftrightarrow (a, x) \in (A \times B - A \times C)$	정의 5(제2장)

그러므로 $(a, x) \in A \times (B-C) \Longleftrightarrow (a, x) \in (A \times B) - (A \times C)$

따라서 $A \times (B-C) = (A \times B) - (A \times C)$

연습문제 3.1

1. 좌표평면에 다음 점의 집합 그래프를 그려라.

(a) $\{(x, y) \in \mathbb{R} \times \mathbb{R} \mid x = y\}$ (b) $\{(x, y) \in \mathbb{R} \times \mathbb{R} \mid x > y\}$

(c) $\{(x, y) \in \mathbb{R} \times \mathbb{R} \mid |x+y| \leqq 1\}$

2. 집합 A, B에 대한 어떤 조건하에 $A \times B = B \times A$이겠는가?

3. 정리 1(b) $A \times (B \cup C) = (A \times B) \cup (A \times C)$를 증명하여라.

4. $A \times B = \phi \Longleftrightarrow A = \phi \vee B = \phi$임을 증명하여라.

5. 집합 A, B, C에 대하여 $A \subseteq B$이면 $A \times C \subseteq B \times C$임을 증명하여라.

6. 집합 A, B의 원소가 각각 m, n개일 때 $A \times B$의 원소는 몇 개인가?

7. 원소가 모두 아홉 개인 데카르트곱 $A \times A$에는 $(-1, 0)$, $(0, 1)$이 속해 있다. 이 집합의 나머지 원소, 집합 A를 각각 구하여라.

8. 다음 각 명제를 증명 또는 반증(반례 하나를 들음으로써)하여라.

(a) $A \subseteq C \wedge B \subseteq D \Longleftrightarrow A \times B \subseteq C \times D$

(b) $A \times B$의 멱집합 $\mathscr{P}(A \times B)$는 멱집합 $\mathscr{P}(A)$와 $\mathscr{P}(B)$의 데카르트곱 $\mathscr{P}(A) \times \mathscr{P}(B)$와 같다.

(c) $(A \times B) \cup (C \times D) = (A \cup C) \times (B \cup D)$

9. (a) 임의의 집합 A, B, C, D에 대하여 다음이 성립함을 증명하여라.

$$(A \times C) \cap (B \times D) = (A \cap B) \times (C \cap D)$$

(b)* (a)의 네 집합이 모두 집합 $\mathbb{R}$의 부분집합일 때 등식에 대한 그림을 좌표평면에 나타내어라[그림 7(131쪽) 참조].

* 옮긴이가 추가한 것이다.

10. Let $A_1, A_2, \cdots, A_n$ be sets. Can you generalize Definition 1 to the Cartesian product $A_1 \times A_2 \times A_3$ of three sets? Can you generalize this further to the Cartesian product $A_1 \times A_2 \times \cdots \times A_n$ of n sets?

11. Define the ordered pair (x, y) to be the set $\{\{x\}, \{x, y\}\}$. Use this definition to prove that $(a, b)=(c, d)$ if and only if $a=b$ and $b=d$.

3.2 Relations

Given two sets A and B, not necessarily distinct, when we say that an element a of A is related to another element b in B by a relation $\mathscr{R}$ we are making a statement about the ordered pair (a, b) in the Cartesian product $A \times B$.

Therefore, a mathematical definition of a relation can be precisely given in terms of ordered pairs in the Cartesian product of sets.

Definition 2. A relation $\mathscr{R}$ from A to B is a subset of the Cartesian product $A \times B$. It is customary to write $\boldsymbol{a \mathscr{R} b}$ for $(a, b) \in \mathscr{R}$. The symbol $a \mathscr{R} b$ is read "a is $\mathscr{R}$-related to b."

Often A and B are the same set, say X. In that case, we shall say that $\mathscr{R}$ is a relation in X instead of "from X to X." For example, in a community X,[2] to say that a (for Albert) is the husband of b (for Bonita), is to consider Albert and Bonita as an (ordered) pair (a, b) in the relation $\mathscr{H}$ (of being the husband of ...). The symbol $a \mathscr{H} b$ or $(a, b) \in \mathscr{H}$ may be read: "a is the husband of b."

It is not necessary to put Bonita behind Albert in the ordered pair (a, b). We may say that Bonita is the wife of Albert, or that the ordered pair (b, a) is in the relation $\mathscr{W}$ (of being the wife of $\cdots$). The symbol $b \mathscr{W} a$ or $(b, a) \in \mathscr{W}$ may be read: "b is the wife of a." In this example, the relation $\mathscr{W}$ is called the inverse of the relation $\mathscr{H}$.

[2] Here, X is the set of all members of the community.

10. 데카르트곱을 세 집합 $A_1 \times A_2 \times A_3$으로 확장할 수 있겠는가? 또한 n개의 집합의 데카르트곱 $A_1 \times A_2 \times \cdots \times A_n$으로 일반화할 수 있는가? [$n$에 관한 수학적 귀납법(61쪽)을 이용하여라.]

11. 순서쌍 (a, b)를 집합 $\{\{a\}, \{a, b\}\}$로 정의할 때 다음을 증명하여라.

$$a=c \wedge b=d \Longleftrightarrow (a, b)=(c, d)$$

3.2 관계

두 가정 A, B에는 아들이 각각 4명, 3명씩 있다. 즉

$$A=\{a_1, a_2, a_3, a_4\}, \quad B=\{b_1, b_2, b_3\}$$

지금 집합 A, B 사이에 의형제를 맺다보니

"a_2는 b_1의 형이다." "a_3는 b_3의 형이다."

일 때, 이것을 순서쌍의 집합 $\mathscr{R}=\{(a_2, b_1), (a_3, b_3)\}$로 나타낼 수 있다. 이러한 생각 끝에 우리는 다음 정의를 얻는다.

정의 2 집합 A에서 집합 B로의 관계 $\mathscr{R}$은 데카르트곱 $A \times B$의 하나의 부분집합*을 뜻한다. 여기서 $(a, b) \in \mathscr{R}$를 $\boldsymbol{a \mathscr{R} b}$와 같이 나타내고 이 기호를 "$a$는 $\mathscr{R}$에 따라 b와 관계된다."라고 읽는다.

특히 두 집합 A, B가 모두 X와 같은 경우 "X에서 X로의 관계 $\mathscr{R}$"이라고 하는 대신 "X에서의 관계 $\mathscr{R}$"이라고 한다. 이를테면 지역 X[2)]의 주민 a(김병기를 가리킴)가 b(이경희)의 남편일 때 김병기와 이경희를 (…의 남편)의 관계 $\mathscr{H}$에 따른 순서쌍 (a, b)로서 생각할 수 있다. 이런 경우 기호 $a\mathscr{H}b$ 또는 $(a, b) \in \mathscr{H}$를 "a는 b의 남편이다."로 읽어도 무난하다.

한편 순서쌍 (a, b)에서와 같이 이경희를 김병기 뒤에만 놓을 것이 아니라 그 앞에 놓을 수도 있다. 이때 이경희는 김병기의 아내이다, 또는 순서쌍 (b, a)는 (…의 아내임의) 관계 $\mathscr{W}$에 속한다고 말하고 기호 $b\mathscr{W}a$ 또는 $(b, a) \in \mathscr{W}$를 "b는 a의 아내이다."로 읽는다. 이런 관계 $\mathscr{W}$를 관계 $\mathscr{H}$의 역관계라고 한다.

---

* $A \times B$의 부분집합마다 그에 따른 관계 하나씩이 대응한다.

2) 여기서 X는 그 지역의 주민의 집합이다.

Definition 3. Let A and B be two sets, not necessarily distinct, and let $\mathscr{R}$ be a relation from A to B. Then inverse $\mathscr{R}^{-1}$ of the relation $\mathscr{R}$ is the relation from B to A such that $b\ \mathscr{R}^{-1}a$ if and only if $a\mathscr{R}b$. That is, $\mathscr{R}^{-1}=\{(b, a)\mid(a, b)\in\mathscr{R}\}$

Example 3 (a) Let $A=\{a, b\}$, $B=\{x, y, z\}$, and let $\mathscr{R}\subseteq A\times B$ be given by $\mathscr{R}=\{(a, x), (b, y)\}$. Then $\mathscr{R}^{-1}=\{(x, a), (y, b)\}\subseteq B\times A$.

(b) Let $\mathscr{R}=\{(x, y)\in\mathrm{N}\times\mathrm{N}\mid x \text{ divides } y\}$

Then $\mathscr{R}^{-1}=\{(y, x)\in\mathrm{N}\times\mathrm{N}\mid y \text{ is a multiple of } x\}$

Let $\mathscr{R}$ be a relation from A to B. The domain of the relation $\mathscr{R}$, denoted by **Dom**($\mathscr{R}$), is the set of all those $a\in A$ such that $a\mathscr{R}b$ for some $b\in B$; and the image of $\mathscr{R}$, denoted by **Im**($\mathscr{R}$), is the set of all those $b\in B$ such that $a\mathscr{R}b$ for some $a\in A$. In symbols,

$$\mathrm{Dom}(\mathscr{R})=\{a\in A\mid(a, b)\in\mathscr{R} \text{ for some } b\in B\}$$

and

$$\mathrm{Im}(\mathscr{R})=\{b\in B\mid(a, b)\in\mathscr{R} \text{ for some } a\in A\}$$

In the example of the relations $\mathscr{H}$ (being the husband of ...) and $\mathscr{W}$ (being the wife of ...) in the community X, the domain of $\mathscr{H}$ is the set of all men in X who are married, and the image of $\mathscr{H}$ is the set of all women in X who are married, whereas, the domain of $\mathscr{W}$ is the set of all the wives in X, and the image of $\mathscr{W}$ is the set of all the husbands in X. That is,

$$\mathrm{Dom}(\mathscr{W})=\mathrm{Im}(\mathscr{H}) \quad \text{and} \quad \mathrm{Im}(\mathscr{W})=\mathrm{Dom}(\mathscr{H})$$

Can you make a general conclusion? (see Problem 3 at the end of this section.)

Example 4 In Example 3(a), $\mathrm{Dom}(\mathscr{R})=\{a, b\}$ and $\mathrm{Im}(\mathscr{R})=\{x, y\}$.
In Example 3(b), $\mathrm{Dom}(\mathscr{R})=\mathrm{N}=\mathrm{Im}(\mathscr{R})$.

정의 3 집합 A, B(서로 같더라도 무방함)에 대하여 A에서 B로의 관계 $\mathscr{R}$의 **역관계**를 $\mathscr{R}^{-1}$로 나타낸다. 여기서 $\mathscr{R}^{-1}$은 B에서 A로의 관계로서 $a\mathscr{R}b \Longleftrightarrow b\mathscr{R}^{-1}a$

이 주장을 다음과 같이 나타내면 더욱 분명하다.

$$\mathscr{R}^{-1}=\{(b, a)\,|\,(a, b)\in\mathscr{R}\}$$

예제 3 (a) $A=\{a, b\}$, $B=\{x, y, z\}$에 대하여 $\mathscr{R}=\{(a, x), (b, y)\}$ 라고 놓으면 $\mathscr{R}^{-1}=\{(x, a), (y, b)\}$

(b)* $\mathscr{R}=\{(x, y)\in\mathbb{N}\times\mathbb{N}\,|\,y$는 x로 나누어 떨어진다$\}$일 때 $\mathscr{R}^{-1}=\{(y, x)\in\mathbb{N}\times\mathbb{N}\,|\,y$는 x의 배수이다$\}$.

집합 A에서 집합 B로의 관계 $\mathscr{R}$에서 적당한 $b\in B$에 대한 $a\mathscr{R}b$인 모든 $a\in A$의 집합을 $\mathscr{R}$의 **정의역**이라 하고 **Dom**($\mathscr{R}$)로 나타낸다.

한편 적당한 $a\in A$에 대한 $a\mathscr{R}b$인 모든 $b\in B$의 집합을 $\mathscr{R}$의 **상**이라 하고 **Im**($\mathscr{R}$)로 나타낸다. 이상을 요약하면 다음과 같다.

$$\mathrm{Dom}(\mathscr{R})=\{a\in A\,|\,\text{적당한 } b\in B\text{에 대하여 } (a, b)\in\mathscr{R}\},$$
$$\mathrm{Im}(\mathscr{R})=\{b\in B\,|\,\text{적당한 } a\in A\text{에 대하여 } (a, b)\in\mathscr{R}\}$$

앞에서의 집합 X에서의(…의 남편임의) 관계 $\mathscr{H}$와 (…의 부인임의) 관계 $\mathscr{W}$에 대하여 $\mathscr{H}$의 정의역은 X에 거주하는 모든 기혼남성의 집합, 그 상은 X의 주민으로서 모든 기혼여성의 집합과 각각 같다.

한편 $\mathscr{H}$의 역관계 $\mathscr{W}$의 정의역은 X에 거주하는 모든 부인의 집합 다시 말하면 그 상은 X에 거주하는 모든 남편의 집합과 각각 같다. 다시 말하면

$$\mathrm{Dom}(\mathscr{W})=\mathrm{Im}(\mathscr{H}),\qquad \mathrm{Im}(\mathscr{W})=\mathrm{Dom}(\mathscr{H})$$

우리는 이 결론을 일반화할 수 있을까?(145쪽 연습문제 3.2의 3 참조).

예제 4 예제 3(a)에 대하여 $\mathrm{Dom}(\mathscr{R})=\{a, b\}$, $\mathrm{Im}(\mathscr{R})=\{x, y\}$.
한편 예제 3(b)에서는 $\mathrm{Dom}(\mathscr{R})=\mathbb{N}=\mathrm{Im}(\mathscr{R})$.

* (b)에서의 관계 $\mathscr{R}$, $\mathscr{R}^{-1}$의 원소 몇 개씩을 지적하여 보아라.

Definition 4. Let $\mathscr{R}$ be a relation in a set X. Then we say that

(a) $\mathscr{R}$ is reflexive if and only if $\forall x \in X,\ x\mathscr{R}x$.

(b) $\mathscr{R}$ is symmetric if and only if $x\mathscr{R}y \Longrightarrow y\mathscr{R}x$.

(c) $\mathscr{R}$ is transitive if and only if $x\mathscr{R}y \wedge y\mathscr{R}z \Longrightarrow x\mathscr{R}z$.

(d) $\mathscr{R}$ is an equivalence relation if and only if $\mathscr{R}$ is reflexive, symmetric, and transitive.

The equal relation, $=$, on the set $\mathbb{R}$ of real numbers is clearly an equivalence relaion. Let X be a set of colored balls and let any two balls a and b related by $\mathscr{R}$ if and only if a and b have the same color. Then the relation $\mathscr{R}$ is an equivalence relation.

Equivalence relations are particularly important in modern mathematics. For instance, factor groups in algebra, quotient spaces in topology, and modular number systems in number theory all involve certain kinds of equivalence relations.

Given a nonempty set X, there always exist at least two equivalence relations in X; one of these is the diagonal relation Δ_X (also called the identity relation, defined by

$$\Delta_X = \{(x, x) \mid x \in X\}$$

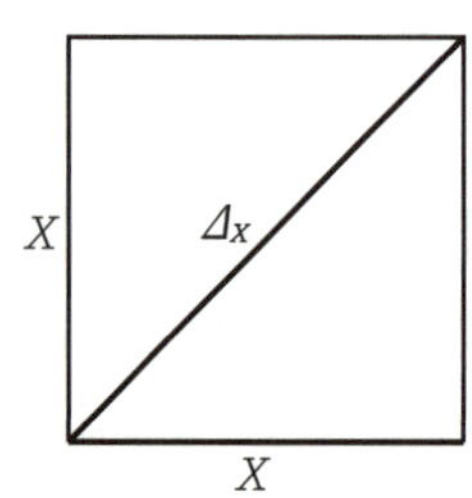

Figure 8

which relates every element with itself. Pictorially, if X is represented as a line interval, then $X \times X$ is a square and Δ_X is the "main" diagonal of the square.

이제 다음 정의에서와 같은 성질을 지닌 관계를 알아보자.

정의 4* 집합 $X(\neq\phi)$에서의 관계 $\mathscr{R}$이 다음 조건을 만족할 때마다 그 왼편의 율이 성립한다고 한다.

(a) 반사율(reflexivelaw) : $\forall x \in X,\ x\mathscr{R}x$

(b) 대칭율(symmetriclaw) : $x\mathscr{R}y \Longrightarrow y\mathscr{R}x$

(c) 추이율(transitivelaw) : $x\mathscr{R}y \wedge y\mathscr{R}z \Longrightarrow x\mathscr{R}z$

(d) 동치관계(equivalencelaw) : $\mathscr{R}$에 대하여 반사율 · 대칭율 · 추이율이 모두 성립한다.

보기 1 $X=\{1, 2, 3, 4\}$에서의 관계

$$\mathscr{R}=\{(1, 1), (1, 4), (2, 2), (2, 4), (4, 1), (4, 2), (4, 4)\}$$

에 대하여 $3\in X$이지만 $(3, 3)\notin\mathscr{R}$. 그러므로 $\mathscr{R}$은 반사적이 아니다.

그러나 $\mathscr{R}$은 대칭적이다(과연?). 한편 $(1, 4)\neq\mathscr{R}\wedge(4, 2)\in\mathscr{R}$이지만 $(1, 2)\notin\mathscr{R}$이므로 추이적이 아니다.

보기 2** 실수의 집합 $\mathbb{R}$에서의 상등관계(등식) $=$는 다름아닌 동치관계임을 증명하여 보자.

[증명] (i) $\forall x\in\mathbb{R}\ \ x-x=0 \Longrightarrow x=x$(등식 $=$의 정의)

(ii) $x=y \Longrightarrow x-y=0 \Longrightarrow -x+y=0 \Longrightarrow y-x=0 \Longrightarrow y=x$(각 단계마다 확인하여라). 그러므로 $x=y \Longrightarrow y=x$

(iii) $x=y \wedge y=z \Longrightarrow x=z$(이것의 증명은 독자에게 맡긴다).

이상으로 $=$은 동치관계이다.

실제로 어느 집합 $X(\neq\phi)$에서나 동치관계는 적어도 둘 존재한다. 그 중의 하나는 $\Delta x=\{(x, x)\mid x\in X\}$로 정의되는 항등관계(대각관계)이다.

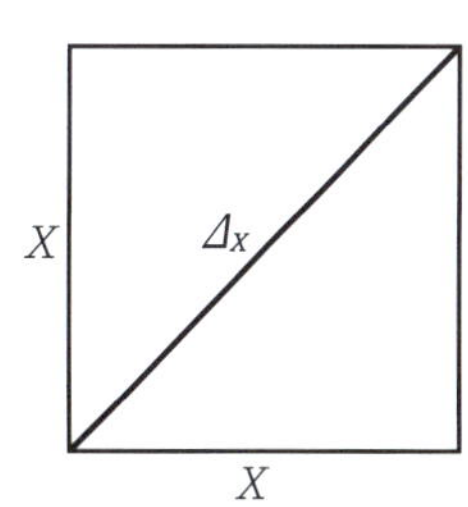

그림 8

여기서 X의 각 원소는 Δx에 따라 다름아닌 자기자신에 관계지어졌음을 뜻한다. 이를테면 오른쪽 그림 8에서와 같이 선분의 점을 원소로 하는 집합 X에 대하여 $X\times X$는 하나의 정사각형(그 내부가 포함된 점의 집합)을 이루며 Δ_X는 그것의 주대각선인 것을 알 수 있다.

* 원서에서와 같이 반사율, 대칭율, 추이율을 차례로 반사적, 대칭적, 추이적이라고도 한다. 특히 반사율은 동일율(identity law)로도 불리운다.

** 보기 2의 자리의 본문에서는 동차관계의 중요성과 그것의 활용범위를 소개하고 있다.

There is, at the other extreme, always another equivalence relation $\mathscr{R}=X\times X$ on[3] X. The relation Δ_X is the smallest of all the equivalence relations among the subsets of $X\times X$ that can be defined on X, whereas $X\times X$ is the largest.

Example 5 Let m be an arbitrary fixed positive integer. The congruence relation $\equiv$ **modulo *m*** on the set $\mathbb{Z}$ of integers is defined by $x\equiv y$ (modulo m) if and only if $x-y=km$ for some $k\in\mathbb{Z}$. The congruence relation is an equivalence relation on $\mathbb{Z}$.

[Proof] (a) For each $x\in\mathbb{Z}$, since $x-x=0\cdot m$, we have $x\equiv x \pmod{m}$. Hence it is reflexive.

(b) If $x\equiv y \pmod{m}$, then $x-y=km$ for some $k\in\mathbb{Z}$. Consequently, $y-x=(-k)m$ and $-k\in\mathbb{Z}$, or $y\equiv x \pmod{m}$. Hence it is symmetric.

(c) If $x\equiv y \pmod{m}$ and $y\equiv z \pmod{m}$, then $x-y=k_1m$ and $y-z=k_2m$ for some k_1 and k_2 in $\mathbb{Z}$. Hence $x-z=(x-y)+(y-z)=(k_1+k_2)m$ and $k_1+k_2\in\mathbb{Z}$, which shows that $x\equiv z \pmod{m}$. Hence it is transitive.

Therefore, we have proved that the congruence relation (modulo m) is an equivalence relation on $\mathbb{Z}$.

As a special case for Example 5, let $m=2$. Then $x\equiv y \pmod{2}$ if and only if $x-y$ is an even integer. Consequently, $x\equiv y \pmod{2}$ if and only if either both x and y are even or both x and y are odd.

---

[3] When the domain of a relation $\mathscr{R}$ in X is obviously X itself, most mathematicians prefer to say "relation $\mathscr{R}$ *on* X" instead of "relation $\mathscr{R}$ *in* X."

사실 동치관계는 대수학, 위상수학, 수론 인자군, 상공간, 모듈수계를 수반하고 있으리만큼 현대수학에서 큰 역할을 한다.

사실 집합 X 위의[3] 동치관계 중 $\mathscr{R}=X\times X$를 들 수 있다. 여기서 항등관계 Δ_X는 $X\times X$의 부분집합으로서 X 위에서의 가장 작은 것(집합의 관점에서)인데 반하여 $X\times X$는 가장 큰 동치관계이다.

예제 5 m이 임의의 정해진 양의 정수일 때 집합 $\mathbb{Z}$ 위에서의 법 $\boldsymbol{m}$에 관한 합동관계(또는 간단히 합동 $\equiv$)이란 것은 적당한 $k\in\mathbb{Z}$에 대하여

$$x-y=km \Longleftrightarrow x\equiv y \pmod{m}$$

로 정의된 것을 말한다. 이 합동은 집합 $\mathbb{Z}$ 위의 동치관계임을 보여라.

[증명] (a) $x\in\mathbb{Z}$에 대하여 $x-x\equiv 0\cdot m$이므로 $x\equiv x \pmod{m}$.

(b) $x\equiv y \pmod{m}$이면 적당한 $k\in\mathbb{Z}$에 대하여

$$x-y=km \text{이므로} \quad y-x=(-k)m, \text{ 여기서 } -k\in\mathbb{Z}.$$

따라서 $y\equiv x \pmod{m}$.

(c) $x\equiv y \pmod{m}$, $y\equiv z \pmod{m}$이면 적당한 $k_1, k_2\in\mathbb{Z}$에 대하여

$$x-y=k_1m, \quad y-z=k_2m.$$

그러므로

$$x-z=(x-y)+(y-z)=k_1m+k_2m=(k_1+k_2)m,$$

즉 $x-z=(k_1+k_2)m$, $k_1+k_2\in\mathbb{Z}$. 따라서 $x\equiv z \pmod{m}$.

이상으로 합동 $\equiv$는 $\mathbb{Z}$ 위의 동치관계이다.

예제 5*에서 특히 $m=2$인 경우

$$x-y\text{가 짝수이면 그리고 그때에만 } x\equiv y \pmod{2}.$$

즉 x, y가 모두 짝수이거나 모두 홀수이면 그리고 그때에만 $x\equiv y \pmod{2}$ [예제 6(147쪽) 참조].

[3] 집합 X에서의 $\mathscr{R}$의 정의역이 X 자신과 같을 때 대부분의 수학자들은 "X에서의 관계 $\mathscr{R}$" 대신 "X 위의 관계 $\mathscr{R}$"이라고 부르기도 한다.

* 주어진 평면 P 위에서의 삼각형 t의 집합 T를 생각하자. 지금 T 위의 임의의 두 삼각형 s, t가 합동, 즉 $s\equiv t$일 때 이 관계 $\equiv$은 동치관계인지 알아보아라. 한편 대칭 $\backsim$에 관해서는 어떠한가? 내친김에 T 위에서의 관계이지만 동치관계가 아닌 것의 보기를 한두 개 들어 보아라[연습문제 3.2의 문제 4, 5, 6(145쪽) 참고].

Exercise 3.2

1. Let $\mathscr{R}$ be a relation from A to B. Prove that

$$(\mathscr{R}^{-1})^{-1}=\mathscr{R}$$

2. Let $A=\{a,b,c\}$ and let $\mathscr{R}=\{(a,c),\ (c,b),\ (a,b)\}$. Find the domain of $\mathscr{R}$ and the image of $\mathscr{R}$.

3. Let $\mathscr{R}$ be a relation from A to B. Prove that

(a) $\mathrm{Dom}(\mathscr{R}^{-1})=\mathrm{Im}(\mathscr{R})$ (b) $\mathrm{Im}(\mathscr{R}^{-1})=\mathrm{Dom}(\mathscr{R})$.

4. Let $A=\{a,b,c\}$ and let

$$\mathscr{R}=\{(a,a),\ (b,b),\ (c,c),\ (a,b),\ (b,a),\ (c,a),\ (a,c)\}$$

Prove that $\mathscr{R}$ is reflexive and symmetric, but not transitive.

5. Give an example of a relation that is reflexive and transitive, but not symmetric.

6. Give an example of a relation that is symmetric and transitive, but not reflexive.

7. Let $\mathscr{R}$ be a relation in a set X. Prove that

(a) $\mathscr{R}$ is reflexive if and only if $\mathscr{R}\supseteq\Delta_X$

(b) $\mathscr{R}$ is symmetric if and only if $\mathscr{R}=\mathscr{R}^{-1}$

(c) $\mathscr{R}$ is reflexive if and only if $\mathscr{R}^{-1}$ is reflexive

(d) $\mathscr{R}$ is symmetric if and only if $\mathscr{R}^{-1}$ is symmetric

(e) $\mathscr{R}$ is transitive if and only if $\mathscr{R}^{-1}$ is transitive

(f) $\mathscr{R}$ is an equivalence relation if and only if $\mathscr{R}^{-1}$ is an equivalence relation.

8. Let $X=\mathbb{Z}\times(\mathbb{Z}-\{0\})$. Define a relation $\sim$ on X by declaring that

$$(a,b)\sim(c,d) \text{ if and only if } ad=bc$$

Prove that the relation $\sim$ is an equivalence relation.

연습문제 3.2

1. 집합 A에서 집합 B로의 관계 $\mathscr{R}$에 대하여 $(\mathscr{R}^{-1})^{-1}=\mathscr{R}$임을 증명하여라.

2. $A=\{a, b, c\}$, $\mathscr{R}=\{(a, b), (c, b), (a, b)\}$일 때 $\mathscr{R}$의 정의역, 상을 각각 구하여라.

3. 집합 A에서 집합 B로의 관계 $\mathscr{R}$에 대하여 다음을 증명하여라.

(a) $\mathrm{Dom}(\mathscr{R}^{-1})=\mathrm{Im}(\mathscr{R})$ (b) $\mathrm{Im}(\mathscr{R}^{-1})=\mathrm{Dom}(\mathscr{R})$

4. (a) $A=\{a, b, c\}$, $\mathscr{R}=\{(a, a), (b, b), (c, c), (a, b), (b, a), (c, a), (a, c)\}$ 일 때 $\mathscr{R}$은 반사적이고 대칭적이지만 추이적이 아님을 보여라.

(b)* 집합 $\mathbb{R}$ 위의 관계 $\mathscr{R}$이 $x<y$로 정의될 때 $\mathscr{R}$은 반사적 · 대칭적 · 추이적 중 어느 경우인가?

5. 반사적이고 추이적이지만 대칭적이 아닌 관계의 보기 하나를 들어라.

6. 대칭적이고 추이적이지만 반사적이 아닌 관계의 보기 하나를 들어라.

7.** 집합 X 위의 관계 $\mathscr{R}$에 대하여 다음을 증명하여라.

(a) $(\Delta_X\subseteq\mathscr{R})\Longleftrightarrow$($\mathscr{R}$은 반사적이다.)

(b) $(\mathscr{R}=\mathscr{R}^{-1})\Longleftrightarrow$($\mathscr{R}$은 대칭적이다.)

(c) ($\mathscr{R}^{-1}$이 반사적이다.)$\Longleftrightarrow$($\mathscr{R}$은 반사적이다.)

(d) ($\mathscr{R}^{-1}$이 대칭적이다.)$\Longleftrightarrow$($\mathscr{R}$은 대칭적이다.)

(e) ($\mathscr{R}^{-1}$이 추이적이다.)$\Longleftrightarrow$($\mathscr{R}$은 추이적이다.)

(f) ($\mathscr{R}^{-1}$이 동치관계이다.)$\Longleftrightarrow$($\mathscr{R}$은 동치관계이다.)

8. 다음은 동치관계이다. 이를 증명하여라.

(a)* 집합 X 위의 항등관계 Δ_X(141쪽 그림 8)

(b) 아래와 같이 정의된 집합 $X=\mathbb{Z}\times(\mathbb{Z}-\{0\})$ 위의 관계 $\sim$

$$ad=bc\Longrightarrow(a, b)\sim(c, d)$$

---

* 옮긴이가 추가한 것이다. 한편 문제 8(b)를 참고로 집합 $X=\mathbb{Q}\times(\mathbb{Q}-\{0\})$ 위의 동치관계 아닌 하나를 생각하여 보아라.

** (a)-(f)에 따른 보기 하나씩을 들어라.

3.3 Partitions and Equivalence Relations

Definition 5. Let X be a nonempty set. By a partition $\mathscr{P}$ of X we mean a set of nonempty subsets of X such that
(a) If $A, B \in \mathscr{P}$ and $A \neq B$, then $A \cap B = \phi$ (b) $\bigcup_{C \in \mathscr{P}} C = X$

Intuitively, a partition of X is a "cutting up" of X into (nonempty disjoint) pieces.

Example 6 Let m be any fixed positive integer. For each integer j, $0 \leqq j < m$, let $\mathbb{Z}_j = \{x \in \mathbb{Z} \mid x - j = km \text{ for some } k \in \mathbb{Z}\}$. Then the set

$$\{\mathbb{Z}_0, \mathbb{Z}_1, \mathbb{Z}_2, \cdots, \mathbb{Z}_{m-1}\}$$

forms a partition of $\mathbb{Z}$. In particular, let $m=2$. Then the set of sets

$$\mathbb{Z}_0 = \{x \in \mathbb{Z} \mid x \text{ is even}\}$$

and $$\mathbb{Z}_1 = \{x \in \mathbb{Z} \mid x - 1 \text{ is even}\} = \{x \in \mathbb{Z} \mid x \text{ is odd}\}$$

forms a partition of $\mathbb{Z}$. (See also Problem 4, Exercise 3.3).

There is a very close connection between the partition of a nonempty set and an equivalence relation on that set. In order to understand this connection, we shall first need the following definition.

Definition 6. Let $\mathscr{E}$ be an equivalence relation on a nonempty set X. For each $x \in X$, we define

$$\boldsymbol{x/\mathscr{E}} = \{y \in X \mid y \mathscr{E} x\}$$

which is called the equivalence class determined by the element x.

The set of all such equivalence classes on X is denoted by $X/\mathscr{E}$; that is, $X/\mathscr{E} = \{x/\mathscr{E} \mid x \in X\}$. The symbol $\boldsymbol{X/\mathscr{E}}$ is read "***X*** **modulo** $\mathscr{E}$," or simply, "***X*** **mod** $\mathscr{E}$."

3.3 분할과 동치관계

정의 5 집합 $X(X \neq \phi)$의 임의의 부분집합 A, B, C에 대하여 다음이 모두 성립할 때 집합 $\mathscr{P}$를 X의 **분할**이라고 한다.

(a) $A, B \in \mathscr{P} \wedge A \neq B \Longrightarrow A \cap B = \phi$ (b) $\bigcup_{C \in \mathscr{P}} C = X$

쉽게 말하면 집합 X를 서로 소인 부분집합(공집합이 아님)들로 "나누어 놓은 것"이 X의 분할이라고 볼 수 있다.

예제 6 m이 임의의 양의 정수일 때 각 정수 j, $0 \leqq j < m$에 대하여 $\mathbb{Z}_j = \{x \in \mathbb{Z} \mid$ 적당한 $k \in \mathbb{Z}$에 대하여 $x - j = km\}$라고 놓으면, 첨수집합족

$$\{\mathbb{Z}_0, \mathbb{Z}_1, \mathbb{Z}_2, \cdots, \mathbb{Z}_{m-1}\} \quad \cdots\cdots ①$$

은 집합 $\mathbb{Z}$의 하나의 분할이다. 특히 $m=2$*인 경우 다음 두 집합은 $\mathbb{Z}$의 분할을 이룬다(143쪽 예제 5 하단 및 연습문제 3.3의 4 참조).

$$\mathbb{Z}_0 = \{x \in \mathbb{Z} \mid x\text{는 짝수이다}\} = \{\cdots, -4, -2, 0, 2, 4, \cdots\}$$

$$\mathbb{Z}_1 = \{x \in \mathbb{Z} \mid x\text{는 홀수이다}\} = \{\cdots, -3, -1, 1, 3, \cdots\}$$

집합의 분할과 그에 따른 집합 위의 동치관계 사이에는 매우 밀접한 관련이 있다. 그 관련을 이해하려면 우선 다음 정의부터 익혀야 한다.

정의 6 집합 $X(X \neq \phi)$ 위의 하나의 동치관계를 $\mathscr{E}$라고 할 때 각 $x \in X$에 대하여 $\boldsymbol{x/\mathscr{E}} = \{y \in X \mid y \mathscr{E} x\}$
로 정의된 집합을 x에 따른 **동치류**라고 한다. 또 이같은 X에서의 동치류의 집합을 $\boldsymbol{X/\mathscr{E}}$와 같이 나타낸다. 다시 말하면

$$X/\mathscr{E} = \{x/\mathscr{E} \mid x \in X\}.$$

이 기호 $X/\mathscr{E}$를 "**X 법 $\mathscr{E}$(엑스법이)**"라고 읽는다.

* m=3인 경우 $\mathbb{Z}_0$, $\mathbb{Z}_1$, $\mathbb{Z}_2$를 차례로 구하고, 특히 m=1인 경우도 생각하여 보아라.

Theorem 3. Let $\mathscr{E}$ be an equivalence relation on a nonempty set X. Then

(a) Each $x/\mathscr{E}$ is a nonempty subset of X.

(b) $x/\mathscr{E} \cap y/\mathscr{E} \neq \phi$ if and only if $x\mathscr{E}y$.

(c) $x\mathscr{E}y$ if and only if $x/\mathscr{E} = y/\mathscr{E}$.

[Proof] (a) Since $\mathscr{E}$ is reflexive, for each $x \in X$, we have $x\mathscr{E}x$. By Definition 6, $x \in x/\mathscr{E}$ and hence $x/\mathscr{E}$ is a nonempty subset of X.

(b) Since $\mathscr{E}$ is an equivalence relation and $X \neq \phi$, we have

$$\begin{aligned} x/\mathscr{E} \cap y/\mathscr{E} \neq \phi &\Longleftrightarrow \exists z(z \in x/\mathscr{E} \wedge z \in y/\mathscr{E}) \\ &\Longleftrightarrow z\mathscr{E}x \wedge z\mathscr{E}y && \text{De. 6} \\ &\Longleftrightarrow x\mathscr{E}z \wedge z\mathscr{E}y && \mathscr{E} \text{ is symmetric} \\ &\Longleftrightarrow x\mathscr{E}y && \mathscr{E} \text{ is transitive} \end{aligned}$$

(c) It follows immediately from (a) and (b) above that $x/\mathscr{E} = y/\mathscr{E} \Longrightarrow x\mathscr{E}y$. We now need to prove that

$$x\mathscr{E}y \Longrightarrow x/\mathscr{E} = y/\mathscr{E}.$$

Let $x\mathscr{E}y$. Then

$$\begin{aligned} z \in x/\mathscr{E} &\Longrightarrow z\mathscr{E}x && \text{Def. 6} \\ z\mathscr{E}x \wedge x\mathscr{E}y &\Longrightarrow z\mathscr{E}y && \mathscr{E} \text{ is transitive} \\ &\Longrightarrow z \in y/\mathscr{E} && \text{Def. 6} \end{aligned}$$

Since z is arbitrary, it follows that $x/\mathscr{E} \subseteq y/\mathscr{E}$. A similar argument gives $y/\mathscr{E} \subseteq x/\mathscr{E}$; hence $x/\mathscr{E} = y/\mathscr{E}$.

Theorem 4. Let $\mathscr{E}$ be an equivalence relation on a nonempty set X. Then $X/\mathscr{E}$ is a partition of X.

정리 3 집합 $X(X \neq \phi)$ 위의 동치관계 $\mathscr{E}$에 대하여 다음이 성립한다.

(a) 각 동치류 $x/\mathscr{E}$는 X의 부분집합(공집합이 아님)이다.

(b) $x\mathscr{E}y \Longleftrightarrow x/\mathscr{E} \cap y/\mathscr{E} \neq \phi$ (c) $x/\mathscr{E} = y/\mathscr{E} \Longleftrightarrow x\mathscr{E}y$

[증명] (a) 동치관계 $\mathscr{E}$는 반사적이므로 각 $x \in X$에 대하여 $x\mathscr{E}x$. 따라서 정의 6에 의하여 $x \in x/\mathscr{E}$. 그러므로 $x/\mathscr{E} \neq \phi$.

(b) $\mathscr{E}$는 집합 X 위의 동치관계이므로

$$
\begin{aligned}
x/\mathscr{E} \cap y/\mathscr{E} \neq \phi &\Longleftrightarrow \exists z (z \in x/\mathscr{E} \wedge z \in y/\mathscr{E}) \\
&\Longleftrightarrow z\mathscr{E}x \wedge z\mathscr{E}y && \text{정의 6} \\
&\Longleftrightarrow x\mathscr{E}z \wedge z\mathscr{E}y && \mathscr{E}\text{는 대칭적이다.} \\
&\Longleftrightarrow x\mathscr{E}y && \mathscr{E}\text{는 추이적이다.}
\end{aligned}
$$

그러므로 $x/\mathscr{E} \cap y/\mathscr{E} \neq \phi \Longleftrightarrow x\mathscr{E}y$

따라서 $x\mathscr{E}y \Longleftrightarrow x/\mathscr{E} \cap y/\mathscr{E} \neq \phi$

(c) (a)와 (b)에 따라 $x/\mathscr{E} = y/\mathscr{E} \Longrightarrow x\mathscr{E}y$

다음에 $x\mathscr{E}y \Longrightarrow x/\mathscr{E} = y/\mathscr{E}$

임을 증명할 필요가 있다. 지금 $x\mathscr{E}y$라고 가정하면

$$
\begin{aligned}
z \in x/\mathscr{E} &\Longrightarrow z\mathscr{E}x && \text{정의 6} \\
z\mathscr{E}x \wedge x\mathscr{E}y &\Longrightarrow z\mathscr{E}y && \mathscr{E}\text{는 추이적이다.} \\
&\Longrightarrow z \in y/\mathscr{E} && \text{정의 6}
\end{aligned}
$$

따라서 $z \in x/\mathscr{E} \Longrightarrow z \in y/\mathscr{E}$

그런데 z는 집합 X의 임의의 원소이므로 $x/\mathscr{E} \subseteq y/\mathscr{E}$.

마찬가지로 $y/\mathscr{E} \subseteq x/\mathscr{E}$. 그러므로 $x/\mathscr{E} = y/\mathscr{E}$

정리 4* 집합 X 위의 하나의 동치관계 $\mathscr{E}$에 대하여 $X/\mathscr{E}$는 X의 분할이다.

*(a) 앞으로 특별한 언급이 없는 한 $X \neq \phi$인 것으로 본다.

(b) 149·151·153쪽에 대한 이해는 그리 쉽지 않은 것이다. 우선 예제 6(147쪽)에서의 첨수집합족 ①을 참고로 정리 3,4부터 이해하여라.

[Proof] By Theorem 3(a) and Definition 6, $X/\mathscr{E}=\{x/\mathscr{E} \mid x \in X\}$ is a family of nonempty subsets of X. We next show that

$$x/\mathscr{E} \neq y/\mathscr{E} \Rightarrow x/\mathscr{E} \cap y/\mathscr{E} = \phi$$

by showing its contrapositive: $x/\mathscr{E} \cap y/\mathscr{E} \neq \phi \Rightarrow x/\mathscr{E} = y/\mathscr{E}$. The last assertion is a direct consequence of Theorem 3(b) and (c). Finally, we have to show that $\bigcup_{x \in X} x/\mathscr{E} = X$. This is also trivial, since each x in X belongs to $x/\mathscr{E}$. This completes the proof of the theorem.

We have just seen, in Theorem 4, that an equivalence relation on the nonempty set X gives rise to a partition of X. We shall next show that the converse of Theorem 4 is true, that is, each partition of X gives rise to an equivalence relation on X.

Definition 7. Let $\mathscr{T}$ be a partition of a nonempty set X. We define a relation $X/\mathscr{T}$ on X by $x(X/\mathscr{T})y$ if and only if there exists a set $A \in \mathscr{T}$ such that $x, y \in A$.

Warning! The reader should read and compare Definitions 6 and 7 carefully to understand the delicate differences among these similar notations: $x/\mathscr{E}$, $X/\mathscr{E}$, and $X/\mathscr{T}$.

Theorem 5. Let $\mathscr{T}$ be a partition of a nonempty set X. Then the relation $X/\mathscr{T}$ is an equivalence relation on X, and the equivalence classes induced by the equivalence relation $X/\mathscr{T}$ are precisely the sets in $\mathscr{T}$. Symbolically,

$$X/(X/\mathscr{T}) = \mathscr{T}$$

[증명] 정리 3(a), 정의 6에 따라 $X/\mathscr{E}=\{x/\mathscr{E} | x\in X\}$는 X의 부분집합의 족이다.

다음에는 $x/\mathscr{E}\neq y/\mathscr{E} \Longrightarrow x/\mathscr{E}\cap y/\mathscr{E}=\phi$인 것을 밝히기 위하여 그 대우

$$x/\mathscr{E}\cap y/\mathscr{E}\neq\phi \Longrightarrow x/\mathscr{E}=y/\mathscr{E}$$

를 알아본다. 여기서 마지막 주장은 정리 3(b)와 (c)로부터 직접 얻는 결과에 불과하다. 끝으로 $\bigcup_{x\in X} x/\mathscr{E}=X$인 것을 밝히면 된다. 그런데 $\forall x\in X$에 대하여 $x\in x/\mathscr{E}$이므로 $X\subseteq\bigcup_{x\in X} x/\mathscr{E}$. 한편 $\bigcup_{x\in X} x/\mathscr{E}\subseteq X$. 따라서 $\bigcup_{x\in X} x/\mathscr{E}=X$. 이상으로 증명이 끝난다.

정리 4를 통하여 집합 X 위의 하나의 동치관계에 따라 그때마다 X의 분할 하나씩이 대응하는 것을 보아왔다. 한편 정리 4의 역이 참인 것도 보일 수 있다. 즉 집합 X의 하나의 분할에 따라 X 위의 동치관계 하나씩을 얻는다.

정의 7 집합 X의 분할 $\mathscr{T}$에 대하여 X 위의 관계 $X/\mathscr{T}$를 다음과 같이 정의한다. 즉 집합 $A\in\mathscr{T}$가 존재함으로써

$$\forall x,\ y\in A \Longleftrightarrow x(X/\mathscr{T})y$$

주의 ! 독자는 정의 6, 7을 주의 깊게 비교함으로써 세 기호

$$x/\mathscr{E},\quad X/\mathscr{E},\quad X/\mathscr{T}$$

사이의 미묘한 차이를 분명히 익혀야 한다.*

정리 5 집합 X의 분할 $\mathscr{T}$에 따른 관계 $X/\mathscr{T}$는 X 위의 하나의 동치관계이다. 이때 $X/\mathscr{T}$에 따라 유도된 X의 동치류들은 바로 분할 $\mathscr{T}$를 이룬다. 이 사실을 기호로 나타내면 다음과 같다.**

$$X/(X/\mathscr{T})=\mathscr{T}$$

* 집합 $X=\{a, b, c\}$ 위의 관계 $\mathscr{E}=\{(a, a), (b, b), (c, c), (a, b), (b, a)\}$는 동치관계임을 확인하여라. 여기서 $a/\mathscr{E}=\{a, b\}$, $b/\mathscr{E}=\{a, b\}$, $c/\mathscr{E}=\{c\}$. 이 경우 $X/\mathscr{E}=\{a/\mathscr{E}, b/\mathscr{E}, c/\mathscr{E}\}$는 X의 분할이다. 각자 보기 하나를 더 들어 보아라(연습문제 3.3 문제 3 참조).

** 이 등식은 0이 아닌 두 실수 x, f에 대한 식 $x/(x/f)=x\div\frac{x}{f}=f$와 유사하다.

[Proof] Since every element x of X is contained in some $A \in \mathscr{T}$, $x(X/\mathscr{T})x$; that is, $X/\mathscr{T}$ is reflexive. The symmetry of $X/\mathscr{T}$ is a clear consequence of Definition 7. To show that the relation $X/\mathscr{T}$ is transitive, let x, y, and z be three elements of X satisfying

$$x(X/\mathscr{T})y \quad \text{and} \quad y(X/\mathscr{T})z$$

Then, by definition 7, there exist A and B in $\mathscr{T}$ such that $x, y \in A$ and $y, z \in B$. Consequently, $y \in A \cap B \neq \phi$. It follows, by the definition of a partition, that $A = B$. Hence, $x, z \in A$ and hence $x(X/\mathscr{T})z$. Thus, $X/\mathscr{T}$ is an equivalence relation on X.

To show the remainder of the theorem, let x be an arbitrary element of X. There exists one and only one set A in $\mathscr{T}$ such that $x \in A$ (Why?). Consequently, by Definition 7, we have

$$x/(X/\mathscr{T}) = A$$

We have just proved that each equivalence class modulo $X/\mathscr{T}$ is a set in the family $\mathscr{T}$. Conversely, let A be any set in the partition $\mathscr{T}$. Since $A \neq \phi$, there exists an element x in X that belongs to A. By our previous argument, $x/(X/\mathscr{T}) = A$. This proves that $X/(X/\mathscr{T}) = \mathscr{T}$. The proof of the theorem is now complete.

Any equivalence relation $\mathscr{E}$ on the nonempty set X gives rise to a partition $X/\mathscr{E}$ (Theorem 4); this partition in turn determines an equivalence relation $X/(X/\mathscr{E})$(Theorem 5). The crucial fact is that $X/(X/\mathscr{E}) = \mathscr{E}$ (see Problem 6). This together with $X/(X/\mathscr{E}) = \mathscr{E}$ establishes the intimate connection between equivalence relations and partitions.

Let us illustrate Theorem 5 by a concrete example. Let $\mathbb{Z}_0$ and $\mathbb{Z}_1$ be the set of even integers and the set of odd integers, respectively. Then $\mathscr{T} = \{\mathbb{Z}_0, \mathbb{Z}_1\}$ forms a partition of the set $\mathbb{Z}$ of integers. By definition of the relation $\mathbb{Z}/\mathscr{T}$, we have $a(\mathbb{Z}/\mathscr{T})b$ if and only if either $a, b \in \mathbb{Z}_0$ or $a, b \in \mathbb{Z}_1$. That is, $a(\mathbb{Z}/\mathscr{T})b$ if and only if either both a and b are even or both a and b are odd. It is easy to verify that this relation $\mathbb{Z}/\mathscr{T}$ is indeed an equivalence relation. In effect, $a(\mathbb{Z}/\mathscr{T})b$ if and only if $a \equiv b (\text{mod } 2)$. Therefore, the relation $\mathbb{Z}/\mathscr{T}$ is really

[증명] 집합 X의 각 원소 x는 적당한 $A\in\mathscr{T}$에 속하므로 $x(X/\mathscr{T})x$, 즉 $X/\mathscr{T}$는 반사적이다. 또 정의 7의 결과에 따라 $X/\mathscr{T}$는 대칭적이다. 한편 $X/\mathscr{T}$가 추이적임을 보이기 위하여 임의의 $x, y, z\in X$에 대하여 $x(X/\mathscr{T})y \wedge y(X/\mathscr{T})z$라고 가정하자. 그러면 적당한 $A, B\in\mathscr{T}$가 존재함으로써 $x, y\in A \wedge y, z\in B$. 따라서 $y\in A\cap B$, 즉 $A\cap B\neq\phi$. 한편 분할의 정의[정의 5(a)의 대우]에 따라 $A=B$. 그러므로 $x, z\in A$. 마침내 $x(X/\mathscr{T})z$.

이상으로 $X/\mathscr{T}$는 X 위의 동치관계이다. 여기서 이 정리의 나머지 부분을 증명하고자 집합 X의 임의의 원소를 x라고 하면 집합 $A\in\mathscr{T}$가 유일하게 존재함으로써 $x\in A$(왜?). 따라서 정의 7에 의하여

$$x/(X/\mathscr{T})=A$$

이것으로써 법 $X/\mathscr{T}$인 각 동치류는 바로 족 $\mathscr{T}$의 하나의 집합이란 것을 알 수 있다.

역으로 분할 $\mathscr{T}$의 임의의 원소인 집합을 A라고 하면 $A\neq\phi$이므로 집합 X의 원소로서 A에 속하는 x가 존재한다. 그러므로 지금까지의 논증에 의하여 $x/(X/\mathscr{T})=A$. 따라서 $X/(X/\mathscr{T})=\mathscr{T}$.

마침내 정리 5의 증명이 끝난다.

이렇듯 집합 X 위의 하나의 동치관계 $\mathscr{E}$는 분할 $X/\mathscr{E}$를 낳고(149쪽 정리 4), 이 분할에 따라 동치관계 $\mathscr{E}$를 얻는다(151쪽 정리 5). 이 경우 다음 등식이 성립한다(155쪽 연습문제 3.3 문제 6 참조).

$$X/(X/\mathscr{E})=\mathscr{E} \qquad \cdots\cdots ②$$

여기서 ②는 집합 X 위의 하나의 동치관계와 그에 따른 분할 사이에 밀접한 관련이 있음을 말해주고 있다.

보기 3 짝수, 홀수의 집합을 각각 $\mathbb{Z}_0, \mathbb{Z}_1$이라 하면 $\mathscr{T}=\{\mathbb{Z}_0, \mathbb{Z}_1\}$는 정수의 집합 $\mathbb{Z}$의 분할을 이룬다. 이 경우 관계 $\mathbb{Z}/\mathscr{T}$의 정의에 따라 $a, b\in\mathbb{Z}_0$ 또는 $a, b\in\mathbb{Z}_1$이면 그리고 그때에만 $a(\mathbb{Z}/\mathscr{T})\,b$, 다시 말해 a, b가 모두 짝수 또는 모두 홀수이면 그리고 그때에만 $a(\mathbb{Z}/\mathscr{T})\,b$.

여기서 $\mathbb{Z}/\mathscr{T}$가 바로 $\mathbb{Z}$ 위의 동치관계임을 쉽게 밝힐 수 있다. 실제로

$$a\equiv b \pmod 2 \Longleftrightarrow a(\mathbb{Z}/\mathscr{T})\,b.$$

[결국 관계 $\mathbb{Z}/\mathscr{T}$는 mod 2에 따른 정수의 집합 $\mathbb{Z}$ 위의 합동관계 $\equiv$에 불과하다(143쪽 예제 5* 참조)].

* 이 보기는 143쪽 예제 5*의 하단 및 147쪽 예제 6을 통해 쉽게 이해될 것이다.

the familiar congruence relation$\equiv$(mod 2). (See Example 5.)

Conversely, given the set $\mathbb{Z}$ together with the equivalence relation $\mathscr{E}$ such that $x\mathscr{E}y$ if and only if $x\equiv y \pmod 2$, then

$$a/\mathscr{E}=\{x\in\mathbb{Z}\mid x\equiv a \pmod 2\}=\begin{cases}\mathbb{Z}_0 & \text{if } a \text{ is even}\\ \mathbb{Z}_1 & \text{if } a \text{ is odd}\end{cases}$$

Therefore, $\mathbb{Z}/\mathscr{E}=\{\mathbb{Z}_0,\mathbb{Z}_1\}$, which is clearly a partition of $\mathbb{Z}$.

Exercise 3.3

1. Let $\mathscr{T}$ be a partition of the nonempty set X. Prove that the equivalence relation $X/\mathscr{T}$ is equal to $\bigcup_{A\in\mathscr{T}} A\times A$.

2. In Problem 1, let X be a finite set and let $\mathscr{T}=\{A_1, A_2, \cdots, A_k\}$ where the set A_j contains n_j elements for $j=1, 2, \cdots, k$. Prove that the number of ordered pairs in the equivalence relation $X/\mathscr{T}$ is equal to

$$n_1^2+n_2^2+\cdots+n_k^2$$

3. Let $X=\{a, b, c, d, e\}$ and let $\mathscr{T}=\{\{a,b\},\{c\},\{d,e\}\}$.
(a) Show that $\mathscr{T}$ is a partition of X.
(b) Find the equivalence relation $X/\mathscr{T}$ on X explicitly as a set of ordered pairs.
(c) Denote $\mathscr{E}=X/\mathscr{T}$ and find $a/\mathscr{E}$, $b/\mathscr{E}$, $c/\mathscr{E}$, $d/\mathscr{E}$, and $e/\mathscr{E}$ explicitly.

4. Verify Example 6 for $m=3$.

5. Let X be the set $\mathbb{Z}$ of integers and let $\mathscr{E}$ be a relation on X defined by $x\mathscr{E}y$ if and only if $x-y=5k$ for some integer k.
(a) Prove that the relation $\mathscr{E}$ is an equivalence relation on X.
(b) Find the partition $X/\mathscr{E}$ of X.
(c) Verify that the equivalence relation $X/(X/\mathscr{E})$ is indeed the equivalence relation $\mathscr{E}$.

6. Let $\mathscr{E}$ be an equivalence relation on the nonempty set X. Prove that

$$X/(X/\mathscr{E})=\mathscr{E}.$$

이 주장의 역을 알아보고자 집합 $\mathbb{Z}$ 위의 동치관계 $\mathscr{E}$에 대하여 $x \equiv y \pmod 2$이면 그리고 그때에만 $x\mathscr{E}y$라고 할 때 다음을 얻는다.

$$a/\mathscr{E} \equiv \{x \in \mathbb{Z} \mid x \equiv a \pmod 2\} = \begin{cases} \mathbb{Z}_0, & a\text{가 짝수인 경우} \\ \mathbb{Z}_1, & a\text{가 홀수인 경우} \end{cases}$$

그러므로 $\mathbb{Z}/\mathscr{E} = \{\mathbb{Z}_0, \mathbb{Z}_1\}$. 이것은 분명히 $\mathbb{Z}$에서의 하나의 분할이다.

연습문제 3.3

1. 집합 X의 분할 $\mathscr{T}$에 따른 동치관계 $X/\mathscr{T}$에 대하여 다음을 증명하여라.

$$X/\mathscr{T} = \bigcup_{A \in \mathscr{T}} A \times A$$

2. 문제 1에서 X가 유한집합이고 그 분할이 다음과 같다고 한다.

$$\mathscr{T} = \{A_1, A_2, \cdots, A_k\}$$

이 때, $j=1, 2, \cdots, k$에 대한 A_j가 n_j개의 원소를 지니면 동치관계 $X/\mathscr{T}$에는 $n_1^2 + n_2^2 + \cdots + n_k^2$개의 순서쌍이 존재함을 증명하여라.

3. $X = \{a, b, c, d, e\}$, $\mathscr{T} = \{\{a, b\}, \{c\}, \{d, e\}\}$에 대하여 다음 물음에 답하여라.

(a) $\mathscr{T}$는 집합 X의 분할임을 보여라.

(b) 집합 X 위의 동치관계 $X/\mathscr{T}$를 순서쌍의 집합으로서 나타내어라.

(c) $\mathscr{E} = X/\mathscr{T}$로 놓고 $a/\mathscr{E}$, $b/\mathscr{E}$, $c/\mathscr{E}$, $d/\mathscr{E}$, $e/\mathscr{E}$를 각각 원소나열법으로 나타내어라.

4. $m=3$일 때의 예제 6을 확인하여라(147쪽 각주 * 참조).

5. X를 정수의 집합 $\mathbb{Z}$라 하고 X 위의 관계 $\mathscr{E}$를 다음과 같이 정의한다.

적당한 정수 k에 대하여 $x - y = 5k$이면 그리고 그때에만 $x\mathscr{E}y$이다.

(a) 관계 $\mathscr{E}$는 X 위의 동치관계임을 증명하여라.

(b) 집합 X의 분할 $X/\mathscr{E}$를 구하여라.

(c) 동치관계 $X/(X/\mathscr{E})$는 다름아닌 동치관계 $\mathscr{E}$와 같음을 확인하여라.

6. 집합 X 위의 임의의 동치관계 $\mathscr{E}$에 대하여 다음 등식을 증명하여라.*

$$X/(X/\mathscr{E}) = \mathscr{E}.$$

---

* 이 등식은 0이 아닌 실수 x, e에 대한 다음 식과 유사하다.

$$x/(x/e) = x \div \frac{x}{e} = x \times \frac{e}{x} = e, \quad \text{즉} \quad x/(x/e) = e$$

3.4 Functions

Unquestionably, the concept of a function is one of the most basic ideas in every branch of mathematics. The reader may have learned the following definition : a function is a rule of correspondence that assigns to each element x of a certain set (called the domain of the function) one and only one element y in another set (called the range* of the function). This definition is cloudy. What is meant precisely by a "rule" ? In order to avoid ambiguities, mathematicians have devised a precise definition of a function using the language of sets.

Definition 8. Let X and Y be sets. A function from X to Y is a triple (f, X, Y), where f is a relation from X to Y satisfying
(a) $\mathrm{Dom}(f) = X$.
(b) If $(x, y) \in f$ and $(x, z) \in f$, then $y = z$.

Let (f, X, Y) be a function from X to Y. In what follows, we shall adhere to the custom of writing $\boldsymbol{f : X \to Y}$ instead of (f, X, Y), and $\boldsymbol{y = f(x)}$ instead of $(x, y) \in f$. The reason that "$y = f(x)$" is a meaningful substitute for "$(x, y) \in f$" is

"Every element $x \in X$ has a uniquely determined $y \in Y$ such that $(x, y) \in f$".

To see that this assertion is true, let $x \in X$. Then by condition (a) of Definition 8, there exists an element $y \in Y$ such that $(x, y) \in f$; if there is another element $z \in Y$ with $(x, z) \in f$, then according to condition (b), $z = y$. This shows that y is uniquely determined by $x \in X$.

Let $f : X \to Y$ be a function. If $y = f(x)$, we say that y is the image of x under f and that x is a preimage of y under f. The reader may picture this as illustrated in Figures 9 and 10.

3.4 함수

수학의 어느 분야에서나 첫째로 꼽을 수 있는 것으로는 "함수"를 들 수 있다. 함수의 개념에 대한 다음 표현은 독자 누구나 알고 있을 것이다.

> "함수란 집합(정의역)의 각 원소 x에 또 하나의 집합(정의역과 같더라도 무방)(공역*)의 원소 y가 하나씩 대응하는 규칙을 말한다."

그러나 이 정의는 어딘가 애매하다. 여기서 언급된 "규칙"이란 무엇을 뜻하는가? 수학자들은 이런 애매한 점을 피하려고 집합의 용어를 빌려 함수의 정의를 정확히 내리려고 힘써왔다. 다음 정의부터 살펴보자.

정의 8 공집합이 아닌 두 집합 X, Y에 대하여 X에서 Y로의 함수란 하나의 세 짝 (f, X, Y)**를 뜻한다. 여기서 f는 다음 두 조건을 만족하는 X에서 Y로의 관계이다.

(a) $\mathrm{Dom}(f)=X$ (b) $(x, y)\in f \wedge (x, z)\in f \Longrightarrow y=z$

함수를 나타내는 기호로서 (f, X, Y) 대신 $\boldsymbol{f : X \to Y}$로 나타내거나 $(x, y)\in f$ 대신으로는 $\boldsymbol{y=f(x)}$와 같이 쓰도록 한다. 이렇게 바꿔 쓰기로 한 이유는 정의 8(b)에 의하여 각 원소 $x\in X$에 관계 f에 따른 $y\in Y$가 하나씩 대응하기 때문이다. 다시 말하면 함수 f***에 따라 각 $x\in X$에 대한 $y\in Y$가 하나씩 정해짐을 알 수 있다.

함수 $f : X \to Y$에 대하여 $y=f(x)$일 때 y를 f에 따른 x에서의 상(또는 함숫값)이라고 하고, x는 f에 따른 y의 원상이라고 한다.

---

* 원서에서의 "*range*"는 공역(*codomain*)을 저자가 그렇게 적고 있다. 사실 함수에 따라 치역은 공역과 같거나 그것의 진부분집합이다[다음 중간 설명 및 예제 7(159쪽) 참조].

** 순서 세 짝 (a, b, c)는 순서쌍 (a, b)에 버금가는 용어이다. 그러나 여기서 말하는 세 짝은 그런 것과는 다른 뜻을 지닌다.

*** 정의 8에 따라 집합 X에서 집합 Y로의 함수 f는 하나의 관계이다. 역으로 X에서 Y로의 임의의 관계 $\mathscr{R}$은 함수일까? 반례를 한두 가지 찾아보아라.

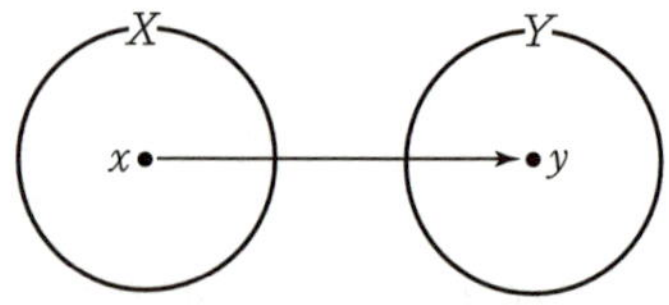

Figure 9 y is the image of x

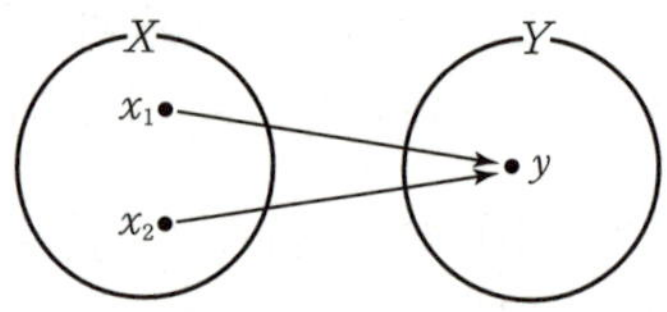

Figure 10 x_1 and x_2 are preimages of y

We call the set Y, in $f: X \longrightarrow Y$, the codomain of the function. The reader should notice that the range of a function need not be the same as the image of the function[4] (see Example 7, below). We call the reader's attention to the fact that some authors use the word "codomain" to mean "image," but for a technical reason, which will be apparent in Section 6, we make a distinction between "image" and "codomain" of a function. In general, the image of a function is a subset of the codomain of that function.

Example 7 Let $f: \mathbb{R} \longrightarrow \mathbb{R}$ be defined by $f(x)=[x]$ for all x in $\mathbb{R}$, where $[x]$ denotes the greatest integer $\leqq x$, e.g., $[\sqrt{2}\,]=1$, $\left[-\frac{1}{2}\right]=-1$. Here, the codomain of f is $\mathbb{R}$, whereas the image of f is $\mathbb{Z}$, a proper subset of $\mathbb{R}$.

It is possible to alter the codomain of a function without otherwise changing the function. For instance, for the same relation f as in Example 7 above, $f: \mathbb{R} \rightarrow \mathbb{Z}$ and $g: \mathbb{R} \rightarrow \mathbb{Q}$ are functions, because Definition 8 is satisfied.

[4] The image of the function $f: X \rightarrow Y$ is equal to the image, $\mathrm{Im}(f)$, of the relation f. Consequently, $\{f(x) \mid x \in X\} = \mathrm{Im}(f)$.

다음 그림 9, 10은 함수 f에 대한 상, 원상을 각각 나타내고 있다.

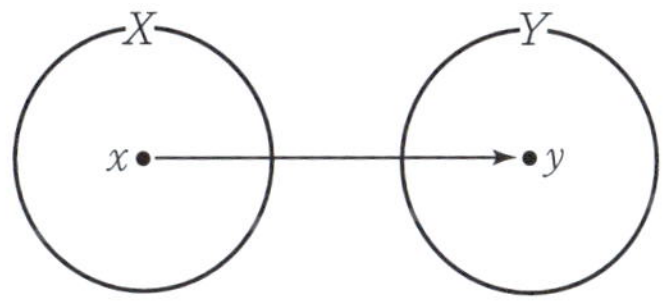

그림 9 y(하나)는 x의 상이다.

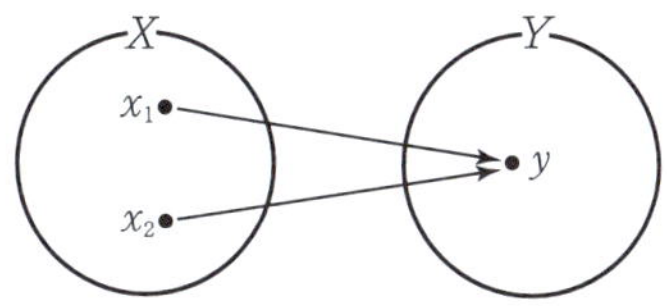

그림 10 x_1, x_2(하나씩)는 각각 y(하나)의 원상이다.

함수 $f: X \to Y$에 대하여 집합 Y를 f의 **공역**(157쪽 * 참조)이라고 한다. 대체로 함수의 공역은 함수의 **치역**(또는 **상**)$\{f(x) \in y \mid x \in X\}$[4]과 같지 않을 수도 있음을 주의해야 한다(다음 예제 7 참조). 저자에 따라서는 두 용어 "함수의 공역"과 "함수의 상"을 같은 뜻으로 사용하지만 우리는 3.6절(181쪽)에서 보듯이 그 두 용어를 구분해서 쓰기로 한다. 사실 함수의 치역이 그 공역의 진부분집합인 경우가 흔하다.

예제 7 각 $x \in \mathbb{R}$에 대한 $[x]$는 x보다 크지 않은 실수 중 가장 큰 정수를 나타내는 기호이다. 이를테면

$$[\sqrt{2}]=1, \quad \left[-\frac{1}{2}\right]=-1, \quad [-5]=-5, \quad [0]=0, \quad [7]=7$$

여기서 $f(x)=[x]$*로 정의된 함수 $f: \mathbb{R} \longrightarrow \mathbb{R}$의 공역은 실수의 집합 $\mathbb{R}$이지만 그 치역은 공역 $\mathbb{R}$의 진부분집합으로서 0 이상의 정수의 집합과 같다.

주어진 함수에 대하여 그 공역을 바꿀 때마다 새로운 함수를 얻는다. 이를테면 예제 7의 함수 f에 대하여 $g: \mathbb{R} \to \mathbb{Z}$, $h: \mathbb{R} \to \mathbb{Q}$인 경우를 생각하면 이들은 제각기 다른 함수**이다.

---

[4] 함수 $f: X \to Y$의 치역은 관계 f의 상 $\mathrm{Im}(f)$과 같다. 즉 $\mathrm{Im}(f)=\{f(x) \mid x \in X\}$.

* 이 함수 f를 **가우스 함수**라고 한다.

** 비록 $\mathrm{Dom}(f)=\mathrm{Dom}(g)=\mathrm{Dom}(h)=\mathbb{R}$, $\mathrm{Im}(f)=\mathrm{Im}(g)=\mathrm{Im}(h)=\{x \in \mathbb{Z} \mid 0 \le x\}$이지만 f, g, h의 공역은 차례로 $\mathbb{R}$, $\mathbb{Z}$, $\mathbb{Q}$로서 제각기 다르다고 본다.

In general, we have the following theorem.

Theorem 6. Let $f: X \longrightarrow Y$ be a function and let W be a set containing the image of f. Then $f: X \longrightarrow W$ is a function.

[Proof] We first prove that f is a relation from X to W:

$(x, y) \in f \Rightarrow x \in X \wedge y \in (\text{Im } f)$	Def. of Im
$\Rightarrow x \in X \wedge y \in W$	$\text{Im}(f) \subseteq W$
$\Rightarrow (x, y) \in X \times W$	Def. 1

Therefore, $(x, y) \in f \Rightarrow (x, y) \in X \times W$ (Why?).

This proves that $f \subseteq X \times W$; in other words, f is a relation from X to W. Now since $f: X \longrightarrow Y$ is a function. $\text{Dom}(f) = X$ and condition (b) of Definition 8 is satisfied. Therefore, $f: X \longrightarrow W$ is a function.

Theorem 7. Let $f: X \longrightarrow Y$ and $g: X \longrightarrow Y$ be functions. Then

$$f = g \text{ if and only if } f(x) = g(x), \ \forall x \in X.$$

[Proof] (1) Suppose that $f = g$ and that x is an arbitrary element in X.

Then, $y = f(x) \Leftrightarrow (x, y) \in f$	Notation
$\Leftrightarrow (x, y) \in g$	Def. of $f = g$
$\Leftrightarrow g(x) = y$	Notation

Hence, $f(x) \Leftrightarrow g(x)$.

Therefore, $f(x) = g(x)$.

이제 우리는 다음 정리를 알아보자.

정리 6 함수 $f: X \to Y$에 대하여 W가 f의 공역 Y를 포함하는 집합일 때 X, W를 각각 정의역, 공역으로 하는 함수 $g^*: X \to W$를 얻는다.

[증명] 우선 g는 X에서 W로의 관계임을 증명한다.

$(x, y) \in g \Rightarrow x \in X \wedge y \in \mathrm{Im}(g)$	$\mathrm{Im}(f)$의 정의
$\Rightarrow x \in X \wedge y \in W$	$\mathrm{Im}(f) \subseteq W$
$\Rightarrow (x, y) \in X \times W$	정의 1
그러므로 $(x, y) \in g \Rightarrow (x, y) \in X \times W$	$\Rightarrow$의 추이율

따라서 $g \subseteq X \times W$. 즉 f는 X에서 W로의 관계이다. 그런데 가정에서 $f: X \to Y$는 함수이므로 $\mathrm{Dom}(f) = X = \mathrm{Dom}(g)$. 한편 함수 g는 정의 8의 조건 (b)도 만족한다(과연?). 따라서 함수 $g: X \to W$를 얻는다.

정리 7** 두 함수 $f: X \to Y$, $g: X \to Y$에 대하여

$$\forall_{x \in X}\, x \in X \quad f(x) = g(x) \Longleftrightarrow f = g.$$

[증명] ($\Rightarrow$: 필요조건) 임의의 $x \in X$에 대하여 $f(x) = g(x)$이라면

$(x, y) \in f \Longleftrightarrow y = f(x)$	기호 f
$\Longleftrightarrow y = g(x)$	가정 $f(x) = g(x)$
$\Longleftrightarrow (x, y) \in g$	기호 g

그러므로 $(x, y) \in f \Longleftrightarrow (x, y) \in g$

따라서 $f = g$

* 원서는 두 함수를 모두 f로 나타내지만 공역이 다른 (여기서는 Y 대신 W) 두 번째의 f는, 이를테면, g로 적어 f와 구분함이 옳다(159쪽 각주 ** 및 아래 각주 ** 참조). 한편 정리 6에서의 "$f: x \to y$는 함수이다."와 같은 표현에 있어서 "함수는 함수이다"로 읽힌 염려가 있다.

** 이 정리의 증명에서 함수를 일종의 관계로 생각하여 순서쌍의 집합으로 보고 있다. 책에 따라서 정리 7을 "$f = g$에 대한 정의"로 내세우는 경우도 있다.

(2) Suppose that $\forall_{x \in X}\, f(x)=g(x)$. Then

$$\begin{aligned} (x, y) \in f &\Longleftrightarrow y = f(x) && \text{Notation} \\ &\Longleftrightarrow y = g(x) && f(x) = g(x) \\ &\Longleftrightarrow (x, y) \in g && \text{Notation} \end{aligned}$$

This proves that $f=g$.

If the domain and the codomain of a function are subsets of the set of real numbers, then, as in analytic geometry, the graph of the function may be sketched on the Cartesian plane. For example, the function in Example 7 has the following graph.

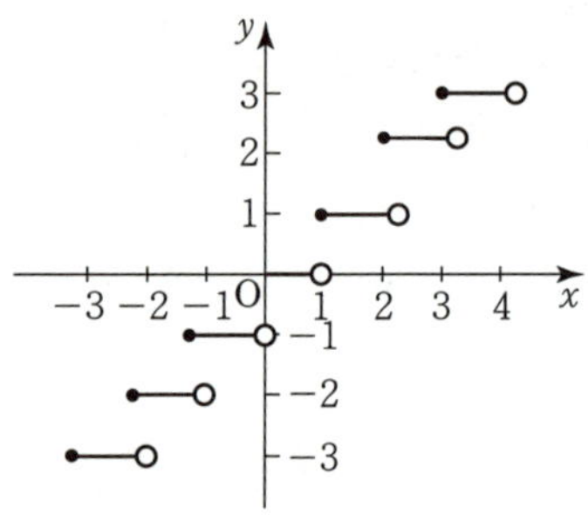

Figure 11

Example 8 Let A be a subset of a nonempty set X. Then the relation

$$\{(x, y) \in X \times \{0, 1\} \mid y=1 \text{ if } x \in A \text{ and } y=0 \text{ if } x \notin A\}$$

gives rise to a function from X to $\{0,1\}$, known as the characteristic function of A in X. This function is usually denoted by the Greek letter chi with a subscript A, χ_A. That is,

$$\chi_A : X \longrightarrow \{0,1\}$$

is defined by

$$\chi_A(x) = \begin{cases} 1 & \text{if } x \in A \\ 0 & \text{if } x \in X - A \end{cases}$$

(⇐: 충분조건) $f=g$이라면 임의의 $x \in X$에 대하여

$$
\begin{array}{lll}
 & y=f(x) \Longleftrightarrow (x, y) \in f & \text{기호 } f \\
 & \quad\quad\quad\ \ \Longleftrightarrow (x, y) \in g & \text{가정 } f=g \\
 & \quad\quad\quad\ \ \Longleftrightarrow y=g(x) & \text{기호 } g \\
\text{그러므로} & y=f(x) \Longleftrightarrow y=g(x) & \Longleftrightarrow \text{의 추이율} \\
\text{따라서} & \quad\quad\quad f(x)=g(x) &
\end{array}
$$

함수의 정의역과 공역이 실수의 집합 ℝ의 부분집합인 경우 우리는 좌표평면에 그 함수의 개형을 그려왔다. 이를테면 보기 7의 함수 $y=[x]$의 그래프의 일부를 다음 그림 11과 같이 그릴 수 있다.

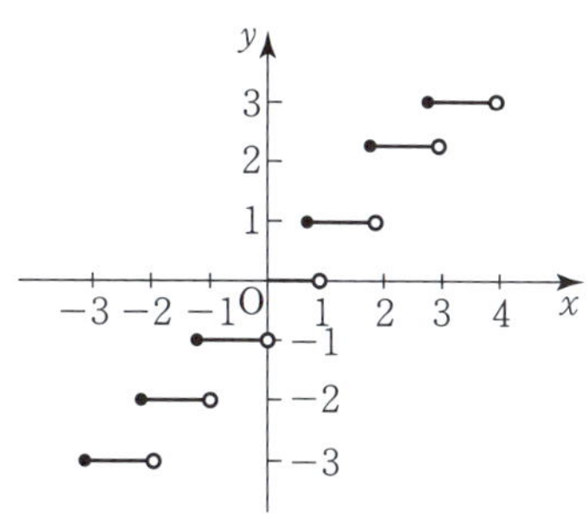

그림 11

예제 8 집합 X의 부분집합 A에 대한 순서쌍의 집합

$$\{(x, y) \mid x \in A \text{이면 } y=1,\ x \notin A \text{이면 } y=0\}$$

에 따라 X에서 $\{0, 1\}$로의 함수 하나를 얻거니와 이 함수를 X에서의 A의 **특성함수**라 하고 그리스 문자 χ(chi: 카이라고 읽는다)에 첨자 A를 적어 χ_A로 나타낸다. 다시 말해 $\chi_A : X \longrightarrow \{0, 1\}$은 다음과 같이 정의된 함수이다.

$$\chi_A(x)=\begin{cases} 1, & x \in A \\ 0, & x \in X-A \end{cases}$$

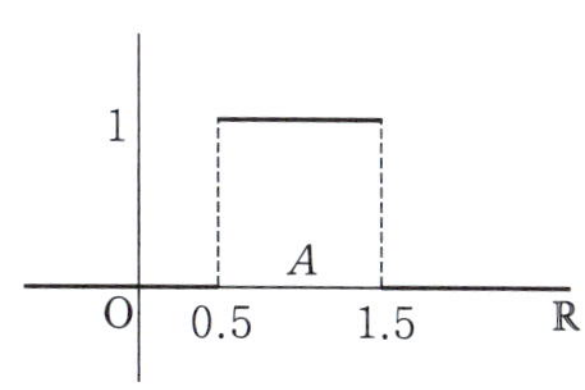

특히 $X=\mathbb{R}$에서의 $A=[0.5, 1.5]$의 특성함수 $\chi_{[0.5, 1.5]} : \mathbb{R} \longrightarrow \{0, 1\}$의 그래프는 오른쪽 그림과 같다.

Although a function is, by definition, written (f, X, Y) or $f : X \to Y$, it is often a nuisance to have to write the domain and the range of the function explicitly when they are implicitly clear from the context. Therefore, we shall denote a function by f when the domain and the range of f are clearly understood, without explicitly giving the domain and the range of f.

Example 9 Let X be a set. The diagonal relation Δ_X on X defined on page 140 is a function from X to X. When we wish to stress that the relation Δ_X is a function, we use the alternative notation $1_X : X \to X$, where $1_X(x) = x$ for all x in X. The function 1_X is called the identity function on X.

Example 10 Let X and Y be two nonempty sets and let b be a fixed element of Y. The relation

$$C_b = \{(x, b) \mid x \in X\}$$

gives rise to the function $C_b : X \to Y$ given by $C_b(x) = b$ for all x in X. The function C_b is called a constant function.

In calculus, we have often seen a function defined by two (or more) rules of correspondence: for example, $h : \mathrm{R} \to \mathrm{R}$ defined by

$$h(x) = \begin{cases} 1-2x, & \text{if } x \leqq 0 \\ x^2+1, & \text{if } x \geqq 0 \end{cases}$$

This function may be considered as the union of the following two functions:

(1) $f : (-\infty, 0] \to R$ defined by $f(x) = 1-2x$, $\forall x \in (-\infty, 0]$
(2) $g : [0, \infty) \to R$ defined by $g(x) = x^2+1$, $\forall x \in [0, \infty)$

The reader should notice that here

$$\mathrm{Dom}(f) \cap \mathrm{Dom}(g) = \{0\} \text{ and that } f(0) = g(0).$$

함수는 정의에 따라 (f, X, Y) 또는 $f: X \to Y$와 같이 나타내지만 정의역과 공역의 언급이 없더라도 문맥상 분명할 때 그 함수를 단지 f로 나타낸다.

예제 9 141쪽에서 정의된 집합 X 위의 항등관계 Δ_X는 X에서 그 자신으로의 함수이다. 여기서 함수 Δ_X를 강조하기 위하여 새로운 기호 $\mathrm{I}_X : X \to X$를 쓰기로 한다. 그러고 보면 모든 $x \in X$에 대하여 $\mathrm{I}_X(x) = x$. 이 경우의 I_X를 집합 X 위의 **항등함수***(이것을 $f(x)=x$ 또는 $y=x$로 나타내 왔다)라고 한다.

예제 10 두 집합 X, Y가 각각 공집합이 아닌 집합일 때 Y의 일정한 원소 b에 대한 순서쌍의 집합 $C_b = \{(x, b) \mid x \in X\}$는 집합 X에서 집합 Y로의 하나의 관계이다. 이 관계에 따라 임의의 $x \in X$에 대하여 $C_b(x) = b$**로 정의된 함수 $C_b : X \to Y$를 얻는다. 이 때의 함수 C_b를 **상수함수**라고 한다.

우리는 미적분학에서 둘 또는 그 이상의 대응규칙에 따라 정의된 함수를 가끔 보아왔다. 이를테면

$$h(x) = \begin{cases} 1-2x, & x \leqq 0 \\ x^2+1, & 0 \leqq x \end{cases}$$

로 정의된 함수 $h : \mathbb{R} \to \mathbb{R}$을 다음 두 함수 (1), (2)의 합집합(이것을 간단히 **합*****이라고 부른다)으로 볼 수 있다.

(1) 임의의 $x \in (-\infty, 0]$에 대하여 $f(x) = 1-2x$로 정의된 함수

$$f : (-\infty, 0] \to \mathbb{R}$$

(2) 임의의 $x \in [0, \infty)$에 대하여 $g(x) = x^2+1$로 정의된 함수

$$g : [0, \infty) \to \mathbb{R}$$

여기서 다음과 같은 사실을 주목하자.

$$\mathrm{Dom}(f) \cap \mathrm{Dom}(g) = \{0\}, \qquad f(0) = g(0) = 1$$

(독자는 두 함수 f, g의 그래프를 같은 좌표평면에 그려 보아라.)

---

* 예제 9의 X 대신 $\mathbb{R}$로 그려 놓고 위의 항등함수 $y=x$의 그래프를 그려 보아라. 한편 $X=\mathbb{Z}$일 때의 그래프도 그려라.

** 상수함수 $C_b(x)=b$를 흔히 $f(x)=b$ 또는 $y=b$(b는 상수)로 나타낸다(이런 함수의 그래프를 그려 보아라).

*** 초보자에게는 낯설지 모르지만 순서쌍의 집합으로서 함수를 생각할 때 두 함수의 합도 나름대로 뜻을 지닌다. 여기서 주어진 $h(x)$의 그래프를 하나의 좌표평면에 그려 보아라.

The last example motivates the following general theorem.

Theorem 8. Let $f: A \to C$ and $g: B \to D$ be two functions such that $f(x) = g(x),\ \forall x \in A \cap B$. Then the union h of f and g defines the function

$$h = f \cup g : A \cup B \to C \cup D$$

where
$$h(x) = \begin{cases} f(x), & \text{if } x \in A \\ g(x), & \text{if } x \in B \end{cases}$$

[Proof] Since f and g are relations, $f \subseteq A \times C$ and $g \subseteq B \times D$, and we have

$$\begin{aligned} h = f \cup g &\subseteq (A \times C) \cup (B \times D) \\ &\subseteq (A \cup B) \times (C \cup D) \end{aligned}$$

because both $A \times C$ and $B \times D$ are subsets of $(A \cup B) \times (C \cup D)$. Thus, h is a relation from $A \cup B$ to $C \cup D$. We leave it to the reader to verify that

$$\begin{aligned} \text{Dom}(h) &= \text{Dom}(f) \cup \text{Dom}(g) \\ &= A \cup B \end{aligned}$$

This shows that the relation h satisfies Definition 8(a).

For each element $x \in A - B$, we may consider the following three cases:

$$(1)\ x \in A - B, \quad (2)\ x \in B - A, \quad \text{and} \quad (3)\ x \in A \cap B$$

Since $f: A \to C$ and $g: B \to D$ satisfy Definition 8(b) and $f(x) = g(x)\ \forall x \in A \cap B$, we have that $h(x)$ is uniquely defined in each of the three cases. Thus, the relation h satisfies Definition 8(b) as well. Hence,

$$h: A \cup B \to C \cup D$$

is indeed a function.

앞의 마지막 함수 $h: \mathbb{R} \longrightarrow \mathbb{R}$은 다음 정리에 대한 동기가 되고 있다.

정리 8* 두 함수 $f: A \to C$, $g: B \to D$에서 정의역의 교집합 $A \cap B$의 임의의 원소 x에 대하여 $f(x)=g(x)$일 때 f, g의 **합 $f \cup g$**은 다음과 같이 정의된 함수 h와 같다.

$$h = f \cup g: A \cup B \to C \cup D$$

여기서 $$h(x) = \begin{cases} f(x), & x \in A\text{일 때} \\ g(x), & x \in B\text{일 때} \end{cases}$$

[증명] 함수 f, g는 각각 관계 $f \subseteq A \times C$, $g \subseteq B \times D$이므로

$$h = f \cup g \subseteq (A \times C) \cup (B \times D)$$

여기서 $A \times C$, $B \times D$는 모두 $(A \cup B) \times (C \cup D)$의 부분집합이다.

그러므로 $\quad (A \times C) \cup (B \times D) \subseteq (A \cup B) \times (C \cup D)$

따라서 $\quad h \subseteq (A \cup B) \times (C \cup D)$

즉 h는 $A \cup B$에서 $C \cup D$로의 하나의 관계이다.

한편 $\quad \mathrm{Dom}(h) = \mathrm{Dom}(f \cup g) = \mathrm{Dom}(f) \cup \mathrm{Dom}(g) = A \cup B$

곧 $\quad \mathrm{Dom}(h) = A \cup B$

(독자는 이것에 대한 보기 하나를 나름대로 들어 보아라.)

이상으로 관계 h는 정의 8(a)를 만족함을 알 수 있다. 이제 남은 과제로 각 원소 $x \in A \cup B$에 대한 다음 세 경우를 알아보는 일이다.

(1) $x \in A - B$ (2) $x \in B - A$ (3) $x \in A \cap B$

여기서 $f: A \to C$, $g: B \to D$는 모두 정의 8(b)를 만족하며 임의의 $x \in A \cap B$에 대해서는 $f(x) = g(x)$. 그러므로 위의 어느 경우나 $h(x)$는 유일하게 정의된다. 이처럼 관계 h 또한 정의 8(b)를 만족하므로 h는 함수 $h: A \cup B \to C \cup D$임을 알 수 있다.

---

* 이를테면 $f(x) = \sin x$, $g(x) = \cos x - 1$로 각각 정의된

$$f: [-\pi, 0] \to [0, 1], \quad g(x): [0, \pi] \to [-2, 0]$$

에 대하여 $f(0) = g(0) = 0$이므로 우리는 함수 $h: [-\pi, \pi] \to [-2, 0]$을 얻는다[$y = h(x)$의 그래프를 그려 보아라].

Exercise 3.4

1. Test whether or not each of the following diagrams defines a function from $X=\{x, y, z\}$ to $Y=\{u, v, w\}$.

(a)

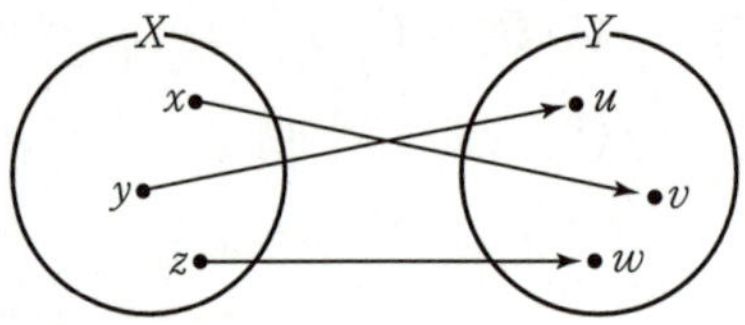

(b)

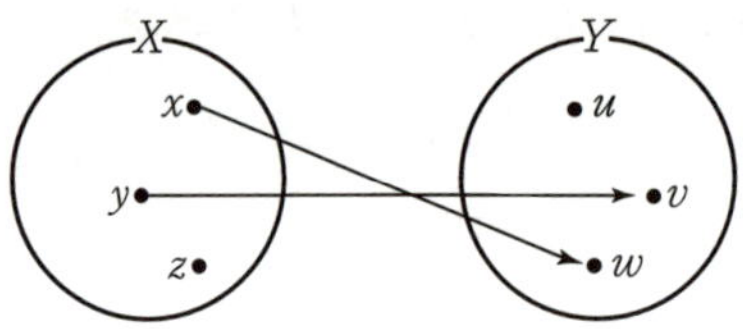

(c)

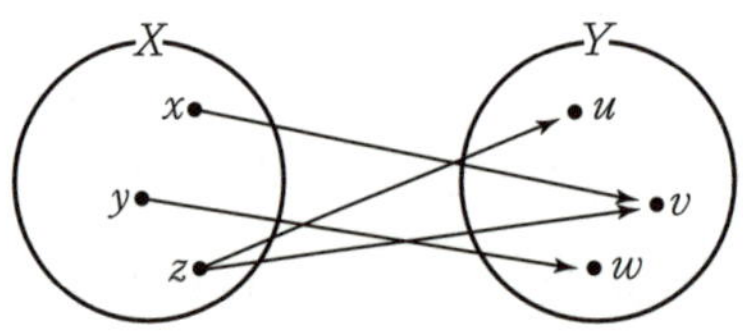

2. Let $f: \mathbb{R} \longrightarrow \mathbb{R}$ be the function given by

$$f(x)=\begin{cases} 5 & \text{if } x \text{ is rational} \\ -3 & \text{if } x \text{ is irrational} \end{cases}$$

Find $f(1/3)$, $f(7)$, and $f(1.323232\cdots)$.

3. Let the function $f: \mathbb{R} \longrightarrow \mathbb{R}$ be given by

$$f(x)=\begin{cases} 4x+3 & \text{if } x>5 \\ x^2-2 & \text{if } -6 \leqq x \leqq 5 \\ 4-5x & \text{if } x<-6 \end{cases}$$

Find $f(-7)$, $f(3)$, and $f(6)$.

연습문제 3.4

1. 다음 그림을 보고 $X=\{u,y,z\}$에서 $Y=\{u,v,w\}$로의 함수인 것을 가리고 함수 아닌 것에 대해서는 그 이유를 말하여라.

(a)

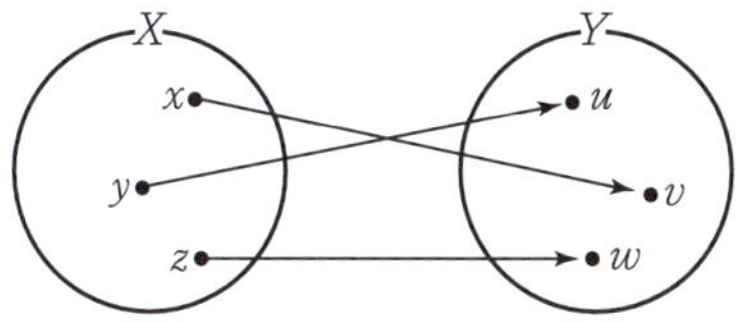

(b)

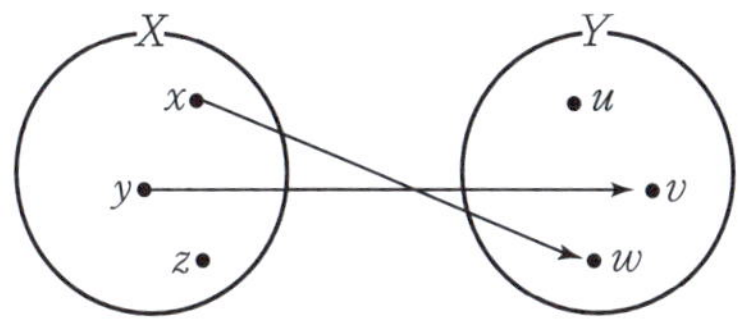

(c)

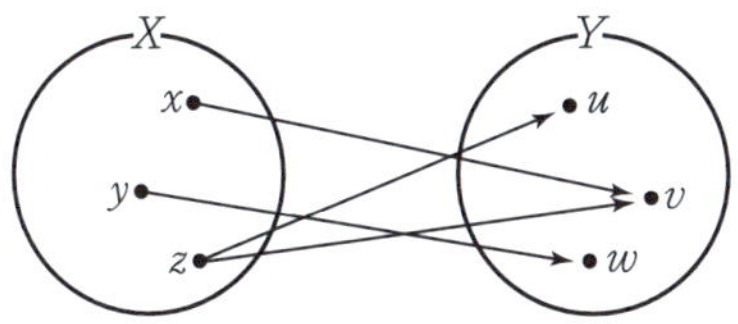

2. 다음과 같이 정의된 함수 $f:\mathbb{R}\longrightarrow\mathbb{R}$가 있다.

$$f(x)=\begin{cases} 5, & x\text{가 유리수일 때} \\ -3, & x\text{가 무리수일 때} \end{cases}$$

아래의 함숫값을 구하여라.

$$f(1/3),\ f(7),\ f(1.323232\cdots),\ f(\sqrt{2}),\ f(-2)$$

3. 다음과 같이 정의된 함수 $g:\mathbb{R}\longrightarrow\mathbb{R}$가 있다.

$$g(x)=\begin{cases} 4x+3, & x>5 \\ x^2-2, & -6\leqq x\leqq 5 \\ 4-5x, & x<-6 \end{cases}$$

아래의 함숫값을 구하여라.

$$g(-7),\ g(6),\ g(-\pi),\ g(0),\ g\left(-\frac{1}{5}\right),\ g(3.141414\cdots)$$

4. Let the function $f: X \to Y$ be defined by the diagram. What is the image of this function?

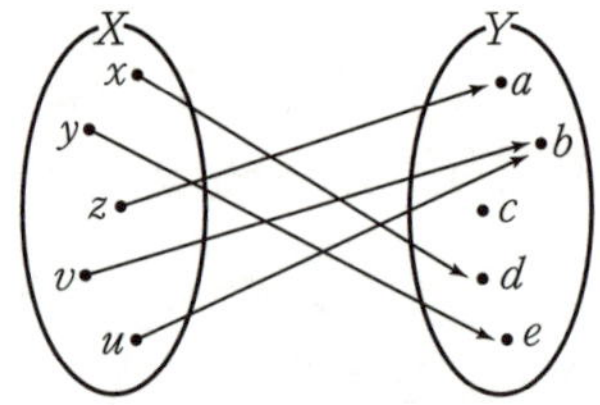

5. Let the function $f: X \to \mathbb{R}$ be defined by $X = \{-2, -1, 0, 1, 2\}$ and $f(x) = x^2 - 3$ for all $x \in X$. Find the image of the function f.

6. Each of the following expressions defines a function from $\mathbb{R}$ to $\mathbb{R}$. Find the image of each function.
(a) $f(x) = 2x^2 + 5$ (b) $g(x) = \cos x$ (c) $h(x) = x^3 - 1$

7. Let $X \subseteq Y$ and $f = \{(x, x) \mid x \in X\}$. Prove that $f: X \to Y$ is a function. (*Remark.* This function is called an inclusion function which may be denoted by $f: X \subseteq Y$.)

8. Let $X = \{x, y, z\}$ and $Y = \{1, 2, 3\}$. Which of the following constitute functions from X to Y? If they do not, give the reason.
(a) $f = \{(x, 1), (y, 2), (z, 3)\}$ (b) $g = \{(x, 2), (y, 3), (z, 2)\}$
(c) $h = \{(x, 2), (y, 1)\}$ (d) $i = \{(x, 1), (x, 2), (y, 1), (z, 3)\}$

9. If $X = \{x, y, z\}$ and $Y = \{1, 2\}$, how many functions from X to Y exist? More generally, if the set X has m elements and if Y has n elements, how many functions from X to Y exist?

10. How many of the functions in Problem 9 are constant functions?

11. Let $f: X \to Y$ be a function. Prove that every subset g of f gives rise to a function.

12. Let $f: X \to Y$ be a function from X to X that is also a reflexive relation on X. Prove that f must be the identity function $1_X: X \to X$.

13. Let X be the unit interval $[0, 1]$. Find a function $f: X \to X$ that is a symmetric relation on X.

14. Let $f: X \to Y$ and $g: X \to Y$ be two functions with the same domain and the same codomain. Prove that if $f \subseteq g$, then $f = g$.

다음 함수 4~6의 치역을 구하여라.

4. 다음 벤 다이어그램*으로 정의된 함수 $f: X \to Y$

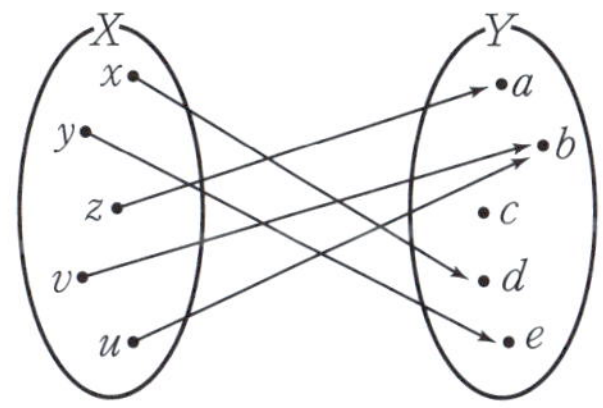

5. 정의역 $X=\{2, -1, 0, 1, 2\}$, 공역 $\mathbb{R}$인 함수 $f(x)=x^2-3$

6. 아래와 같은 $\mathbb{R}$에서 $\mathbb{R}$로의 함수

(a) $f(x)=2x^2+5$ (b) $g(x)=\cos x$ (c) $h(x)=x^3-1$

7. 두 집합 X, Y에 대하여 $X\subseteq Y$**, $f=\{(x, x) \mid x\in X\}$일 때 f는 X에서 Y로의 함수임을 증명하여라(이 함수를 **포함함수**라 하고 $f: X\subseteq Y$와 같이 나타내기도 한다).

8. $X=\{x, y, z\}$, $Y=\{1, 2, 3\}$에 대하여 다음 중 X에서 Y로의 함수인 것을 지적하고 그렇지 않은 것에 대해서는 그 이유를 대라.

(a) $f=\{(x, 1), (y, 2), (z, 3)\}$ (b) $g=\{(x, 2), (y, 3), (z, 2)\}$

(c) $h=\{(x, 2), (y, 1)\}$ (d) $i=\{(x, 1), (x, 2), (y, 1), (z, 3)\}$

9. $X=\{x, y, z\}$, $Y=\{1, 2\}$일 때 X에서 Y로의 함수는 몇 개인가? 집합 X, Y의 원소가 각각 m, n일 때의 X에서 Y로의 함수의 수를 구하여라.

10. 문제 9의 함수 중 상수함수인 것의 수를 말하여라.

11. 함수 $f: X\to Y$에 대하여 f의 각 부분집합 g(f, g를 각각 순서쌍의 집합으로 생각함)마다 함수를 얻는다. 이것을 증명하여라.

12. 집합 X에서 X로의 함수 $f: X\to X$가 X 위의 반사적 관계이면 f는 항등함수 $I_X: X\to X$와 같다. 이것을 증명하여라.

13. 집합 X가 단위폐구간 $[0, 1]$일 때 이 구간 위의 대칭적 관계로서 함수 $f: X\to X$인 것을 구하여라.

14. 정의역과 공역이 각각 같은 두 함수 $f: X\to Y$, $g: X\to Y$에 대하여 $f\subseteq g$이면 $f=g$. 이것을 증명하여라.

---

* 벤 다이어그램으로서 원이 아니더라도 무방하다.

3.5 Images and Inverse Images of Sets

Recall that if $f : X \longrightarrow Y$ is a function and if x and y are elements of X and Y, respectively, such that $y = f(x)$, then y is the image of x under f, and x is a preimage of y under f.

This concept can be extended naturally from elements to subsets as follows :

Definition 9. Let $f : X \longrightarrow Y$ be a function, and let A and B be subsets of X and Y, respectively.

(a) The image of A under f, which we denote $\boldsymbol{f(A)}$, is the set of all images $f(x)$ such that $x \in A$.

(b) The inverse image of B under f, which we denote $\boldsymbol{f^{-1}(B)}$, is the set of all preimages of y in B.

Using set builder notation, we have the following expressions :

$$f(A) = \{f(x) \mid x \in A\}, \qquad f^{-1}(B) = \{x \mid f(x) \in B\}$$

Theorem 9. Let $f : X \longrightarrow Y$ be a function. Then

(a) $f(\phi) = \phi$.

(b) $f(\{x\}) = \{f(x)\} \forall x \in X$.

(c) If $A \subseteq B \subseteq X$, then $f(A) \subseteq f(B)$.

(d) If $C \subseteq D \subseteq Y$, then $f^{-1}(C) \subseteq f^{-1}(D)$.

[Proof] Theorem 9 follows easily from Definition 9 ; therefore the proof is left to the reader.

Theorem 10. Let $f : X \longrightarrow Y$ be a function and let $\{A_\gamma \mid \gamma \in \Gamma\}$ be a family of subsets of X. Then

(a) $f(\bigcup_{\gamma \in \Gamma} A_\gamma) = \bigcup_{\gamma \in \Gamma} f(A_\gamma)$. (b) $f(\bigcap_{\gamma \in \Gamma} A_\gamma) \subseteq \bigcap_{\gamma \in \Gamma} f(A_\gamma)$.

3.5 집합의 상과 역상

함수 $f: X \to Y[y=f(x)]$에 대하여 y는 f에 따른 x의 상이고 x는 f에 따른 y의 원상(157쪽 아래 구절)임을 알고 있다. 집합의 원소로서 이 개념을 우리는 다음과 같이 집합의 경우로 확장한다.

정의 9* 함수 $f: X \to Y$에 대하여 A, B가 각각 X, Y의 부분집합($A=X$ 또는 $B=Y$인 경우도 포함)일 때

(a) 각 $x \in A$의 상 $f(x)$의 집합을 f에 따른 A의 **상**이라 하고 이것을 $\boldsymbol{f(A)}$로 나타낸다.

(b) 원소 $y \in B$의 원상의 집합을 f에 따른 B의 **역상**이라 하고 이것을 $\boldsymbol{f^{-1}(B)}$로 나타낸다.

집합 $f(A)$, $f^{-1}(B)$을 각각 다음과 같이 조건제시법으로 나타낼 수 있다.

$$f(A)=\{f(x) \mid x \in A\}, \qquad f^{-1}(B)=\{x \mid f(x) \in B\}$$

정리 9 함수 $f: X \to Y$에 대하여 다음이 성립한다.

(a)** $f(\phi)=\phi$　　　(b) $\forall x \in X \; f(\{x\})=\{f(x)\}$

(c) $A \subseteq B \wedge B \subseteq X$이면　$f(A) \subseteq f(B)$

(d) $C \subseteq D \wedge D \subseteq Y$이면　$f^{-1}(C) \subseteq f^{-1}(D)$

[증명] 정리 9는 정의 9로부터 쉽게 유도되므로 그 증명은 독자에게 돌린다(힘이 든다면 증명 대신 보기라도 하나 들어 확인하여 보아라).

정리 10* ** 함수 $f: X \to Y$에 있어서 정의역 X의 부분집합의 족 $\{A_\gamma \mid \gamma \in \Gamma\}$에 대하여 다음이 성립한다.

(a) $f(\bigcup_{\gamma \in \Gamma} A_\gamma)=\bigcup_{\gamma \in \Gamma} f(A_\gamma)$　　　(b) $f(\bigcap_{\gamma \in \Gamma} A_\gamma) \subseteq \bigcap_{\gamma \in \Gamma} f(A_\gamma)$

---

* 해당하는 보기부터 하나 둘 들어보아라.

** $X \neq \phi$을 감안할 때 이 등식은 별로 뜻이 없다.

*** 우선 진부분집합 $\subset$, 등호 $=$(정리 13, 185쪽)인 두 경우에 대한 보기를 하나씩 들어라.

[Proof] (a) By repeated use of Definition 9 and Definition 6 of Chapter 2, we have

$$\begin{aligned} y\in f(\bigcup_{\gamma\in\Gamma} A_\gamma) &\Longleftrightarrow y=f(x) \quad \text{for some } x\in \bigcup_{\gamma\in\Gamma} A_\gamma \\ &\Longleftrightarrow y=f(x) \quad \text{for some } x\in A_\gamma, \text{ for some } \gamma\in\Gamma \\ &\Longleftrightarrow y\in f(A_\gamma) \quad \text{for some } \gamma\in\Gamma \\ &\Longleftrightarrow y\in \bigcup_{\gamma\in\Gamma} f(A_\gamma) \end{aligned}$$

Therefore, $f(\bigcup_{\gamma\in\Gamma} A_\gamma) = \bigcup_{\gamma\in\Gamma} f(A_\gamma)$.

(b) Since $\bigcap_{\gamma\in\Gamma} A_\gamma \subseteq A_\gamma$ for every $\gamma\in\Gamma$, by Theorem 9(c), we have

$$f(\bigcap_{\gamma\in\Gamma} A_\gamma)\subseteq f(A_\gamma) \quad \text{for every } \gamma\in\Gamma.$$

It follows from Definition 7 of Chapter 2 that

$$f(\bigcap_{\gamma\in\Gamma} A_\gamma)\subseteq \bigcap_{\gamma\in\Gamma} f(A_\gamma).$$

The inclusion symbol $\subseteq$ in Theorem 10(b) may not be replaced by an equal sign, as the next example shows.

Example 11 Let $X=\{a,b\}$, $Y=\{c\}$, $\Gamma=\{1,2\}$, $A_1=\{a\}$, $A_2=\{b\}$, and let $f: X\longrightarrow Y$ be the constant function, $f(a)=f(b)=c$. Then

$$f(A_1\cap A_2)=f(\phi)=\phi,$$

whereas $f(A_1)\cap f(A_2)=\{c\}$. This shows that

$$f(\bigcap_{\gamma\in\Gamma} A_\gamma) = \bigcap_{\gamma\in\Gamma} f(A_\gamma)$$

is not true in general.

Theorem 11. Let $f: X\longrightarrow Y$ be a function and let $\{B_\gamma \mid \gamma\in\Gamma\}$ be a family of subsets of Y. Then

(a) $f^{-1}(\bigcup_{\gamma\in\Gamma} B_\gamma) = \bigcup_{\gamma\in\Gamma} f^{-1}(B_\gamma)$, (b) $f^{-1}(\bigcap_{\gamma\in\Gamma} B_\gamma) = \bigcap_{\gamma\in\Gamma} f^{-1}(B_\gamma)$.

[증명] (a) 정의 9, 정의 6(111쪽)을 반복 인용함으로써 다음을 얻는다.

$y\in f(\bigcup_{\gamma\in\Gamma}A_\gamma)\Longleftrightarrow$ 적당한 $x\in\bigcup_{\gamma\in\Gamma}A_\gamma$에 대하여 $y=f(x)$

$\Longleftrightarrow$ 적당한 $\gamma\in\Gamma$에 대한 $x\in A_\gamma$가 존재함으로써 $y=f(x)$

$\Longleftrightarrow$ 적당한 $\gamma\in\Gamma$에 대하여 $y\in f(A_\gamma)$

$\Longleftrightarrow y\in\bigcup_{\gamma\in\Gamma}f(A_\gamma)$ 기호 $\cup$

그러므로 $y\in f(\bigcup_{\gamma\in\Gamma}A_\gamma)\Longleftrightarrow y\in\bigcup_{\gamma\in\Gamma}f(A_\gamma)$

따라서 $f(\bigcup_{\gamma\in\Gamma}A_\gamma)=\bigcup_{\gamma\in\Gamma}f(A_\gamma)$*

(b) 각 $\gamma\in\Gamma$에 대하여 $\bigcap_{\gamma\in\Gamma}A_\gamma\subseteq A_\gamma$이므로 정리 9(c)에 위하여

$$f(\bigcap_{\gamma\in\Gamma}A_\gamma)\subseteq f(A_\gamma)$$

그러므로 제 2 장 정의 7에 따라 $f(\bigcap_{\gamma\in\Gamma}A_\gamma)\subseteq\bigcap_{\gamma\in\Gamma}f(A_\gamma)$.

다음 예제에서 보듯이 정리 10(b)에서의 포함기호 ⊆를 등호 =로 바꿔 놓을 수 없는 경우가 있다.

예제 11 $X=\{a,b\}$, $Y=\{c\}$, $\Gamma=\{1,2\}$, $A_1=\{a\}$, $A_2=\{b\}$,

상수함수 $f:X\to Y$, $f(a)=f(b)=c$

에 대하여 $f(A_1\cap A_2)=f(\phi)=\phi$

즉 $f(A_1\cap A_2)=\phi$

한편 $f(A_1)\cap f(A_2)=\{c\}\cap\{c\}=\{c\}$, 곧 $f(A_1)\cap f(A_2)=\{c\}$.

따라서 $f(A_1\cap A_2)\neq f(A_1)\cap f(A_2)$.

대체로 함수 f에 대하여 다음과 같이 등식이 성립하지 않는다.

$$f(\bigcap_{\gamma\in\Gamma}A_\gamma)\neq\bigcap_{\gamma\in\Gamma}f(A_\gamma)^*$$

정리 11 함수 $f:X\to Y$에서의 치역 Y의 부분집합족 $\{B_\gamma\mid\gamma\in\Gamma\}$에 대하여 다음이 성립한다.

(a) $f^{-1}(\bigcup_{\gamma\in\Gamma}B_\gamma)=\bigcup_{\gamma\in\Gamma}f^{-1}(B_\gamma)$ (b) $f^{-1}(\bigcap_{\gamma\in\Gamma}B_\gamma)=\bigcap_{\gamma\in\Gamma}f^{-1}(B_\gamma)$

* $f(A_1\cup A_2)=f(A_1)\cup f(A_2)$에 대한 보기 하나를 들어라.

[Proof] (a) By repeated applications of Definition 9, and Definition 6 of Chapter 2, we have

$$\begin{aligned} x\in f^{-1}(\bigcup_{\gamma\in\Gamma} B_\gamma) &\Leftrightarrow f(x)\in \bigcup_{\gamma\in\Gamma} B_\gamma \\ &\Leftrightarrow f(x)\in B_\gamma, \quad \text{for some} \quad \gamma\in\Gamma \\ &\Leftrightarrow x\in f^{-1}(B_\gamma), \quad \text{for some} \quad \gamma\in\Gamma \\ &\Leftrightarrow x\in \bigcup_{\gamma\in\Gamma} f^{-1}(B_\gamma) \end{aligned}$$

Thus, we have proved $f^{-1}(\bigcup_{\gamma\in\Gamma} B_\gamma)=\bigcup_{\gamma\in\Gamma} f^{-1}(B_\gamma)$.

(b) Replacing $\cup$ by $\cap$ and the phrase "for some" by "for all" in the proof of part (a) yields a proof for part (b). The student should write down each change, step by step, until he is fully convinced.

Theorem 12. Let $f: X\rightarrow Y$ be a function and let B and C be any subsets of Y. Then

$$f^{-1}(B-C)=f^{-1}(B)-f^{-1}(C)$$

[Proof] Let us examine the following equivalences :

$$\begin{aligned} x\in f^{-1}(B-C) &\Leftrightarrow f(x)\in B-C && \text{Def. 9} \\ &\Leftrightarrow f(x)\in B\wedge f(x)\notin C && \text{Def. 5(Ch. 2)} \\ &\Leftrightarrow x\in f^{-1}(B)\wedge x\notin f^{-1}(C) && \text{Def. 9} \\ &\Leftrightarrow x\in[f^{-1}(B)-f^{-1}(C)] && \text{Def. 5(Ch. 2)} \end{aligned}$$

This proves that $f^{-1}(B-C)=f^{-1}(B)-f^{-1}(C)$

Exercise 3.5

1. In Problem 2, Exercise 3.4, find

(a) $f(\{-1, 0, 1\})$, $f(\{\sqrt{2}, \pi\})$ and $f(\{2, \log 2\})$

(b) $f^{-1}(\{0, 1\})$, $f^{-1}(\{-3, 3\})$, $f^{-1}(\{4, 5\})$ and $f^{-1}(\{-3, 4, 5\})$.

[증명] (a) 정의 9, 제 2 장 정의 6을 반복 인용함으로써 다음을 얻는다.

$$x \in f^{-1}(\bigcup_{\gamma\in\Gamma} B_\gamma) \iff f(x) \in \bigcup_{\gamma\in\Gamma} B_\gamma$$

$\iff$ 적당한 $\gamma\in\Gamma$에 대하여 $f(x)\in B_\gamma$

$\iff$ 적당한 $\gamma\in\Gamma$에 대하여 $x\in f^{-1}(B_\gamma)$

$\iff x\in \bigcup_{\gamma\in\Gamma} f^{-1}(B_\gamma)$ 기호 $\cup$

그러므로 $x\in f^{-1}(\bigcup_{\gamma\in\Gamma} B_\gamma) \iff x\in \bigcup_{\gamma\in\Gamma} f^{-1}(B_\gamma)$

따라서 $f^{-1}(\bigcup_{\gamma\in\Gamma} B_\gamma) = \bigcup_{\gamma\in\Gamma} f^{-1}(B_\gamma))$

(b) (a)의 증명에서 기호 $\cup$를 $\cap$로, 또 "적당한 … 에 대하여"를 "모든 … 에 대하여"로 각각 바꿔 놓음으로써 (b)의 증명을 얻는다. (독자는 위 증명을 충분히 납득할 때까지 각 단계마다의 과정을 일일이 살펴보기 바란다).

정리 12 함수 $f: X \to Y$에서의 공역 Y의 임의의 부분집합 B, C에 대하여 다음이 성립한다.

$$f^{-1}(B-C) = f^{-1}(B) - f^{-1}(C)$$

[증명] $x\in f^{-1}(B-C) \iff f(x)\in(B-C)$ 정의 9

$\iff f(x)\in B \wedge f(x)\notin C$ 정의 5(제 2 장)

$\iff x\in f^{-1}(B) \wedge x\notin f^{-1}(C)$ 정의 9

$\iff x\in[f^{-1}(B)-f^{-1}(C)]$ 정의 5(제 2 장)

그러므로 $x\in f^{-1}(B-C) \iff x\in[f^{-1}(B)-f^{-1}(C)]$

따라서 $f^{-1}(B-C) = f^{-1}(B) - f^{-1}(C)$

연습문제 3.5

1. 연습문제 3.4 문제 2(169쪽)의 함수 f에 대하여 다음을 구하여라.

(a) $f(\{-1,0,1\})$, $f(\{\sqrt{2}, \pi\})$, $f(\{2, \log 2\})$

(b) $f^{-1}(\{0,1\})$, $f^{-1}(\{-3,3\})$, $f^{-1}(\{4,5\})$, $f^{-1}(\{-3,4,5\})$

2. In Problem 3, Exercise 3.4, find

(a) $f(\{-7, 3, 6\})$, $f(\{-8, 2, 7\})$, and $f(\{-9, 1, 8\})$

(b) $f^{-1}(\{0, 1\})$, $f^{-1}(\{-3, 3\})$, and $f^{-1}(\{1, 2, 3\})$.

3. In Problem 4, Exercise 3.4, find $f(\{u, v\})$, $f^{-1}(\{c\})^*$, and $f^{-1}(\{a, b\})$.

4. Let $f : X \to Y$ be a function, and let $A \subseteq X$, $B \subseteq Y$. Prove that

(a) $A \subseteq f^{-1}(f(A))$ (b) $f(f^{-1}(B)) \subseteq B$.

5. Let $f : X \to Y$ be a function, and let $A \subseteq X$, $B \subseteq Y$. Find examples which show that the following statements are false.

(a) If $B \neq \phi$, then $f^{-1}(B) \neq \phi$ (b) $f^{-1}(f(A)) = A$

(c) $f(f^{-1}(B)) = B$ (d) $f(X) = Y$

6. Prove that Problem 5(c) is true if $f(X) = Y$.

7. Let $f : X \to Y$ be a function such that $f(X) = Y$, and let B and C be subsets of Y. Prove that $B = C$ if $f^{-1}(B) = f^{-1}(C)$. Give an example which shows that the assertion is false if $f(X) \neq Y$.

8. Let X and Y be two sets, and let $p_X : X \times Y \to X$ and $p_Y : X \times Y \to Y$ be two functions given respectively by $p_X(x, y) = x$, $p_Y(x, y) = y$ for all $(x, y) \in X \times Y$ (p_X and p_Y are called the X-projection and the Y-projection, respectively). Prove that if $\mathscr{R}$ is a relation from X to Y, that is, $\mathscr{R} \subseteq X \times Y$, then $p_X(\mathscr{R}) = \mathrm{Dom}(\mathscr{R})$ and $p_Y(\mathscr{R}) = \mathrm{Im}(\mathscr{R})$.

9. Let $f : X \to Y$ be a function, and let $A \subseteq X$, $B \subseteq Y$. Prove that

(a) $f(A \cap f^{-1}(B)) = f(A) \cap B$ (b) $f(f^{-1}(B)) = f(X) \cap B$.

10. Let $f : X \to Y$ be a function, and let $B \subseteq Y$. Prove that

$$f^{-1}(Y - B) = X - f^{-1}(B)$$

11. Let $f : X \to Y$ be a function, and let A and B be subsets of X. Give an example which shows that it is not true that

$$f(A - B) = f(A) - f(B).$$

12. Prove Theorem 9.

2. 연습문제 3.4의 3(169쪽)의 함수 f에 대하여 다음을 구하여라.
(a) $f(\{-7, 3, 6\})$, $f(\{-8, 2, 7\})$, $f(\{-9, 1, 8\})$
(b) $f^{-1}(\{0, 1\})$, $f^{-1}(\{-3, 3\})$, $f^{-1}(\{1, 2, 3\})$

3. 연습문제 3.4의 4(171쪽)의 함수 f에 대하여 다음을 구하여라.

$$f(\{u, v\}),\ f^{-1}(\{e\})^{*},\ f^{-1}(\{a, b\})$$

4. 함수 $f: X \to Y$에 대하여 $A \subseteq X$, $B \subseteq Y$일 때 다음을 증명하여라.
(a) $A \subseteq f^{-1}(f(A))$
(b) $f(f^{-1}(B)) \subseteq B$

5. 함수 $f: X \to Y$에 있어서 $A \subseteq X$, $B \subseteq Y$일 때 다음 명제에 대한 반례 하나씩을 찾아라.
(a) $B \neq \phi$이면 $f^{-1}(B) \neq \phi$
(b) $f^{-1}(f(A)) = A$
(c) $f(f^{-1}(B)) = B$
(d) $f(X) = Y$

6. 문제 5(c)에서 $f(X) = Y$이면 그 등식이 성립함을 증명하여라.

7. 함수 $f: X \to Y$에 대하여 $f(X) = Y$이고 B, C가 각각 집합 Y의 부분집합일 때 $f^{-1}(B) = f^{-1}(C)$이면 $B = C$. 이것을 증명하여라. 그러나 이 주장은 $f(X) \neq Y$이면 거짓이다. 그런 보기를 하나 들어라.

8. 집합 X, Y의 데카르트곱 $X \times Y$의 각 원소 (x, y)에 대하여

$$p_X(x, y) = x, \qquad p_Y(x, y) = y$$

(p_X, p_Y를 각각 X-사영, Y-사영이라고 한다)와 같이 정의한다. 지금 $\mathscr{R}$이 X에서 Y로의 관계, 즉 $\mathscr{R} \subseteq X \times Y$일 때 다음을 증명하여라.

$$p_X(\mathscr{R}) = \mathrm{Dom}(\mathscr{R}), \qquad p_Y(\mathscr{R}) = \mathrm{Im}(\mathscr{R})$$

9. 함수 $f: X \to Y$에 대하여 $A \subseteq X$, $B \subseteq Y$일 때 다음을 증명하여라.
(a) $f(A \cap f^{-1}(B)) = f(A) \cap B$
(b) $f(f^{-1}(B)) = f(X) \cap B$.

10. 함수 $f: X \to Y$에 대하여 $B \subseteq Y$일 때 다음이 성립함을 증명하여라.

$$f^{-1}(Y - B) = X - f^{-1}(B)$$

11. 함수 $f: X \to Y$에 대하여 A, B가 각각 집합 X의 부분집합일 때 등식 $f(A - B) = f(A) - f(B)$** 에 대한 반례를 하나 들어라.

12. 정리 9를 증명하여라.

* 원서의 $f^{-1}(\{c\})$은 공집합 ϕ와 같아 별 뜻이 없으므로 c를 e로 바꿔 놓았다.
** 그러나 이 등식이 성립할 경우가 있지 않을까? 있으면 그런 보기도 하나 들어라.

3.6 Injective, Surjective, and Bijective Functions

In the study of functions, we find it convenient to give names to three important types of functions.

Definition 10. A function $f: X \to Y$ is said to be injective or one-to-one provided that if $x_1, x_2 \in X$ with $f(x_1) = f(x_2)$, then $x_1 = x_2$. An injective function is called an injection.

By the Contrapositive Laws in logic, we may say equivalently that the function $f: X \to Y$ is an injection if and only if $x_1, x_2 \in X$ with $x_1 \neq x_2$ implies that $f(x_1) \neq f(x_2)$. For example, the inclusion function in Problem 7, Exercise 3.4, is an injection.

Definition 11. A function $f: X \to Y$ is said to be surjective or onto provided that if $y \in Y$, then there exists at least one $x \in X$ such that $f(x) = y$. A surjective function is called a surjection. In other words, $f: X \to Y$ is a surjection if and only if $f(X) = Y$.

The function in Example 7, Section 4, for instance, is not surjective.

3.6 단사 · 전사 · 전단사

함수를 다룰 때 다음 세 가지 형태인 것에 일일이 특별한 명칭을 붙여 두면 편리하다.

정의 10* 함수 $f : X \to Y$에서의 $x_1, x_2 \in X$에 대하여

$$f(x_1) = f(x_2) \Longrightarrow x_1 = x_2$$

일 때, 이 함수를 **일대일** 또는 **단사적**이라고 한다. 단사적 함수를 간단히 **단사**라고 부른다.

논리에서 대우에 따라 단사를 다음과 같이 정의할 수 있다.

함수 $f : X \to Y$에서 $x_1, x_2 \in X$에 대하여

$$x_1 \neq x_2 \Longrightarrow f(x_1) \neq f(x_2)$$

일 때 이 함수는 단사이다.

이를테면 연습문제 3.4의 7(171쪽)에서의 포함함수 f는 단사이다.

정의 11 함수 $f : X \to Y$에 있어서 임의의 $y \in Y$에 대하여 적어도 하나의 $x \in X$가 존재함으로써 $y = f(x)$일 때 이 함수를 **위로** 또는 **전사적**이라고 한다. 전사적 함수를 간단히 **전사**라고 부른다.

다시 말하면 함수 $f : X \to Y$에 대하여

$$f(X) = Y$$

일 때 그리고 그 때에만 f는 전사이다.

예제 7(159쪽)의 가우스함수 $f(x) = [x]$는 전사가 아니다(왜?). 전사가 되려면 그 함수의 공역 R을 어떻게 바꿔 놓아야 하는가? 이때 공역이 바뀐 함수는 처음 함수 f와 다르다고 본다.

* 단사이지만 전사가 아닌 것과 전사이지만 단사가 아닌 보기를 하나씩을 들어라. 한편 그 어느 것도 아닌 함수의 보기도 들어 보아라. 이렇듯 정의, 정리의 가정을 독자 나름대로 바꿔가며 그에 따른 결론을 음미하는 습관을 붙이면 수학 공부에 큰 도움이 되리라 믿는다.

Example 12 The sine function $f : \mathbb{R} \longrightarrow [-1, 1]$ given by $f(x) = \sin x$ is a surjection; but if the codomain $[-1, 1]$ is replaced by $\mathbb{R}$, then the function $g : \mathbb{R} \longrightarrow \mathbb{R}$ is not surjective.

Definition 12. A function $f : X \longrightarrow Y$ is called a bijection or said to be bijective if it is both injective and surjective. A bijection is also called a one-to-one correspondence.

For example, the identity function in Example 9, Section 4, is a bijection. Definition 10, 11 and 12 are illustrated in three diagrams below. Sets X and Y are represented as sets of dots within the circles. In each picture, every dot in X is paired with some dot in Y by an arrow drawn between them. The set of pairs so obtained gives rise to a function $f : X \longrightarrow Y$.

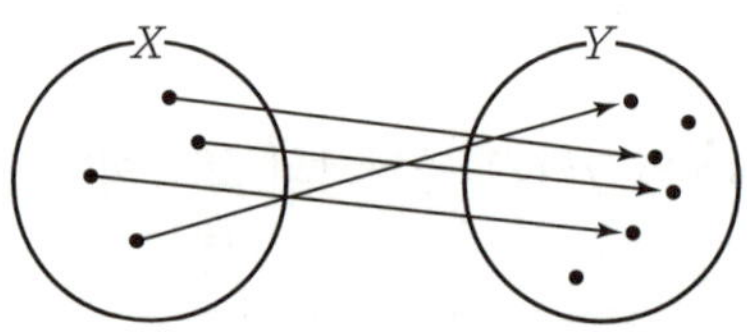

Figure 12 $f : X \longrightarrow Y$ is injective

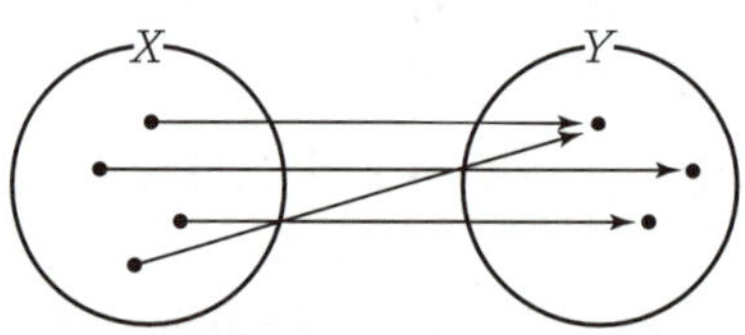

Figure 13 $f : X \longrightarrow Y$ is surjective

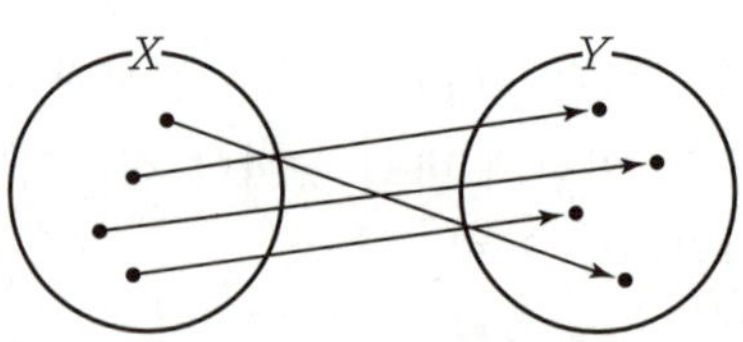

Figure 14 $f : X \longrightarrow Y$ is bijective

예제 12 $f(x)=\sin x$로 정의된 함수 $f:\mathbb{R}\longrightarrow[-1,1]$은 전사*이지만 그 공역 $[-1,1]$을 집합 $\mathbb{R}$로 바꿔 놓은 함수 $g:\mathbb{R}\longrightarrow\mathbb{R}$은 전사가 아니다.

정의 12 함수 $f:X\rightarrow Y$가 단사적이고 전사적일 때 이 함수를 **전단사적**이라 하고 그 같은 함수를 간단히 **전단사*** 또는 **일대일 대응**이라고 한다.

이를테면 3.4절 예제 9의 항등함수 I_X(165쪽)는 전단사이다. 다음 그림 12, 13, 14는 정의 10, 11, 12에 따른 함수의 보기 하나씩을 차례로 나타낸 것이다. 각 그림에서 X, Y는 원의 내부에 표시된 점들의 집합을 나타내며 집합 X의 각 점은 화살표에 따라 집합 Y의 어느 점과 짝지어져 있다. 이처럼 얻어진 순서쌍의 집합 $\{(x,y)\mid x\in X, y\in Y\}$으로서 우리는 하나의 함수 $f:X\rightarrow Y$를 생각하여 왔다.

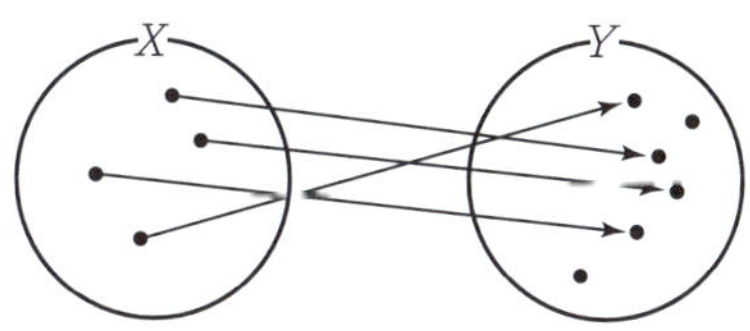

그림 12 $f:X\rightarrow Y$는 단사이다.

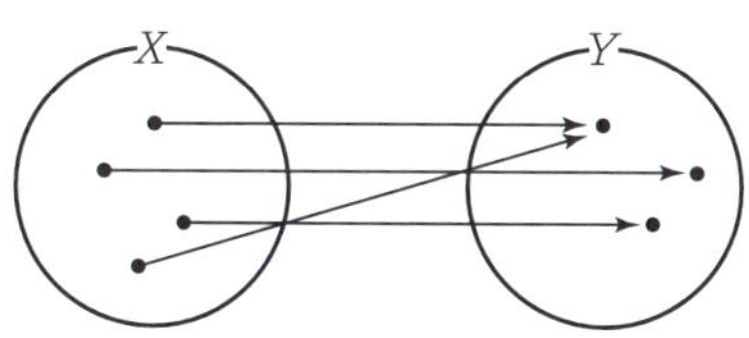

그림 13 $f:X\rightarrow Y$는 전사이다.

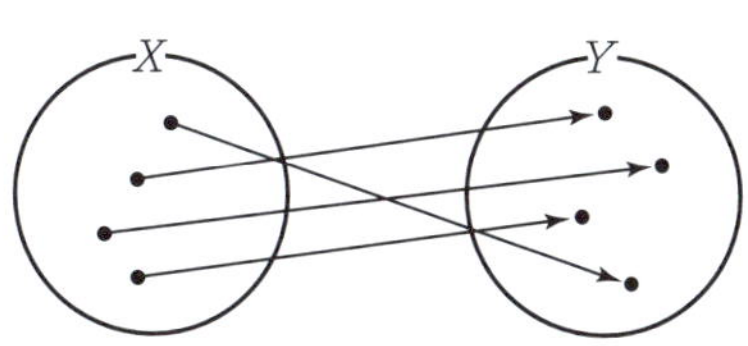

그림 14 $f:X\rightarrow Y$는 전단사이다.

* 함수 f가 전단사이면 그 함수는 전사이다($p\wedge q\Rightarrow p$). 이 주장의 역은 참인가? 거짓이면 그에 대한 보기(반례) 하나를 들어라.

For an injection, the result of Theorem 10(b) may be improved.

Theorem 13. Let $f: X \longrightarrow Y$ be a injection and let $\{A_\gamma \mid \gamma \in \Gamma\}$ be a family of subsets of X. Then

$$f(\bigcap_{\gamma \in \Gamma} A_\gamma) = \bigcap_{\gamma \in \Gamma} f(A_\gamma)$$

[Proof] By Definition 9, and Definition 7 of Chapter 2, we have

$$\begin{aligned} y \in \bigcap_{\gamma \in \Gamma} f(A_\gamma) &\Longleftrightarrow \forall_{\gamma \in \Gamma}\ y \in f(A_\gamma) \\ &\Longleftrightarrow \forall_{\gamma \in \Gamma}\ [\exists x_\gamma \in A_\gamma \text{ such that } y = f(x_\gamma)] \end{aligned}$$

Since $f: X \rightarrow Y$ is injective, all these x_γ's are the same; we denote this element by x_0. Then we have

$$\begin{aligned} y \in \bigcap_{\gamma \in \Gamma} f(A_\gamma) &\Longleftrightarrow \forall_{\gamma \in \Gamma}\ [\exists x_0 \in A_\gamma \text{ such that } y = f(x_0)] \\ &\Longleftrightarrow \exists x_0 \in \bigcap_{\gamma \in \Gamma} A_\gamma \text{ such that } y = f(x_0) \\ &\Longleftrightarrow y \in f(\bigcap_{\gamma \in \Gamma} A_\gamma) \end{aligned}$$

Therefore, $f(\bigcap_{\gamma \in \Gamma} A_\gamma) = \bigcap_{\gamma \in \Gamma} f(A_\gamma)$.

Recall that if $\mathscr{R}$ is a relation from X to Y, then its inverse

$$\mathscr{R}^{-1} = \{(y, x) \mid (x, y) \in \mathscr{R}\}$$

is a relation from Y to X. Since a function $f: X \longrightarrow Y$ is a (particular kind of) relation from X to Y, f^{-1} is at least a relation from Y to X. It is natural to inquire when f^{-1} becomes a function. This question is considered in the following theorem.

앞의 정리 10(173쪽)에서 함수 f가 단사라면 (b)에서 포함기호 ⊆는 등호 =로 바뀐다. 즉 다음 정리를 얻는다.

정리 13 주어진 함수 $f: X \to Y$가 단사이면 그 정의역 X의 부분집합족 $\{A_\gamma \mid \gamma \in \Gamma\}$에 대하여 다음이 성립한다.

$$f\left(\bigcap_{\gamma\in\Gamma} A_\gamma\right) = \bigcap_{\gamma\in\Gamma} f(A_\gamma) \qquad \cdots\cdots ①^*$$

[증명] 정의 9와 제 2 장 정의 7에 따라 다음이 성립한다.

$$\begin{aligned} y \in \bigcap_{\gamma\in\Gamma} f(A_\gamma) &\Longleftrightarrow \forall_{\gamma\in\Gamma}\ y \in f(A_\gamma) \\ &\Longleftrightarrow \forall_{\gamma\in\Gamma}\ [\exists x_\gamma \in A_\gamma\ \ y = f(x_\gamma)] \end{aligned}$$

가정에서 $f: X \to Y$는 단사이므로 이들 $x_\gamma \in A_\gamma$는 모두 같다. 여기서 그것을 x_0이라고 놓으면

$$\begin{aligned} y \in \bigcap_{\gamma\in\Gamma} f(A_\gamma) &\Longleftrightarrow \forall_{\gamma\in\Gamma}\ [\exists x_0 \in A_\gamma,\ y = f(x_0)] \\ &\Longleftrightarrow \exists x_0 \in \bigcap_{\gamma\in\Gamma} A_\gamma,\ y = f(x_0) \\ &\Longleftrightarrow y \in f\left(\bigcap_{\gamma\in\Gamma} A_\gamma\right) \end{aligned}$$

그러므로 $y \in \bigcap_{\gamma\in\Gamma} f(A_\gamma) \Longleftrightarrow y \in f\left(\bigcap_{\gamma\in\Gamma} A_\gamma\right)$

따라서 $\bigcap_{\gamma\in\Gamma} f(A_\gamma) = f\left(\bigcap_{\gamma\in\Gamma} A_\gamma\right)$ 곧 $f\left(\bigcap_{\gamma\in\Gamma} A_\gamma\right) = \bigcap_{\gamma\in\Gamma} f(A_\gamma)$

집합 X에서 집합 Y로의 관계 $\mathscr{R}$에 대하여 그 역관계

$$\mathscr{R}^{-1} = \{(y, x) \mid (x, y) \in \mathscr{R}\}$$

은 Y에서 X로의 하나의 관계일 것으로 짐작된다. 하물며 함수 $f: X \to Y$는 X에서 Y로의 (특별한) 관계**이므로 f^{-1} 또한 Y에서 X로의 관계일 것이다. 더 나아가 그 f^{-1}가 어떤 조건하에 함수가 되는가를 다음 정리가 말해준다.

---

* 정리 13에서 가정 "함수 $f: x \to y$는 단사임"을 제거하면 등식 ①은 $f(\bigcap_{\gamma\in\Gamma} A_\gamma) \subseteq \bigcap_{\gamma\in\Gamma} f(A_\gamma)$로 바뀐다. 이를 증명할 수 있겠는가?(우선 $\forall_{\gamma\in\Gamma} \bigcap_{\gamma\in\Gamma} A_\gamma \subseteq A_\gamma$임에 착안하여라).

** "함수 $f: A \to B$는 A에서 B로의 하나의 관계 $\mathscr{R}$이다."는 참이다. 이 주장의 역을 말하고 그 것의 참, 거짓을 가려라. 거짓인 경우의 보기는 쉽게 들 수 있을 것이다.

Theorem 14. Let $f: X \to Y$ be a bijection. Then $f^{-1}: Y \to X$ is a bijection.

[Proof] We shall first prove that the relation f^{-1} from Y to X forms a function. Since $f: X \to Y$ is surjective, by Problem 3(a), Exercise 3.2, we have $\mathrm{Dom}(f^{-1}) = \mathrm{Im}(f) = Y$. Thus, condition (a) of Definition 8 is satisfied. To show that f^{-1} satisfies the other condition, we let $(y, x_1) \in f^{-1}$ and $(y, x_2) \in f^{-1}$. Then we have $(x_1, y) \in f$ and $(x_2, y) \in f$. Consequently $f(x_1) = y = f(x_2)$. Now, because $f: X \to Y$ is injective, the last equality implies that $x_1 = x_2$. Thus we have established that $f^{-1}: Y \to X$ is a function.

To show that the function $f^{-1}: Y \to X$ is injective, let $y_1, y_2 \in Y$ with $f^{-1}(y_1) = f^{-1}(y_2) = x$ (say). Then we have $f(x) = y_1$ and $f(x) = y_2$, and hence $y_1 = y_2$. This proves that f^{-1} is injective.

Finally, it remains to be shown that $f^{-1}: Y \to X$ is surjective. By Problem 3(b) of Exercise 3.2, we have

$$\mathrm{Im}(f^{-1}) = \mathrm{Dom}(f) = X,$$

which proves that f^{-1} is surjective. Thus the proof is complete.

If $f: X \to Y$ is a bijection, the function $f^{-1}: Y \to X$ is called the inverse function of f (see also Problem 14, Exercise 3.6).

By virtue of Theorem 14, if $f: X \to Y$ is a bijection, we shall say that f is a one-to-one correspondence between the sets X and Y.

Exercise 3.6

1. Which of the function in Problem 2, 3, and 4 of Exercise 3.4 are injective? surjective?

2. Which of the functions in Problems 5 and 6 of Exercise 3.4 are injective? surjective? bijective?

정리 14 전단사 $f: X \to Y$에 대하여 $f^{-1}: Y \to X$도 전단사*이다.

[증명] 우선 집합 Y에서 집합 X로의 관계 f^{-1}는 함수임을 증명하자. 여기서 $f: X \to Y$는 전사이므로 145쪽 연습문제 3.2의 3(a)에 따라

$$\mathrm{Dom}\,(f^{-1}) = \mathrm{Im}(f) = Y$$

그러므로 f^{-1}는 157쪽 정의 8(a)의 조건을 만족한다. 또 f^{-1}는 정의 8(b)의 조건도 만족함을 보이기 위하여 $(y, x_1) \in f^{-1}$, $(y_1, x_2) \in f^{-1}$라고 각각 놓으면 $(x_1, y) \in f$, $(x_2, y) \in f$이므로 $y = f(x_1) = f(x_2)$. 그런데 $x_1 = x_2$. 그러므로 $f^{-1}: Y \to X$는 하나의 함수이다.

이제 함수 $f^{-1}: Y \to X$가 단사임을 보이기 위해 $y_1, y_2 \in Y$에 대하여 $f^{-1}(y_1) = f^{-1}(y_2) = x$라고 놓으면 $f(x) = y_1$, $f(x) = y_2$. 역시 가정에 따라 함수 f는 단사이므로 $y_1 = y_2$. 따라서 f^{-1}는 단사이다. 끝으로 f^{-1}가 전사임을 보여야 한다. 여기서 연습문제 3.2의 3(b)에 따르면

$$\mathrm{Im}\,(f^{-1}) = \mathrm{Dom}\,(f) = X$$

따라서 f^{-1}는 전사이다. 이상으로 증명이 끝난다.

전단사 $f: X \to Y$에 대하여 $f^{-1}: Y \to X$를 f의 **역함수****라고 부른다 [190쪽 연습문제 3.6의 14(190쪽) 참조].

정리 14를 감안하여 함수 $f: X \to Y$가 전단사일 때 이 f를 집합 X, Y **사이**의 일대일 대응이라고도 한다.

연습문제 3.6

1. 연습문제 3.4의 2, 3, 4의 함수 중 단사, 전사인 것을 지적하여라.

2. 연습문제 3.4의 5, 6의 함수 중 단사, 전사, 전단사인 것을 지적하여라.

---

* 전단사는 함수로서 강한 조건을 지니고 있다. 그런 보기를 둘 들어 보아라.

** 전단사가 아닌 함수의 역함수도 존재하는가? 존재하지 않으면 그에 따른 반례를 들어라.

3. Let $f : \mathbb{R} \longrightarrow \mathbb{R}$ be the function defined by $f(x)=3x-2$, for all $x \in \mathbb{R}$.

(a) Prove that the function f is a bijection.

(b) Find the inverse f^{-1} of f.

4. Let $g : (-\pi/2, \pi/2) \longrightarrow \mathbb{R}$ be the function given by $g(x)=\tan x$, for all $-\pi/2<x<\pi/2$. Is this function bijective? If so, describe its inverse function.

5. Prove that the characteristic function $\chi_A : X \longrightarrow \{0,1\}$ in Example 8, Section 4, is surjective if and only if $\phi \neq A \subset X$.* When does the function χ_A become an injection?

6. Prove that the constant function $C_b : X \longrightarrow Y$ is surjective if and only if $Y=\{b\}$. When does $C_b : X \longrightarrow Y$ become an injection?

7. Prove that the X-projection $p_X : X \times Y \longrightarrow X$ and the Y-projection $p_Y : X \times Y \longrightarrow Y$ in Problem 8, Exercise 3.5, are surjective. When is the X-projection an injection?

8. Prove that there is a one-to-one correspondence between the set $\mathbb{N}$ of natural numbers and the set $\mathbb{N}_e$ of all even natural numbers.

9. Prove that there is a one-to-one correspondence between the set $\mathbb{Z}$ of integers and the set $\mathbb{Z}_o$ of odd integers.

10. Let X be a finite set with m elements and let Y be a finite set with n elements. Prove that

(a) if $m>n$, then there can be no injection $f : X \longrightarrow Y$.

(b) If $m \leq n$, there exist exactly $n!/(n-m)!$ injections. (See also Problem 9, Exercise 3.4.)

11. Let X be a finite set with m elements. How many bijections from X onto X exist? (Remark : A bijection from a finite set onto itself is sometimes called a permutation.)

3. $\forall_{x\in\mathbb{R}}\ f(x)=3x-2$로 정의된 함수 $f:\mathbb{R}\longrightarrow\mathbb{R}$에 대하여
(a) f는 전단사임을 증명하여라. (b) f의 역함수 f^{-1}*를 구하여라.

4. $\forall x\in(-\pi/2,\pi/2)\ g(x)=\tan x$로 정의된 함수
$$g:(-\pi/2,\pi/2)\longrightarrow\mathbb{R}$$
은 전단사인가? 그렇다면 함수 g의 역함수 g^{-1}를 구하여라.

5. 이 장의 4절 보기 8(163쪽)의 특성함수 $\chi_A: X\longrightarrow\{0,1\}$에 대하여
$$A\subset X\ (A\neq\phi)^{**}\Longleftrightarrow\chi_A\text{는 전사임}$$
을 밝히고 함수 χ_A가 단사인 것은 어느 경우인지 말하여라.

6. 상수함수 $C_b: X\rightarrow Y$는 $Y=\{b\}$이면 그리고 그때에만 전사임을 증명하여라. 한편 $C_b: X\rightarrow Y$는 어느 경우에 단사이겠는가?

7. 우리는 연습문제 3.5의 8에서 다음과 같음 것을 정의한 바 있다.
$$X\text{-사영 } p_X: X\times X\rightarrow X,\qquad Y\text{-사영 } p_Y: X\times X\rightarrow Y$$
여기서 X-사영, Y-사영 모두 전사임을 증명하여라. 한편 어느 경우에 X-사영이 단사이겠는가?

8. 자연수의 집합 $\mathbb{N}$과 짝수의 집합 $\mathbb{N}_e$ 사이에 일대일 대응이 존재한다.** 이것을 증명하여라.

9. 집합 $\mathbb{Z}$과 홀수인 정수의 집합 $\mathbb{Z}_o$ 사이에 일대일 대응이 존재한다.*** 이를 증명하여라.

10. X, Y는 각각 m, n개의 원소를 지닌 집합일 때 다음을 밝혀라.
(a) $m>n$이면 함수 $f: X\rightarrow Y$로서 단사인 것은 존재하지 않는다.
(b) $m\leqq n$이면 $n!/(n-m)!$개의 단사가 존재한다(연습문제 3.4의 9).

11. 원소가 m개인 집합 X에 대하여 X에서 X로의 전단사는 모두 몇 개 존재하는가?(유한집합에서 그 자신으로의 전단사를 순열이라고도 한다).

* f, f^{-1}의 그래프를 같은 좌표평면에 그려 두 그래프 사이의 관계를 말하여라.
** 원서의 문제 5에서 $\phi\neq A\subset X$와 같은 표현이 마음에 들지 않아서 수정하였다.
*** 여기서 $\mathbb{N}$과 $\mathbb{N}_e$, $\mathbb{Z}$과 $\mathbb{Z}_0$이 모두 무한집합이라는 사실에 새삼 주목할 필요가 있다.

12. Let $f : X \to Y$ be a function, and let $A \subseteq X$, $B \subseteq Y$. Prove that

(a) If f is injective, then $f^{-1}(f(A)) = A$.

(b) If f is surjective, then $f(f^{-1}(B)) = B$.

13. Let $f : X \to Y$ be an injection, and let A and B be subsets of X. Prove that

$$f(A-B)=f(A)-f(B).$$

(Compare this with Problem 11 of Exercise 3.5.)

14. Prove the following converse of Theorem 13 :

Let $f : X \to Y$ be a function such that $f^{-1} : Y \to X$ is a bijection. Then $f : X \to Y$ is a bijection.

3.7 Composition of Functions

To a thoughtful reader, a function $f : X \to Y$ may be considered as a machine that takes an arbitrary object x of the set X, operates on it in a certain way, and transforms it into a new object $f(x)$, an output of the machine. This idea is illustrated in Figure 15.

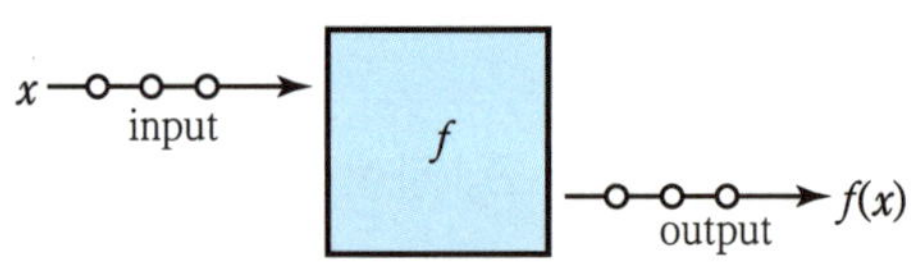

Figure 15

Let $f : X \to Y$ and $g : Y \to Z$ be two given functions, where the domain of the second function is the same as the range of the first function. Imagine these two functions as two machines such as a washer and a dryer. We don't have to be inventors to imagine the possibility of combining these two machines

12.* 함수 $f: X \to Y$에 대하여 $A \subseteq X$, $B \subseteq Y$일 때 다음이 성립한다. 이를 증명하여라(증명에 앞서 그런 보기부터 들어보아라!).

(a) f가 단사이면 $f^{-1}(f(A)) = A$ (b) f가 전사이면 $f(f^{-1}(B)) = B$

13. 함수 $f: X \to Y$가 단사이면 그 정의역 X의 부분집합 A, B에 대하여

$$f(A-B) = f(A) - f(B).$$

이것을 증명하여라(이것과 181쪽 연습문제 3.5의 11을 비교하여 보아라).

14. 다음은 정리 14의 역이다.

함수 $f: X \to Y$에 대하여 $f^{-1}: Y \to X$가 전단사이면 f도 전단사이다. 이것을 증명하여라.

3.7 함수의 합성

우리는 함수 $f: X \to Y$를 하나의 기계로 비유할 수 있다. 다시 말하면 집합 X의 임의의 원소 x에 대하여 한 가지 방법으로 작용**(여기서는 f)을 일으킴으로써 새로운 출력물 $f(x)$를 만들어내는 기계를 상상할 수 있다. 다음 그림 15는 바로 이러한 과정을 상상한 것이다.

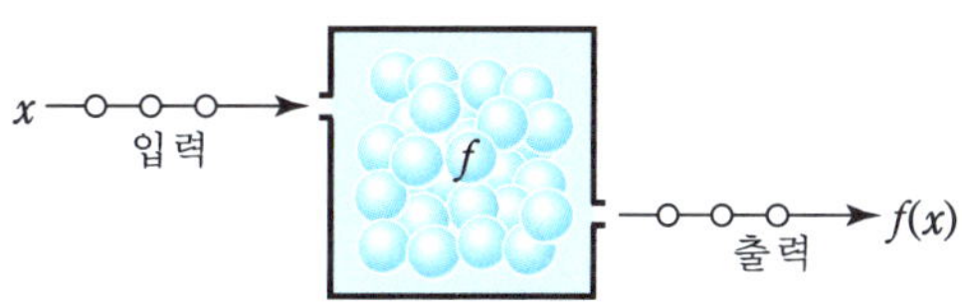

그림 15

지금 두 함수 $f: X \to Y$, $g: Y \to Z$에 대하여 함수 g의 정의역 Y는 함수 f의 공역 Y와 같을 때 이 두 함수를 각각 세탁기, 건조기로 빗대어 보자. 우리는 비록 이 두 기계를 하나의 새로운 기계로 결합시키려는 디자이너가 아니더라도 다음과 같은 일은 쉽게 짐작할 수 있을 것이다.

---

* 함수 $f: \mathbb{R} \to \mathbb{R}$의 정의역 $\mathbb{R}$의 부분집합을 $A = \{3\}$이라 하면 $f^{-1}(f(A)) = \{3, -3\}$이다. 그러므로 $f^{-1}(f(A)) \neq A$. 왜 이런 결과가 생길까? $f(f^{-1}(B)) \neq B$인 경우도 지적하여 보아라.

** 그림 15에서 보듯이 함수 f는 상자(여기서는 정사각형) 속에서 일어나는 일종의 작용(기능)을 연상케 한다. 그런 뜻에서 함수를 영어로 "*function*"이라고 불러왔다. 한편 함수를 한자로 "函數"라고 적는데 函이 상자를 뜻한다니 흥미롭지 않은가?

into one new machine; the result would be a washer-dryer combination that takes a dirty garment x, washes it so that it becomes a clean but wet garment $f(x)$, and then dries it. The outcome is a clean and dry garment $g(f(x))$. The idea is illustrated in Figure 16.

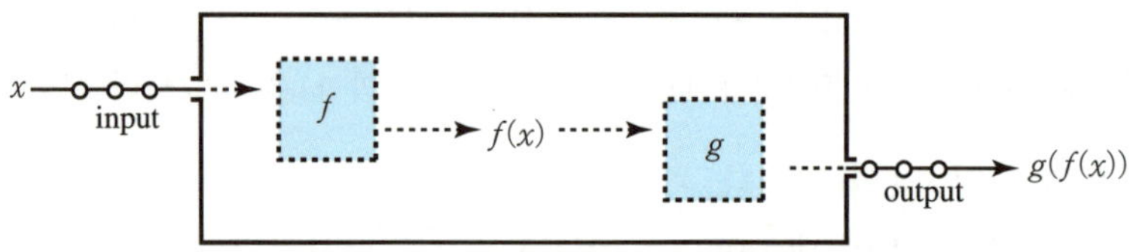

Figure 16

The "combination" of the machines $f: X \to Y$ and $g: Y \to Z$ results in a new machine, denoted by $h: X \to Z$, which takes an arbitrary object x in X and transforms it into the object $h(x)=g(f(x))$ in Z. The traditional notation for h is $\boldsymbol{g \circ f}$ and

$$(g \circ f)(x) = g(f(x))$$

the traditional name for the term "combination" is "composition."

Now we are ready for the following definition.

Definition 13. Let $f: X \to Y$ and $g: Y \to Z$ be two functions. The composition of these two functions is the function $g \circ f: X \to Z$ where $(g \circ f)(x)=g(f(x))$ for all x in X. In another notation

$$g \circ f = \{(x, z) \in X \times Z \mid \exists_y \in Y \text{ such that } (x, y) \in f \wedge (y, z) \in g\}$$

즉 때묻은 옷 x를 세탁기로 빨아(작용 f에 따라) 젖은 상태의 깨끗한 옷 $f(x)$를 얻은 뒤 다시 이것을 건조기로 말림(작용 g에 따라)으로써 얻는 마지막 결과물은 깨끗하고 마른 옷 $g(f(x))$가 될 것이다. 다음은 이러한 과정을 나타낸 그림이다.

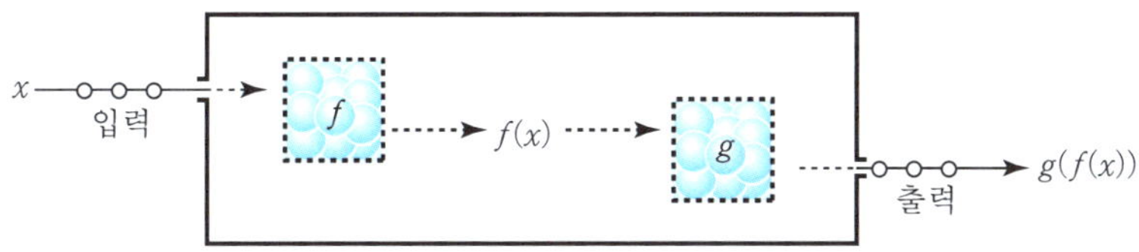

그림 16

대체로 두 기계 $f: X \rightarrow Y$, $g: Y \rightarrow Z$의 결합으로 얻은 새로운 기계를 $h: X \rightarrow Z$라고 생각할 때 이 기계는 X의 임의의 대상 x를 Z의 대상 $h(x)=g(f(x))$로 변환시킨다고 볼 수 있다. 여기서 h에 대한 기호 $\boldsymbol{g \circ f}$를 이용하여

$$\boldsymbol{(g \circ f)(x) = g(f(x))}$$

와 같이 적고 "결합"에 해당하는 h 즉 $g \circ f$를 함수 f와 g의 **합성함수***라고 한다. 이상으로 다음 정의를 얻는다.

정의 13 두 함수 $f: X \rightarrow Y$, $g: Y \rightarrow Z$가 주어졌을 때 각 $x \in X$에 대하여

$(g \circ f)(x) = g(f(x))$로 정의된 함수 $g \circ f: X \rightarrow Z$

를 f와 g의 합성함수라고 한다. 이 합성함수를 순서쌍을 써서 다음과 같이 나타낼 수 있다.

$$g \circ f = \{(x, z) \in X \times Z \mid \exists_y \in Y\ [(x, y) \in f \wedge (y, z) \in g]\}$$

보기 4* 함수 $f: A \rightarrow B$, $g: B \rightarrow C$가 오른쪽 그림과 같이 정의되어 있다. 이때

$$(g \circ f)(a) = g(f(a)) = g(5) = y$$

$$(g \circ f)(b) = g(f(b)) = g(8) = z$$

$$(g \circ f)(c) = g(f(c)) = g(6) = y$$

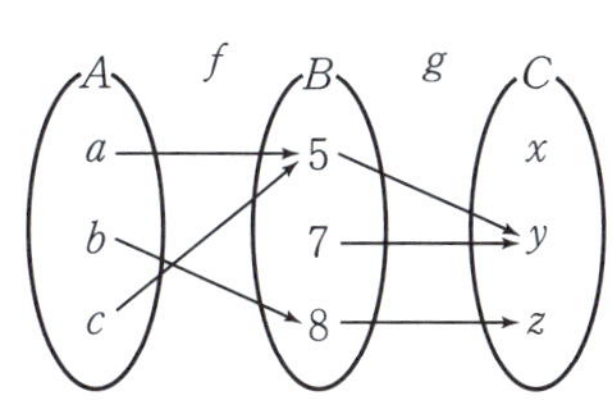

* 합성함수에 관한 더 엄밀한 정의와 문제는 다음 책에서 찾아 볼 수 있다.
대학교재편찬위원회, 《미적분학 제2판》, 경문사, 2005년, 19 및 20쪽 참조.

Example 13 Let $f:\mathbb{R}\rightarrow\mathbb{R}$ and $g:\mathbb{R}\rightarrow\mathbb{R}$ be two functions given respectively by $f(x)=x+1$ and $g(x)=x^2$ for all x in $\mathbb{R}$. Find the composition $g\circ f(x)$ and $f\circ g(x)$.

[Solution] Using Definition 13, we have

(1) $g\circ f(x)=g(f(x))=g(x+1)$
$\quad = (x+1)^2 = x^2+2x+1$

(2) $f\circ g(x)=f(g(x))=f(x^2)=x^2+1$

The result of Example 13 tells us that in general $g\circ f\neq f\circ g$; therefore, functional composition is not commutative.

Theorem 15. Functional composition is associative. That is, if

$$f:X\rightarrow Y,\ g:Y\rightarrow Z,\text{ and } h:Z\rightarrow W,$$

then
$$(h\circ g)\circ f=h\circ(g\circ f)$$

[Proof] We first note that both $h\circ(g\circ f)$ and $(h\circ g)\circ f$ give rise to functions from X to W. Therefore, to show that

$$h\circ(g\circ f)=(h\circ g)\circ f,$$

by Theorem 7 of Section 4, we need only to show that

$$h\circ(g\circ f)(x)=(h\circ g)\circ f(x)$$

for all x in X. We use Definition 13 to derive the following:

$$(h\circ(g\circ f))(x)=h((g\circ f)(x))=h(g(f(x)))$$

and
$$((h\circ g)\circ f)(x)=(h\circ g)(f(x))=h(g(f(x)))$$

for all x in X. This shows that

$$\forall x\in X\ (h\circ(g\circ f))(x)=((h\circ g)\circ f)(x)$$

The proof is now complete.

예제 13 $\forall x \in \mathbb{R}$ $f(x)=x+1$, $g(x)=x^2$으로 정의된 함수

$$f:\mathbb{R}\to\mathbb{R},\quad g:\mathbb{R}\to\mathbb{R}$$

에 대하여 합성함수 $g\circ f(x)$, $f\circ g(x)$를 각각 구하여라.

[풀이] 정의 13에 따라 다음을 얻는다.

$$g\circ f(x)=g(f(x))=g(x+1)=(x+1)^2=x^2+2x+1$$

즉 $g\circ f(x)=x^2+x+1$

한편 $f\circ g(x)=f(g(x))=f(x^2)=x^2+1$

즉 $f\circ g(x)=x^2+1$

여기서 특히 $x=0$이면 $g\circ f(0)=1$, $f\circ g(0)=1$

따라서 $g\circ f(0)=f\circ g(0)$

그러나 $x\neq 0$이면 $x^2+2x+1\neq x^2+1$(왜?)이므로

$g\circ f(x)\neq f\circ g(x)$

결국 $g\circ f\neq f\circ g$.

대체로 두 함수의 합성함수에 있어서는 교환법칙*이 성립하지 않는다.

정리 15 합성함수에 대한 **결합법칙**이 성립한다. 즉

$$f:X\to Y,\quad g:Y\to Z,\quad h:Z\to W$$

에 대하여 $(h\circ g)\circ f=h\circ(g\circ f)$**

[증명] 먼저 $(h\circ g)\circ f$, $h\circ(g\circ f)$는 각각 X에서 W로의 함수인 것을 쉽게 알 수 있다. 그러므로 위의 등식이 성립함을 보이려면 4.4절의 정리 7에 따라 단지

$$((h\circ g)\circ f)(x)=(h\circ(g\circ f))(x)$$

인 것을 밝히면 된다. 그런데 정의 13에 따르면 모든 $x\in X$에 대하여

$$((h\circ g)\circ f)(x)=h\circ g(f(x))=h(g(f(x)))$$

한편 $(h\circ(g\circ f))(x)=h((g\circ f)(x))=h(g(f(x)))$

그러므로 $\forall x\in X\ ((h\circ g)\circ f)(x)=(h\circ(g\circ f))(x)$

따라서 $(h\circ g)\circ f=h\circ(g\circ f)$

---

* 교환법칙이 성립하는 특별한 경우가 있으면 그런 보기를 하나 들어라.

** 세 함수 f, g, h의 보기를 들어 이 등식이 성립하는지 확인하여 보아라.

Theorem 16. Let $f: X \rightarrow Y$ be a function. Then

(a) If there exists a function $g: Y \rightarrow X$ such that $g \circ f = 1_X$ (where $1_X: X \rightarrow X$ is the identity function defined in Example 9, Section 4), then $f: X \rightarrow Y$ is injective.

(b) If there exists a function $h: Y \rightarrow X$ such that $f \circ h = 1_Y$, then $f: X \rightarrow Y$ is surjective.

[Proof] (a) Suppose that there exists a function $g: Y \rightarrow X$ such that $g \circ f = 1_X$. Then for any x_1 and x_2 in X with $f(x_1) = f(x_2)$, we have

$$x_1 = (g \circ f)(x_1) = g(f(x_1)) = g(f(x_2)) = (g \circ f)(x_2) = I_X(x_2) = x_2$$

This proves that $f: X \rightarrow Y$ is injective.

(b) Suppose that there exists a function $h: Y \rightarrow X$ such that $f \circ h = I_Y$. Then for each $y \in Y$, there exists an element

$$x = h(y) \in X$$

such that $$f(x) = f(h(y)) = (f \circ h)(y) = I_Y(y) = y$$

By Definition 11, $f: X \rightarrow Y$ is surjective.

Exercise 3.7

1. Let $f: \mathbb{R} \rightarrow \mathbb{R}$ and $g: \mathbb{R} \rightarrow \mathbb{R}$ be two functions defined by $f(x) = 2x^3 + 1$ and $g(x) = \cot x$, respectively, for all $x \in \mathbb{R}$. Find the compositions.

(a) $g \circ f$ (b) $f \circ g$

2. Let $f: \mathbb{R}_+ \rightarrow \mathbb{R}$ and $g: \mathbb{R} \rightarrow \mathbb{R}_+$ be two functions defined by $f(x) = \log_{10} x$ and $g(x) = 10^x$, respectively, for all $x \in \mathbb{R}$. Find the compositions

(a) $g \circ f: \mathbb{R}_+ \rightarrow \mathbb{R}_+$. (b) $f \circ g: \mathbb{R} \rightarrow \mathbb{R}$.

정리 16 함수 $f: X \to Y$에 대하여

(a) 함수 $g: Y \to X$가 존재함으로써 $g \circ f = 1_X$($1_X : X \to X$는 165쪽 예제 9에서의 항등함수이다)이면 $f: X \to Y$는 단사이다.

(b) 함수 $h: Y \to X$가 존재함으로써 $f \circ h = 1_Y$이면 $f: X \to Y$는 전사이다.

[증명] (a) 임의의 $x_1, x_2 \in X$에 대하여 $f(x_1) = f(x_2)$라고 가정하면

$$\begin{aligned} x_1 &= 1_X(x_1) = (g \circ f)(x_1) = g(f(x_1)) \\ &= g(f(x_2)) = (g \circ f)(x_2) \\ &= 1_X(x_2) = x_2 \end{aligned}$$

그러므로 $x_1 = x_2$

따라서 $f(x_1) = f(x_2)$이면 $x_1 = x_2$.

그러므로 [정의 10(181쪽)에] 따라 $f: X \to Y$는 단사이다.

(b) 가정에 따라 $\forall y \in Y$ $x = h(y)$인 $x \in X$가 존재함으로써

$$f(x) = f(h(y)) = (f \circ h)(y) = 1_Y(y) = y$$

따라서 $y = f(x)$

그러므로 [정의 11(181쪽)에] 따라 $f: X \to Y$는 전사이다.

연습문제 3.7

1. $\forall x \in \mathbb{R}$ $f(x) = 2x^3 + 1$, $g(x) = \cot x$로 정의된 함수

$$f: \mathbb{R} \to \mathbb{R}, \quad g: \mathbb{R} \to \mathbb{R}$$

에 대하여 다음 합성함수를 구하여라.

(a) $g \circ f$ (b) $f \circ g$

2. $\forall x \in \mathbb{R}$ $f(x) = \log_{10} x$, $g(x) = 10^x$로 정의된 함수

$$f: \mathbb{R}_+ \to \mathbb{R}, \quad g: \mathbb{R} \to \mathbb{R}_+$$

에 대하여 다음 합성함수를 구하여라.

(a) $g \circ f: \mathbb{R}_+ \to \mathbb{R}_+$ (b) $f \circ g: \mathbb{R} \to \mathbb{R}$

3. Let f, g and h be the functions given in Problem 6, Exercise 3.4. Find the compositions

(a) $g \circ f$. (b) $h \circ g$. (c) $h \circ (g \circ f)$. (d) $(h \circ g) \circ f$.

(e) Compare your answers for $h \circ (g \circ f)$ and $(h \circ g) \circ f$; are they the same?

4. Let $f: X \to Y$ be a function. Prove that $f \circ 1_X = f = 1_Y \circ f$.

5. Let $f: X \to Y$ be a bijection and let $f^{-1}: Y \to X$ be the inverse function of f. Prove that $f^{-1} \circ f = 1_X$ and $f \circ f^{-1} = 1_Y$

6. Let $f: X \to Y$ be a function. If there exist functions $g: Y \to X$ and $h: Y \to X$ such that $g \circ f = 1_X$ and $f \circ h = 1_Y$, prove that $f: X \to Y$ is bijective and that $g = h = f^{-1}$.

7. Let $f: X \to Y$ and $g: Y \to Z$ be functions. Prove that

(a) If $f: X \to Y$ and $g: Y \to Z$ are injective, then so is $g \circ f: X \to Z$.

(b) If $f: X \to Y$ and $g: Y \to Z$ are surjective, then so is $g \circ f: X \to Z$.

8. Let $\mathscr{R}$ be a relation from X to Y and let $\mathscr{S}$ be a relation from Y to Z. We may, as in the composition of functions, define the composition of these relations by

$$\mathscr{S} \circ \mathscr{R} = \{(x, z) \in X \times Z \mid \exists y[(x, y) \in \mathscr{R} \wedge (y, z) \in \mathscr{S}]\}$$

which is a relation from X to Z. Prove that

(a) $(\mathscr{S} \circ \mathscr{R})^{-1} = \mathscr{R}^{-1} \circ \mathscr{S}^{-1}$.

(b) If furthermore $\mathscr{T}$ is a relation from Z to W, then

$$\mathscr{T} \circ (\mathscr{S} \circ \mathscr{R}) = (\mathscr{T} \circ \mathscr{S}) \circ \mathscr{R}.$$

9. Let $f: X \to Y$ and $g: Y \to Z$ be two bijections. Prove that

$$g \circ f : X \to Z$$

is a bijection, and that the inverse function $(g \circ f)^{-1} : Z \to X$ is the same as the composition $f^{-1} \circ g^{-1} : Z \to X$ of the inverse functions $g^{-1} : Z \to Y$ and $f^{-1} : Y \to X$. That is,

$$(g \circ f)^{-1} = f^{-1} \circ g^{-1}.$$

3. 연습문제 3.4의 6(170쪽)의 함수 $f \cdot g \cdot h$에 대하여 다음 합성함수 (a)~(d)를 구하고 (e)에 답하여라.

(a) $g \circ f$ (b) $h \circ g$ (c) $h \circ (g \circ f)$ (d) $(h \circ g) \circ f$

(e) 두 합성함수 $h \circ (g \circ f)$, $(h \circ g) \circ f$는 같은가?

4. 함수 $f : X \longrightarrow Y$에 대하여 $f \circ 1_X = f = 1_Y \circ f$임을 증명하여라.

5. 함수 $f : X \longrightarrow Y$가 전단사일 때 그 역함수 $f^{-1} : Y \longrightarrow X$에 대하여

$$f^{-1} \circ f = 1_X, \quad f \circ f^{-1} = 1_Y.$$

이것을 각각 증명하여라.

6. 함수 $f : X \longrightarrow Y$에 대하여 두 함수 $g : Y \longrightarrow X$, $h : Y \longrightarrow X$가 존재함으로써 $g \circ f = 1_X$, $f \circ h = 1_Y$이면 $f : X \longrightarrow Y$는 전단사이고 $g = h = f^{-1}$이다. 이를 증명하여라.

7. 두 함수 $f : X \longrightarrow Y$, $g : Y \longrightarrow Z$에 대하여 다음을 증명하여라.

(a) f, g가 모두 단사이면 $g \circ f$도 단사이다.

(b) f, g가 모두 전사이면 $g \circ f$도 전사이다.

8. 집합 X에서 집합 Y로의 관계 $\mathscr{R}$, 집합 Y에서 집합 Z로의 관계 $\mathscr{S}$에 대하여(합성함수의 경우와 같이) 이 두 관계의 **합성**, 즉 X에서 Z로의 관계를 다음과 같이 정의한다.

$$\mathscr{S} \circ \mathscr{R} = \{(x, z) \in X \times Z \mid \exists y\, [(x, y) \in \mathscr{R} \wedge (y, z) \in \mathscr{S}]\}$$

다음을 증명하여라.

(a) $(\mathscr{S} \circ \mathscr{R})^{-1} = \mathscr{R}^{-1} \circ \mathscr{S}^{-1}$.

(b) 집합 Z에서 집합 W로의 관계 $\mathscr{T}$에 대하여

$$\mathscr{T} \circ (\mathscr{S} \circ \mathscr{R}) = (\mathscr{T} \circ \mathscr{S}) \circ \mathscr{R}$$

9. 두 전단사 $f : X \longrightarrow Y$, $g : Y \longrightarrow Z$에 대하여 다음을 증명하여라.
$g \circ f : X \longrightarrow Z$는 전단사이다. 한편 역함수 $(g \circ f)^{-1} : Z \longrightarrow X$는 역함수 $g^{-1} : Z \longrightarrow Y$, $f^{-1} : Y \longrightarrow X$의 합성함수 $f^{-1} \circ g^{-1} : Z \longrightarrow X$와 같다. 즉

$$(g \circ f)^{-1} = f^{-1} \circ g^{-1} *$$

---

*(1) 이것에 해당하는 보기를 하나 들어라.

(2) 대체로 $(g \circ f)^{-1} \neq g^{-1} \circ f^{-1}$임에 주목하고 그런 보기부터 들어 보자. 한편 $(g \circ f)^{-1} = g^{-1} \circ f^{-1}$이 성립하는 특별한 경우가 없을까? 있으면 그 보기를 하나 들어라.

Chapter 4
Denumerable Sets and Nondenumerable Sets

제 4 장
가부번집합과 비가부번집합

4

Denumerable Sets and Nondenumerable Sets

Dedekind's definition of an infinite set is used to discuss properties of infinite sets and of finite sets. It is proved, among other things, that denumerable sets are the smallest in "size" among the infinite sets. Properties and examples of denumerable sets and of nondenumerable sets are given.

4.1 Finite and Infinite Sets

In Section 1, Chapter 2, we mentioned casually that a finite set is a set which contains only finitely many elements; although this concept may be developed into a more precise mathematical definition, we prefer an alternate definition (Definition 1) originated by Dedekind.

It was pointed out in Section 1 of Chapter 2 that the set N of all natural numbers is an infinite set. Let $\mathrm{N}_e=\{2,4,6,\cdots\}$ be the set of all even natural numbers. As the reader has shown in Problem 8, Exercise 3.6, there is a one-to-one correspondence between the set N and its proper subset N_e.

In other words,

A part is as numerous as the whole.[1)]

[1)] A striking difference from Euclid's axiom: "The whole is greater than any of its parts" (325 B.C.).

4 가부번집합과 비가부번집합

먼저 데데킨트가 내린 무한집합에 관한 정의를 써서 무한집합과 유한집합의 성질을 논한다. 무엇보다도 무한집합의 "크기"를 다룰 때 *가부번집합*이 가장 작다는 것이 증명되어 있다.

이 장에서는 가부번집합과 *비가부번집합*의 성질을 다루며 그에 따른 보기들도 함께 제시하고 있다.

4.1 유한집합과 무한집합

2.1절에서 유한집합은 원소의 수를 자연수로 나타낼 수 있는 집합(75쪽)이며 유한집합이 아닌 집합을 무한집합이라고 주장한 이 개념은 점차 수학적으로 정확한 정의로 마무리되어 왔다. 그 중에서도 우리는 데데킨트가 내린 다음의 정의 1을 더욱 선호한다.

또 2.1절에서는 자연수의 집합 $\mathbb{N}$이 무한집합임을 지적하였다. 지금 자연수 중 그 일부로서 짝수의 집합을

$$\mathbb{N}_e = \{2, 4, 6, \cdots\}$$

라고 놓으면 독자가 연습문제 3.6 문제 8에서 다루었듯이 집합 $\mathbb{N}$과 그것의 진부분집합 $\mathbb{N}_e$ 사이에는 일대일 대응이 존재한다. 다시 말하면

부분으로서 전체만큼 여러 개를 지닌 것도 있다.[1)]

[1)] 유클리드의 공리 중의 하나인 "전체는 그 어느 부분보다도 크다"(기원전 325년)와 비교할 때 엄청난 차이가 난다. 무한집합에나 있는 이 같은 사실을 위대한 유클리드조차 깨닫지 못한 것으로 보인다. 독자는 4장을 통하여 "무한"의 개념을 분명히 이해하게 될 것이다.

This strange property (of an infinite set) bothered many mathematicians including Georg Cantor. It was Richard Dedekind (1831 ~ 1916)[2)] who made this property the defining characteristic of an infinite set. The following definition was given by Dedekind in 1888.

Definition 1. A set X is infinite provided that it has a proper subset Y such that there exists a one-to-one correspondence between X and Y. A set is finite if it is not infinite.

In other words, a set X is infinite if and only if there exists an injection $f : X \to X$ such that $f(X)$ is a proper subset of X. Thus, the set $\mathbb{N}$ of natural numbers is an infinite set.

Example 1 The empty set ϕ and the singleton set are finite.

[Solution] (a) Since the empty set has no proper subset, it cannot be infinite. Therefore, the empty set is finite. (b) Let $\{a\}$ be any singleton set. Since the only proper subset of $\{a\}$ is the empty set ϕ and there is no one-to-one correspondence between $\{a\}$ and ϕ, $\{a\}$ must be finite.

Theorem 1. (a) Every superset of an infinite set is infinite.
(b) Every subset of an finite set is finite.

---

2) Richard Dedekind, one of the greatest mathematicians, was born on October 6, 1831, in Brunswick, Germany. At first, Dedekind's interest lay in physics and chemistry; he considered mathematics merely as the servant of the sciences. But this did not continue long; by the age of seventeen he had turned from physics and chemistry to mathematics, whose logic he found more satisfactory. At the age of nineteen he enrolled in the University of Göttingen to study mathematics, and he received his doctor's degree three years later under Gauss. His fundamental contributions to mathematics include the famous "Dedekind Cut," and important concept in the study of irrational numbers, which the reader may have an opportunity to study in a real analysis course.

게오르크 칸토어를 위시하여 많은 수학자들이 무한집합의 낯선 성질에 당혹해 있을 동안 리하르트 데데킨트(Julius Wihelm Richard Dedekind : 1831~1916)[2]는 1888년 다음과 같은 정의를 내렸다.

정의 1 주어진 집합 $X(X \neq \phi)$에 대하여 그 진부분집합 Y가 존재함으로써 X와 Y 사이에 일대일 대응이 존재할 때 X를 **무한집합***이라고 한다. 한편 무한집합이 아닌 집합을 **유한집합**이라고 한다.

다시 말하면 집합 X에 대하여 단사 $f : X \to X$가 존재함으로써 $f(X)$가 X의 진부분집합이면 그리고 그때에만 X는 무한집합이다. 이를테면 자연수의 집합 $\mathbb{N}$은 무한집합*이다.

예제 1 공집합 ϕ과 **한원소집합**(109쪽 윗 구절)은 유한집합**이다.

[풀이] 공집합의 진부분집합이 존재하지 않으므로(과연?) ϕ는 무한집합일 수 없다. 그러므로 ϕ는 유한집합이다.

임의의 한 원소 집합 $\{a\}$에 대하여 이 집합의 진부분집합은 공집합뿐이다. 그런데 $\{a\}$와 ϕ 사이에는 일대일 대응이 존재하지 않는다(왜?). 따라서 $\{a\}$는 유한집합이다.

정리 1 (a) 임의의 무한집합의 초집합은 무한집합이다.
(b) 임의의 유한집합의 부분집합은 유한집합이다.

---

[2] 위대한 수학자 중의 한 사람인 데데킨트는 1831년 11월 6일 독일의 브룬스빅에서 태어났다. 처음에 물리학과 화학에 흥미를 가졌던 그는 과학의 시종으로만 여겼던 수학공부를 하면서 논리에 더욱 흥미를 느낀 나머지 17세 때 전공을 수학으로 바꾸었다. 19세 때 괴팅겐 대학으로 진학한지 3년만에 가우스 지도하에 수학박사학위를 받았다. 수학발전에 크게 이바지한 그의 공로 중에는 유명한 '데데킨트의 절단'과 무리수에 관한 중요한 개념이 있다. 독자는 장차 실해석학을 공부할 때 그런 것을 이해할 수 있을 것이다.

* 두 집합 $\mathbb{N}$, $\mathbb{N}_e$ 이외의 무한집합의 보기를 한두 가지 더 들어 보아라.

** 독자 누구나 짐작할 수 있듯이 좋은 예제가 못된다.

[Proof] (a) Let X be an infinite set and let Y be a superset of X, i.e., $X \subseteq Y$. Then by Definition 1 there exists an injection $f: X \to X$ such that $f(X) \neq X$. Define a function $g: Y \to Y$ by

$$g(y) = \begin{cases} f(y) & \text{if } y \in X \\ y & \text{if } y \in (Y - X) \end{cases}$$

We leave it to the reader to verify that the function $g: Y \to Y$ is injective and that $g(Y) \neq Y$. It now follows by Definition 1 that Y is infinite.

(b) Let Y be a finite set and let X be a subset of Y, i.e., $X \subseteq Y$. To show that X is finite, we suppose the contrary, that X is infinite. Then by (a), the set Y must be infinite. This is a contradiction. Therefore, the set X is finite.

Theorem 2. Let $g: X \to Y$ be a one-to-one correspondence. If the set X is infinite, then Y is infinite.

[Proof] Since X is infinite, by Definition 1 there exists an injection $f: X \to X$ such that $f(X) \neq X$. Since $g: X \to Y$ is a one-to-one correspondence, so is $g^{-1}: Y \to X$ (Theorem 14, chapter 3). We now have the following diagram of injections:

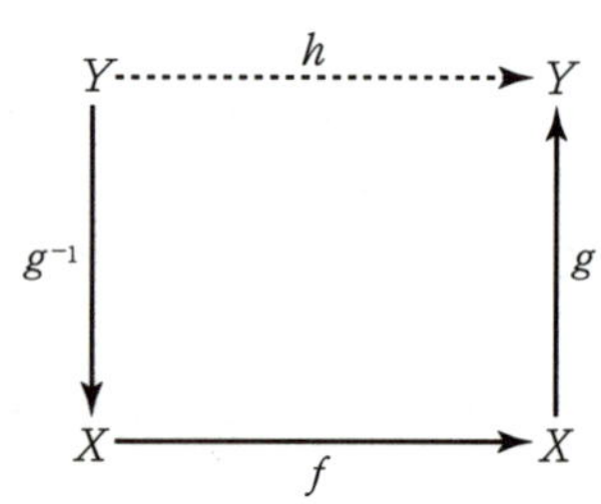

Consequently, the composition

$$h = g \circ f \circ g^{-1}: Y \to Y$$

of injections is an injection (Problem 7, Exercise 3.7).

[증명] (a) 임의의 무한집합을 X라 하고 그 초집합을 Y, 즉 $X \subseteq Y$라고 놓는다. 정의 1에 따라 단사 $f : X \to X$가 존재함으로써 $f(X) \neq X$. 지금 함수 $g : Y \to Y$를

$$g(y) = \begin{cases} f(y), & y \in X \\ y, & y \in (Y-X) \end{cases}$$

로 정의하면 이 함수는 단사이고 $g(Y) \neq Y$(이것의 증명은 독자에게 맡긴다). 그러므로 정의 1에 따라 Y는 무한집합이다.

(b)* 임의의 유한집합을 Y라 하고 그 부분집합을 X, 즉 $X \subseteq Y$라고 놓는다. 여기서 X가 유한집합임을 보이기 위하여 이를 부정함으로써 X가 무한집합이라고 가정한다. 그러면 (a)에 따라 집합 Y는 무한집합이다. 이것은 Y가 유한집합이라는 가정에 모순이다. 따라서 X는 유한집합이다.

정리 2 일대일 대응 $g : X \to Y$에 대하여 정의역 X가 무한집합이면 공역 Y도 무한집합이다.

[증명] X는 무한집합이므로 정의 1에 따라 단사 $f : X \to X$가 존재함으로써 $f(X) \neq X$. 한편 가정에 따라 $g : X \to Y$는 일대일 대응이므로 $g^{-1} : Y \to X$도 일대일 대응이다[제3장 정리 14(187쪽)]. 다음 그림은 여기서의 단사들을 나타낸다.

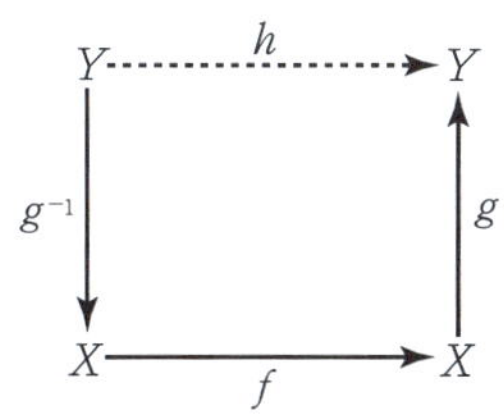

그런데 이들 단사의 합성

$$h^{**} = g \circ f \circ g^{-1} \qquad \text{곧} \qquad h : Y \to Y$$

또한 단사이다[연습문제 3.7 문제 7(199쪽)].

* 지극히 당연한 주장이지만 수학에서는 논리적 증명을 필요로 한다.

** 원서의 표현(합성함수인 h의 오른쪽)이 마음에 들지 않아서 수정한다.

Finally, we have

$$h(Y)=(g\circ f\circ g^{-1})(Y)=(g\circ f)(g^{-1}(Y))$$
$$=(g\circ f)(X)=g(f(X))$$

and $g(f(X))\neq Y$, because $f(X)\neq X$.

Thus, $h(Y)$ is a proper subset of Y and hence Y is infinite.

Corollary. Let $g: X\to Y$ be a one-to-one correspondence. If the set X is finite, then Y is finite.

[Proof] Exercise.

Theorem 3. Let X be an infinite set and let $x_0\in X$. Then $X-\{x_0\}$ is infinite.

[Proof] By Definition 1, there exists an injection $f: X\to X$ such that $f(X)\subset X$. There are two cases to be considered: (1) $x_0\in f(X)$ or (2) $x_0\in X-f(X)$. In each case we must construct an injection $g: X-\{x_0\}\to X-\{x_0\}$ such that $g(X-\{x_0\})\neq X-\{x_0\}$.

Case 1. $x_0\in f(X)$.

There exists an element x_1 in X such that $f(x_1)=x_0$.
A function

$$g: X-\{x_0\}\to X-\{x_0\}$$

may now be defined by

$$g(x)=\begin{cases} f(x) & \text{if } x\neq x_1 \\ x_2 & \text{if } x=x_1\in X-\{x_0\} \end{cases}$$

where x_2 is an arbitrarily fixed element in the nonempty set $X-f(X)$. It follows that $g: X-\{x_0\}\to X-\{x_0\}$ is injective and that

$$g(X-\{x_0\})=f(X-\{x_0, x_1\})\cup\{x_2\}\neq X-\{x_0\}.$$

Hence, $X-\{x_0\}$ is infinite in this case.

마침내 $h(Y)=(g\circ f\circ g^{-1})(Y)=(g\circ f)(g^{-1}(Y))$
$=(g\circ f)(X)=g(f(X))$

즉 $h(Y)=g(f(X))$

한편 $f(X)\neq X$이고 g는 일대일 대응이므로 $g(f(X))\neq Y$이다.

따라서 $h(Y)\neq Y$ 곧 $h(Y)$는 Y의 진부분집합이다. 그러므로 정의 1에 따라 Y는 무한집합이다.

따름정리 일대일 대응 $g: X\to Y$에 있어서 정의역 X가 유한집합이면 공역 Y도 유한집합이다.

[증명] 독자의 연습문제로 돌린다.

정리 3 무한집합 X의 임의의 원소 x_0에 대하여 $X-\{x_0\}$은 무한집합이다.

[증명] 정의 1에 따라 단사 $f: X\to X$가 존재함으로써 $f(X)\subset X$. 이제 두 경우 $x_0\in f(X)$, $x_0\in[X-f(X)]$의 각각에 대하여 다음과 같은 단사를 얻도록 한다.

$$g: X-\{x_0\}\to X-\{x_0\},\quad g(X-\{x_0\})\neq X-\{x_0\}$$

경우 (1) $x_0\in f(X)$

적당한 $x_1\in X$이 존재함으로써 $f(x_1)=x_0$.

지금 함수 $g: X-\{x_0\}\to X-\{x_0\}$를 아래와 같이 정의한다.

$$g(x)=\begin{cases} f(x), & x\neq x_1 \\ x_2, & x=x_1\in X-\{x_0\} \end{cases}$$

여기서 x_2는 공집합이 아닌 집합 $X-f(X)$의 임의로 고정된 원소이다. 그러면 함수 g는 단사이고

$$g(X-\{x_0\})=f(X-\{x_0, x_1\})\cup\{x_2\}\neq X-\{x_0\}\ (\text{왜?})$$

즉 $g(X-\{x_0\})\neq X-\{x_0\}$이므로 $X-\{x_0\}$은 무한집합이다.

Case 2. $x_0 \in X - f(X)$.

Define a function $g : X - \{x_0\} \to X - \{x_0\}$ by $g(x) = f(x)$ for all $x \in \{X - \{x_0\}\}$. Since $f : X \to X$ is injective, so is $g : X - \{x_0\} \to X - \{x_0\}$. Finally,

$$g(X - \{x_0\}) = f(X) - \{f(x_0)\} \neq X - \{x_0\}$$

Therefore, either case, $X - \{x_0\}$ is infinite.

In what follows, let us denote by N_k, $k \in \mathrm{N}$, the set of all natural numbers from 1 to k; that is, $\mathrm{N}_k = \{1, 2, 3, \cdots, k\}$. As an application of Theorem 3, we show in the following example that each N_k is a finite set.

Example 2 For each $k \in \mathrm{N}$, the set N_k is finite.

[Proof] We shall prove this by the principle of mathematical induction. By Example 1, the assertion is true for $k = 1$. Now assume that the set N_k is finite for some natural number k. Consider the set N_{k+1} i.e., $\mathrm{N}_k \cup \{k+1\}$. If N_{k+1} is an infinite set, then by Theorem 3 $\mathrm{N}_{k+1} - \{k+1\}$ i.e., N_k is an infinite set, which contradicts the induction hypothesis. Thus, if N_k is finite then N_{k+1} is finite. Therefore, by the principle of mathematical induction, the set N_k is finite for every $k \in \mathrm{N}$.

In effect, there is a close connection between a nonempty finite set and a set N_k.

Theorem 4. A set X is finite if and only if either $X = \phi$ or X is in one-to-one correspondence with some N_k.

[Proof] If X is either empty or in one-to-one correspondence with some N_k, then by the corollary to Theorem 2 and Example 1 and 2, the set X is finite.

경우 (2) $x_0 \in X - f(X)$

함수 $g : X-\{x_0\} \to X-\{x_0\}$를 다음과 같이 정의한다.

$$각\ x \in \{X-\{x_0\}\}에\ 대하여 \quad g(x)=f(x)$$

여기서 $f : X \to X$가 단사이므로 $g : X-\{x_0\} \to X-\{x_0\}$도 단사이다. 결국 $g(X-\{x_0\})=f(X)-\{f(x_0)\} \neq X_0-\{x_0\}$

즉 $g(X-\{x_0\}) \neq X-\{x_0\}$이므로 $X-\{x_0\}$은 무한집합이다.

이상으로 어느 경우에도 $X-\{x_0\}$은 무한집합이다.*

앞으로는 임의의 $k \in \mathbb{N}$에 대하여 1부터 k까지의 자연수의 집합을 $\mathbb{N}_k$로 나타낸다. 즉 $\mathbb{N}_k=\{1,2,3,\cdots,k\}$. 이때 정리 3의 응용으로써 $\mathbb{N}_k$는 하나의 유한집합임을 다음 예제에서 살펴본다.

예제 2 각 $k \in \mathbb{N}$에 대하여 $\mathbb{N}_k$는 유한집합이다.

[증명] 이것을 수학적 귀납법으로 증명한다.

$k=1$일 때 예제 1에 따라 위의 주장은 참이다. 이제 적당한 자연수 k에 대하여 $\mathbb{N}_k$는 유한집합이라고 가정한다. 지금 집합 $\mathbb{N}_{k+1}$, 즉 $\mathbb{N}_k \cup \{k+1\}$이 무한집합이라면 정리 3에 따라 $\mathbb{N}_{k+1}-\{k+1\}$ 곧 $\mathbb{N}_k$도 무한집합이다. 이것은 $\mathbb{N}_k$가 유한집합이라는 가정에 모순이다. 그러므로 $\mathbb{N}_k$가 유한집합이면 $\mathbb{N}_{k+1}$도 유한집합이다. 따라서 수학적 귀납법에 의하여 각 $k \in \mathbb{N}$에 대한 $\mathbb{N}_k$는 유한집합이다.

사실 다음 정리에서 볼 수 있듯이 공집합이 아닌 유한집합과 집합 $\mathbb{N}_k$ 사이에는 밀접한 관련이 있다.

정리 4 집합 X에 대해서 $X=\phi$ 또는 X와 하나의 $\mathbb{N}_k$ 사이에 일대일 대응이 존재하면 그리고 그때에만 X는 유한집합이다.

[증명] 정리 2의 따름정리(209쪽)와 예제 1(205쪽), 2(바로 위)에 의하여 X는 유한집합이다.

---

* 정리 3은 당연한 것 같지만 그 증명은 까다롭다.

To show the converse, we show, equivalently, its contrapositive: If $X \neq \phi$ and X is not in one-to-one correspondence with any N_k, then X is infinite. Then we can take an element x_1 from X and, again, $X-\{x_1\}$ is not empty; for otherwise we would have $X=\{x_1\}$ in one-to-one correspondence with N_1, a contradiction to the assumption about X. Similarly, we can choose an element x_2 of $X-\{x_1\}$.

Continuing in this manner, suppose that we have chosen elements $x_1, x_2, \cdots, x_k$ of X. Then $X-\{x_1, x_2, \cdots, x_k\}$ is nonempty; otherwise $X=\{x_1, x_2, \cdots, x_k\}$ would be in one-to-one correspondence with N_k, a contradiction to our assumption about X. Thus, we can always choose an element x_{k+1} from $X-\{x_1, x_2, \cdots, x_k\}$. Then, by mathematical induction for every natural number n, there exists a proper subset $\{x_1, x_2, \cdots, x_n\}$ of X. Denote the set of x_n chosen for every natural number n by Y[3]. Then the function

$$f: Y \to Y-\{x_1\} \text{ defined by } f(x_k)=x_{k+1}, \text{ for all } k \in \mathrm{N},$$

establishes a one-to-one correspondence between Y and its proper subset $Y-\{x_1\}$. Therefore, by Definition 1, Y is infinite and hence, by Theorem 1(a), X is infinite.

We mention here that Theorem 4 suggests an alternate definition of finite and infinite sets. We can define a set to be finite if and only if it is either an empty set or in one-to-one correspondence with some N_k, and to be infinite if and only if it is not finite. From this alternate definition, our Definition 1 can be proved as a theorem. However, this would require about the same amount of work as our present approach.

---

[3] Here the authors have implicitly used the "axiom of choice," an important axiom to be discussed in Chapter 6. One form of the axiom of choice may be stated as:

"Let $\mathscr{P}$ be a nonempty set of nonempty subsets of a given set X. Then there exists a set $R \subseteq X$ such that for every $C \in \mathscr{P}$, $C \cap R$ is equal to a singleton set."

this axiom will be used throughout this book without being mentioned explicitly.

한편 정리 4의 역(그것을 말하여 보아라)도 참이다. 여기서 그 사실을 밝히기 위하여 역의 대우, 즉

"공집합이 아닌 집합 X와 임의의 집합 $\mathbb{N}_k$ 사이에 일대일 대응이 존재하지 않으면 X는 무한집합이다."

임을 밝히도록 한다.

지금 집합 X에서 택한 하나의 원소를 x_1이라고 하면 $X-\{x_1\}$은 공집합이 아니다. 왜냐하면 $X-\{x_1\}=\phi$, 즉 $X=\{x_1\}$이라면 X와 $\mathbb{N}_1$ 사이에 일대일 대응이 존재함으로써 X에 대한 가정에 모순이기 때문이다. 같은 방법으로 $X-\{x_1\}$에서 원소 x_2를 택할 수 있다.

이러한 과정을 계속함으로써 X의 원소 $x_1, x_2, \cdots, x_k$를 택하였다고 보자. 그러면 $X-\{x_1, x_2, \cdots, x_k\}$는 공집합이 아니다. 왜냐하면 $X-\{x_1, x_2, \cdots, x_k\}=\phi$이라면 $X=\{x_1, x_2, \cdots, x_k\}$와 $\mathbb{N}_k$ 사이에 일대일 대응이 존재함으로써 역시 X에 대한 가정에 모순이기 때문이다. 그러므로 항상 $X-\{x_1, x_2, \cdots, x_k\}$에서 원소 x_{k+1}을 택할 수 있다. 따라서 수학적 귀납법에 따라 각 자연수 n에 대한 X의 진부분집합 $\{x_1, x_2, \cdots, x_n\}$이 존재한다. 여기서 각 자연수 n에 대하여 X에서 택한 x_n의 집합을 Y[3]라고 놓는다.

이때 각 $k\in\mathbb{N}$에 대한 함수 $f: Y\to Y-\{x_1\}$를 $f(x_k)=x_{k+1}$로 정의하면 이 함수는 집합 Y와 그 진부분집합 $Y-\{x_1\}$ 사이의 일대일 대응에 지나지 않는다(과연?). 그러므로 정의 1(205쪽)에 따라 Y는 무한집합이다. 그런데 X는 Y의 초집합이므로 또한 정리 1(a)(205쪽)에 따라 X는 무한집합이다.

이제 정의 1(205쪽) 대신 정리 4(211쪽)를 유한집합과 무한집합의 정의로 삼을 수 있다. 다시 말하면 집합 X에 대하여 $X=\phi$이거나 X와 적당한 집합 $\mathbb{N}_k$ 사이에 일대일 대응이 존재하면, 그리고 그때에만 X는 유한집합이라고 정의하여 놓고 유한집합이 아니면 무한집합이라고 재차 정의할 수 있다. 그럴 경우 정의 1은 마땅히 증명을 요하는 정리의 구실을 하겠으나 결국 우리가 지금까지 다루어 온 만큼의 작업이 뒤따른다.

---

[3] 여기서 저자들은 제6장에서 다룰 **선택공리***를 암암리에 인용하고 있다. 이 공리의 형식 하나를 들자면 다음과 같다.

"주어진 집합 X의 공집합이 아닌 부분집합의 족을 $\mathscr{P}(\mathscr{P}\neq\phi)$라 하면 X의 부분집합, 이를테면 $R\subseteq X$이 존재함으로써 각 $C\in\mathscr{P}$에 대한 $C\cap R$이 한원소의 집합과 같다."

* 283쪽 하단 [7] 및 289쪽 각주 * 참조.

사실 그것은 무한집합의 경우라야 공리로서 보람이 있다. 독자 나름대로 유한집합, 무한집합을 하나씩 골라 선택공리를 각각 적용하여 보아라.

Exercise 4.1

1. Complete the proof of Theorem 1.

Prove the following statements 2 through 8.

2. If a set X is finite such that $g : X \to Y$ is a one-to-one correspondence, then Y is finite.
3. The sets $\mathbb{Z}$, $\mathbb{Q}$, and $\mathbb{R}$ are infinite.
4. If A is an infinite set, then so is $A \times A$.
5. If A and B are infinite sets, then $A \cup B$ is an infinite set.
6. The union of finitely many finite sets is a finite set.
7. Let A and B be two sets such that $A \cup B$ is infinite. Then at least one of the two sets A and B is infinite.
8. The following generalization of Theorem 3 :
   If Y is a finite subset of the infinite set X, then $X - Y$ is infinite.

4.2 Equipotence of Sets

Two finite sets X and Y have the same number of elements if and only if there exists a one-to-one correspondence $f : X \to Y$. Although the phrase "same number of elements" does not apply here if X and Y are infinite, it seems natural to think that two (infinite) sets that are in one-to-one correspondence are of the same size. We formalize this intuition as follows :

Definition 2. Two sets X and Y are said to be equipotent, symbolized as $X \sim Y$ provided that there exists one-to-one correspondence $f : X \to Y$.

Obviously, every set is equipotent to itself. Since the inverse of a one-to-one correspondence is a one-to-one correspondence (Theorem 14, Chapter 3),

$$X \sim Y \text{ if and only if } Y \sim X.$$

연습문제 4.1

1. 정리 1의 증명을 마무리하여라.

다음 2~8을 증명하여라.

2. 일대일 대응 $g: X \to Y$에서 정의역 X가 유한집합이면 Y도 유한집합이다.

3. 지금까지 다룬 세 집합 $\mathbb{Z}$, $\mathbb{Q}$, $\mathbb{R}$은 모두 무한집합이다.

4. 집합 A가 무한집합이면 $A \times A$도 무한집합이다.

5. 집합 A, B가 각각 무한집합이면 $A \cup B$도 무한집합이다.

6. 유한집합 $A_1, A_2, \cdots, A_k, \cdots, A_m$에 대하여 $\bigcup_{k=1}^{m} A_k$는 유한집합이다.

7. 두 집합 A, B의 합집합 $A \cup B$가 무한집합이면 A, B 중 하나는 무한집합이다.

8. 다음은 정리 3(209쪽)을 일반화한 것이다.
무한집합 X의 부분집합 Y가 유한집합이면 $X-Y$는 무한집합이다.

4.2 집합의 대등

두 유한집합 X, Y에 있어서 일대일 대응 $f: X \to Y$가 존재하면 그리고 그때에만 X와 Y는 같은 수의 원소를 지닌다. 한편 X, Y가 무한집합일 때 "같은 수의 원소"의 뜻을 쉽게 이해할 수 있도록 다음 정의부터 내린다.

> **정의 2** 두 집합 X, Y에 대하여 일대일 대응 $f: X \to Y$가 존재할 때 X와 Y는 **대등**하다고 말하고 이것을 $X \sim Y$로 나타낸다.

임의의 집합 $X(X \neq \phi)$는 분명히 자기자신과 대등하다. 또 일대일 대응의 역함수는 일대일 대응[제 3 장 정리 14(187쪽)]이므로 다음 결과를 얻는다.

$$X \sim Y \Longleftrightarrow Y \sim X.$$

Let us agree that the symbol $f: X \sim Y$ represents "$f: X \longrightarrow Y$ is a one-to-one correspondence and hence $X \sim Y$." Using this convenient notation, the first half of the result of Problem 9, Exercise 3.7 may be restated as:

$$\text{If } f: X \sim Y \text{ and } g: Y \sim Z, \text{ then } g \circ f: X \sim Z$$

We have thus proved the following theorem.

Theorem 5. Let $\mathscr{S}$ be a set of sets and let $\mathscr{R}$ be a relation on $\mathscr{S}$ given by $X \mathscr{R} Y$ if and only if X and Y are members of $\mathscr{S}$ and $X \sim Y$. Then $\mathscr{R}$ is an equivalence relation on $\mathscr{S}$.

In the following example the symbols $(0, 1)$ and $(-1, 1)$ represent open intervals of real numbers, not ordered pairs of integers.

Example 3 (a) $(0, 1) \sim (-1, 1)$ (b) $(-1, 1) \sim \mathbb{R}$ and $(0, 1) \sim \mathbb{R}$.

[Solution] (a) The function $f: (0, 1) \longrightarrow (-1, 1)$ given by $f(x) = 2x - 1$ is a one-to-one correspondence. Hence, $(0, 1) \sim (-1, 1)$.

(b) The trigonometric function $g: (-1, 1) \longrightarrow \mathbb{R}$ given by $g(x) = \tan(\pi x/2)$ is a one-to-one correspondence; therefore $(-1, 1) \sim \mathbb{R}$. The reader should verify this assertion by sketching a graph of $g(x) = \tan(\pi x/2)$.

A rigorous proof may be obtained by verifying the following two observations:

(1) $g: (-1, 1) \longrightarrow \mathbb{R}$ is continuous and unbounded both below and above.

(2) $\forall_{x \in (-1, 1)}\, g'(x) = (\pi/2)\sec^2(\pi x/2) > 0$. Therefore, g is strictly increasing. Since the equipotence "relation" $\sim$ is transitive,[4] $(0, 1) \sim (-1, 1)$ and $(-1, 1) \sim \mathbb{R}$ imply $(0, 1) \sim \mathbb{R}$.

[4] Strictly speaking "$\sim$" is not a relation, because its domain is not a set(see Theorem 10 of Chapter 2). But, we may call it a relation here if we consider it to be defined on any given set of sets $\mathscr{S}$(Theorem 5).

앞으로 기호 $f: X \sim Y$는 "$f: X \to Y$는 일대일 대응이다. 그러므로 $X \sim Y$"를 나타내는 것으로 약속한다. 이 편리한 기호를 써서 연습문제 3.7 문제 9 (199쪽)의 전반부의 결과를 다음과 같이 나타낼 수 있다.

$$f: X \sim Y \wedge g: Y \sim Z \Longrightarrow g \circ f: X \sim Z$$

이로써 다음 정리의 증명이 이미 끝난 셈이다.

정리 5 집합족 $\mathscr{S}$ 위의 관계 $\mathscr{R}$을 다음과 같이 정의한다.
지금 $\mathscr{S}$의 임의의 원소 X, Y에 대하여

$$X \sim Y \Longleftrightarrow X \mathscr{R} Y$$

여기서 $\mathscr{R}$은 $\mathscr{S}$ 위의 하나의 동치관계이다.

다음 예제 3에서 기호 $(0, 1)$, $(-1, 1)$은 각각 순서쌍이 아니라 집합 $\mathbb{R}$의 진부분집합이다.

예제 3 다음이 성립함을 보여라.
(a) $(0, 1) \sim (-1, 1)$ (b) $(-1, 1) \sim \mathbb{R}$, $(0, 1) \sim \mathbb{R}$

[풀이] (a) $f(x) = 2x - 1$*로 주어진 함수 $f: (0, 1) \to (-1, 1)$는 일대일 대응이다(왜?). 그러므로 $(0, 1) \sim (-1, 1)$

(b) $g(x) = \tan(\pi x/2)$로 주어진 삼각함수 $g: (-1, 1) \to \mathbb{R}$은 일대일 대응이다. 그러므로 $(-1, 1) \sim \mathbb{R}$. 독자는 $g(x) = \tan(\pi x/2)$의 그래프의 개형을 그려서 이 주장을 확인하여 보아라.
더욱 엄격한 증명은 다음 두 사실(미분학에서의 연속함수, 도함수의 성질)을 확인함으로써 가능하다.

(1) $g: (-1, 1) \to \mathbb{R}$은 연속함수이고(아래 위로) 유계가 아니다.

(2) $\forall_{x \in (-1,1)}\, g'(x) = (\pi/2)\sec^2(\pi x/2) > 0$.
그러므로 g는 엄밀한 뜻에서 증가한다.
한편 대등 관계 "$\sim$"은 추이적[4]이므로 $(0, 1) \sim (-1, 1)$, $(-1, 1) \sim \mathbb{R}$. $(0, 1) \sim \mathbb{R}$**.

* 이 함수 이외의 일대일 대응인 것이 또 존재한다. 그 중 하나, 둘을 들어보아라.

[4] 엄밀히 말해서 "$\sim$"은 관계가 아니다. 왜냐하면 그것의 정의역은 집합이 아니기 때문이다(제 2장 정리 10 참조). 그러나 여기서는 "$\sim$"을 주어진 집합의 족 $\mathscr{S}$ 위에서 정의된 것으로 여겨 하나의 관계로 인정한다(정리 5).

** 이런 결과에 놀란 독자도 있을 것이다. 무한집합이 지닌 신비라고나 할까.

Theorem 6. Let X, Y, Z, and W be sets with $X \cap Z = \phi = Y \cap W$, and let $f : X \sim Y$ and $g : Z \sim W$. Then $f \cup g : (X \cup Z) \sim (Y \cup W)$.

[Proof] Since $f : X \to Y$ and $g : Z \to W$ are functions with $X \cap Z = \phi$, by Theorem 8 of Chapter 3, $f \cup g : X \cup Z \to Y \cup W$ is a function. We leave it's proof to the reader to prove that the latter function is a one-to-one correspondence.

Theorem 7. Let X, Y, Z, and W be sets such that $X \sim Y$ and $Z \sim W$. Then $X \times Z \sim Y \times W$.

[Proof] Let $f : X \sim Y$ and $g : Z \sim W$. We define the function

$$f \times g : X \times Z \to Y \times W$$

by $\quad (f \times g)(x, z) = (f(x), g(z)) \quad$ for all $(x, z) \in X \times Z$

We ask the reader to verify that the latter function is a one-to-one correspondence.

Examining the vaious finite sets $\mathbb{N}_k = \{1, 2, 3, \cdots, k\}$ as k increases and noting that the infinite sets $\mathbb{Z}$, $\mathbb{Q}$ and $\mathbb{R}$(see Problem 3, Exercise 4.1) are supersets of $\mathbb{N}$, it appears that the "smallest" infinite set is the set $\mathbb{N}$ of all natural numbers, or any set that is equipotent to $\mathbb{N}$. We shall soon learn, in Section 4, that not all infinite sets are equipotent to $\mathbb{N}$.

Definition 3. A set X is said to be denumerable provided that $X \sim \mathbb{N}$. A countable set is a set which is either finite or denumerable.

정리 6 집합 X, Y, Z, W에 대하여 $X \cap Z = \phi \wedge Y \cap W = \phi$* $f : X \sim Y \wedge g : Z \sim W$이면 다음이 성립한다.

$$f \cup g : (X \cup Z) \sim (Y \cup W)$$

[증명] 두 함수 $f : X \longrightarrow Y$, $g : Z \longrightarrow W$에 대하여 $X \cap Z = \phi$이므로 제3장 정리 8(167쪽)에 따라 함수

$$f \cup g : X \cup Z \longrightarrow Y \cup W$$

를 얻는다. 이 함수는 일대일 대응이지만 그 증명은 독자에게 맡긴다.

정리 7 집합 X, Y, Z, W에 대하여 $X \sim Y$, $Z \sim W$이면 다음이 성립한다.

$$X \times Z \sim Y \times W$$

[증명] $f : X \sim Y$, $g : Z \sim W$라 놓고 각 $(x, z) \in X \times Z$에 대하여

$$(f \times g)(x, z) = (f(x), g(z))$$

로 정의된 함수 $f \times g : X \times Z \rightarrow Y \times W$는 일대일 대응이다. 이것의 증명도 독자에게 맡긴다.

짐작컨데 자연수 k가 커짐에 따른 유한집합 $\mathbb{N}_k = \{1, 2, 3, \cdots, k\}$을 비롯하여 무한집합 $\mathbb{Z}$, $\mathbb{Q}$, $\mathbb{R}$(215쪽 연습문제 4.1 문제 3 참조)은 모두 집합 $\mathbb{N}$의 초집합임을 감안할 때 "가장 작은" 무한집합은 $\mathbb{N}$이거나 $\mathbb{N}$과 대등한 임의의 집합일 것으로 여길 것이다. 그러나 독자는 그렇지 않은 무한집합들이 더 많이 존재한다는 사실을 다음 4절(231쪽)에서 발견할 것이다.

정의 3 집합 X에 대하여 $X \sim \mathbb{N}$일 때 X를 **가부번집합****이라고 한다. 한편 유한집합 또는 가부번집합 어느 것이나 **가산집합****이라고 한다.

---

* $X \cap Z \neq \phi$ 또는 $Y \cap W \neq \phi$이면 정리 6의 결론이 거짓인 경우의 보기를 들 수 있을까?

** 가부번집합, 가산집합을 각각 *번호 붙일 수 있는 집합*, *셀 수 있는 집합*이라고 생각하면 편리하다.

Let X be a denumerable set. Then there is a one-to-one correspondence $f: \mathrm{N} \sim X$. If we denote

$$f(1)=x_1,\quad f(2)=x_2,\quad f(3)=x_3,\quad \cdots,\quad f(k)=x_k,\quad \cdots$$

then X may be alternatively denoted as $\{x_1, x_2, x_3, \cdots, x_k, \cdots\}$; the dots are used to indicate that the elements are labeled in a definite order as indicated by the subscripts. An explanation of the term "countable" is now in order. For a finite set, it is theoretically possible to count its elements and the term is suitable. Even though the actual counting of all elements of a denumerable set $X=\{x_1, x_2, x_3, \cdots\}$ is impossible, nevertheless, the set X is in one-to-one correspondence with the natural, or counting numbers.

Theorem 8. Every infinite subset of a denumerable set is denumerable.

[Proof] Let Y be an infinite subset of the denumerable set $X=\{x_1, x_2, x_3, \cdots\}$. Let n_1 be the smallest subscript for which $x_{n_1} \in Y$, and let n_2 be the smallest subscript for which $x_{n_2} \in Y-\{x_{n_1}\}$. Having defined $x_{n_{k-1}} \in Y$, let n_k be the smallest subscript such that $x_{n_k} \in Y-\{x_{n_1}, x_{n_2}, \cdots, x_{n_{k-1}}\}$. Such an x_{n_k} always exists for each $k \in \mathrm{N}$, because Y is infinite, which ensures that $Y-\{x_{n_1}, x_{n_2}, \cdots, x_{n_{k-1}}\} \neq \phi$ for each $k \in \mathrm{N}$. We have thus constructed a one-to-one correspondence $f: \mathrm{N} \sim Y$ where $f(k)=x_{n_k}$ for each $k \in \mathrm{N}$. Therefore, Y is denumerable.

A shorter but less intuitive alternate proof of Theorem 8 is indicated in Problem 10 at the end of this section. The following corollary is an immediate consequence of Definition 3 and Theorem 8.

Corollary. Every subset of a countable set is countable.

More examples and properties of denumerable sets are given in the next section.

우리가 쉽게 알고 있듯이 집합 X가 가부번집합이면 일대일 대응 $f:\mathbb{N}\sim X$가 존재한다. 지금

$$f(1)=x_1,\ f(2)=x_2,\ f(3)=x_3,\ \cdots,\ f(k)=x_k,\ \cdots \qquad \cdots\cdots ①$$

이라고 놓아 집합 X를 $\{x_1, x_2, x_3, \cdots, x_k, \cdots\}$와 같이 나타낼 수 있다. 여기서 첨들은 원소의 순서를 분명히 매기는 데 쓰이고 있다.

이 시점에서 "가산"이란 용어에 대한 설명이 더 필요하다. 먼저 유한집합의 경우 원소들을 셀 수 있다는 것을 쉽게 이해할 수 있으므로 "가산"이란 용어는 적절하다고 본다. 한편 가부번집합 $X=\{x_1, x_2, x_3, \cdots\}$에서는 비록 그 모든 원소들을 일일이 셀 수 없을지라도 ①과 같은 과정을 감안한다면 집합 X와 집합 $\mathbb{N}$ 사이에는 일대일 대응이 존재한다고 볼 수 있다.

정리 8 임의의 가부번집합의 부분집합으로서 무한집합인 것은 역시 가부번집합이다.

[증명] 가부번집합을 $X=\{x_1, x_2, x_3, \cdots\}$으로 놓고 X의 무한부분집합의 하나를 Y라고 한다. 지금 Y의 원소 중 그 첨자가 가장 작은 것을 n_1, 즉 $x_{n_1}\in Y$이라 놓는다. 다음에 집합 $Y-\{x_{n_1}\}$의 원소 중 첨자가 가장 작은 것을 n_2, 즉 $x_{n_2}\in Y-\{x_1\}$. 이와 같이 $x_{n_{k-1}}\in Y$를 정의한 뒤 원소 $x_{n_k}\in Y-\{x_{n_1}, x_{n_2}, \cdots, x_{n_{k-1}}\}$를 택하되 그 첨자 n_k는 가장 작은 경우의 것으로 정의한다. 여기서 Y는 무한집합이므로 각 $k\in\mathbb{N}$에 대한 $x_{n_k}\in Y$는 항상 존재한다. 그러므로 각 $k\in\mathbb{N}$에 대하여 $Y-\{x_{n_1}, x_{n_2}, \cdots, x_{n_{k-1}}\}\neq\phi$. 따라서 각 $k\in\mathbb{N}$에 대하여 $f(k)=x_{n_k}$로서 정의되는 일대일 대응 $f:\mathbb{N}\sim Y$를 얻는다. 그러므로 Y는 가부번집합이다.

다음 연습문제 4.2 문제 10(223쪽)에는 정리 8의 증명으로서 더 짧지만 덜 직관적인 것이 제시되어 있다. 한편 다음 따름정리는 정의 3과 정리 8의 결과로서 다음과 같은 주장을 할 수 있다.

따름정리 가산집합의 임의의 부분집합은 가산집합이다.

가부번집합의 예제와 성질에 관해서는 다음 절에서 더 많이 다룰 것이다.

Exercise 4.2

1. Complete the proof of Theorem 6.

2. Complete the proof of Theorem 7.

Prove the following statements 3 through 10.

3. If X and Y are two sets, then $X \times Y \sim Y \times X$.

4. If $(X-Y) \sim (Y-X)$ then $X \sim Y$.

5. The following generalization of Theorem 6: Let $\{X_\gamma \mid \gamma \in \Gamma\}$ and $\{Y_\gamma \mid \gamma \in \Gamma\}$ be two families of disjoint sets such that $X_\gamma \sim Y_\gamma$ for each $\gamma \in \Gamma$. Then $\bigcup_{\gamma \in \Gamma} X_\gamma \sim \bigcup_{\gamma \in \Gamma} Y_\gamma$.

6. If X is a denumerable set and Y is a finite subset of X, then $X-Y$ is denumerable. (Compare with Problem 8, Exercise 4.1.)

7. If X is a denumerable set and Y is a finite set, then $X \cup Y$ is denumerable.

8. The set N_e of all even natural numbers and the set N_o of all odd natural numbers are denumerable.

9. Let A be a nonempty set, and let 2^A be the set of all functions from the set A to the set $\{0, 1\}$. Then $\mathscr{P}(A) \sim 2^A$.

10. Let X be a denumerable set and Y an infinite subset of X. Let $g : X \sim \mathrm{N}$, and let $h : Y \to \mathrm{N}$ be defined by

$$h(y) = \text{the number of elements in } \{1, 2, 3, \cdots, g(y)\} \cap g(Y)$$

Then, h is a one-to-one correspondence and hence that Y is denumerable.

연습문제 4.2

1. 정리 6의 증명을 마무리 지어라.

2. 정리 7의 증명을 마무리 지어라.

다음 3∼10을 증명하여라.

3. 임의의 집합 X, Y에 대하여 $X\times Y \sim Y\times X$.

4. 임의의 집합 X, Y에 대하여 $(X-Y)\sim(Y-X)$이면 $X\sim Y$.

5. 원소가 서로 소인 집합의 족 $\{X_\gamma \mid \gamma\in\Gamma\}$, $\{Y_\gamma \mid \gamma\in\Gamma\}$에 있어서 각 $\gamma\in\Gamma$에 대하여 $X_\gamma\sim Y_\gamma$이면 $\bigcup_{\gamma\in\Gamma} X_\gamma \sim \bigcup_{\gamma\in\Gamma} Y_\gamma$(정리 6의 일반화).

6. 가부번집합 X의 부분집합 Y가 유한집합이면 $X-Y$는 가부번집합이다(연습문제 4.1 문제 8과 비교하여라).

7. 가부번집합 X, 유한집합 Y에 대하여 $X\cup Y$는 가부번집합이다.

8. 짝수인 자연수의 집합 $\mathbb{N}_e$와 홀수인 자연수의 집합 $\mathbb{N}_o$은 각각 가부번집합이다.

9. 집합 $A(A\neq\phi)$에 대하여 A에서 집합 $\{0, 1\}$로의 함수의 집합을 2^A라고 할 때 $\mathscr{P}(A)\sim 2^A$.

10.* 가부번집합 X의 무한부분집합 Y가 있다. 지금 함수 g, h에 대하여 당연히 $g: X\sim\mathbb{N}$이지만, $h: Y\longrightarrow\mathbb{N}$은 다음과 같이 정의하기로 한다.

$$h(y)=\{1, 2, 3, \cdots, g(y)\}\cap g(Y)\text{의 원소의 개수}$$

그러면 함수 h는 일대일 대응으로써 Y는 가부번집합이다.

---

* 문제 10을 다음과 같이 요약할 수 있다.
"가부번집합의 부분집합으로서 무한집합인 것은 가부번집합이다."

4.3 Examples and Properties of Denumerable Sets

The set N_e of all even natural numbers and the set N_o of all odd natural numbers are denumerable (Problem 8, Exercise 4.2). Since the union $\mathrm{N}_e \cup \mathrm{N}_0$ $(=\mathrm{N})$ of these two denumerable sets is denumerable, the next theorem should be predictable.

Theorem 9. The union of two deunmerable sets is denumerable.

[Proof] Let A and B be any two denumerable sets. We shall show that $A \cup B$ is denumerable in the following two cases:

Case 1. $A \cap B = \phi$.

Since $A \sim \mathrm{N}$ and $\mathrm{N} \sim \mathrm{N}_o$, we have $A \sim \mathrm{N}_o$. Similarly, we have $B \sim \mathrm{N}_e$. Consequently, by theorem 6, we have $(A \cup B) \sim (\mathrm{N}_0 \cup \mathrm{N}_e)$, that is, $(A \cup B) \sim \mathrm{N}$, which shows that $A \cup B$ is denumerable.

Case 2. $A \cap B \neq \phi$.

Let $C = B - A$. Then $A \cup C = A \cup B$ and $A \cap C = \phi$; the set $C \subseteq B$ is either finite or denumerable (corollary to theorem 8). If C is finite, by Problem 7 of Exercise 4.2, $A \cup C$ is denumerable, whereas if C is denumerable, then $A \cup C$ is denumerable by case 1, above. Therefore, the union $A \cup B$ is denumerable.

Corollary. Let $A_1, A_2, \cdots, A_n$ be denumerable sets. Then $\bigcup_{k=1}^{n} A_k$ is denumerable.

[Proof] Left to the reader as an exercise.

4.3 가부번집합의 성질

이미 우리에게 낯익은 짝수인 자연수의 집합 $\mathbb{N}_e$, 홀수인 자연수의 집합 $\mathbb{N}_o$은 각각 가부번집합이다[연습문제 4.2 문제 8(223쪽)]. 이 두 집합의 합집합 $\mathbb{N}_e \cup \mathbb{N}_o (=\mathbb{N})$은 가부번집합이므로 다음 정리를 짐작할 수 있다.

정리 9 임의의 가부번집합 A, B에 대한 $A \cup B$도 가부번집합이다.

[증명] 다음 두 경우로 나누어 증명한다.

(1) $A \cap B = \phi$

$A \sim \mathbb{N}$, $\mathbb{N} \sim \mathbb{N}_o$이므로 $A \sim \mathbb{N}_o$. 마찬가지로 $B \sim \mathbb{N}_e$. 따라서 정리 6(219쪽)에 의하여 $(A \cup B) \sim (\mathbb{N}_e \cup \mathbb{N}_o)$.

여기서 $\mathbb{N}_e \cup \mathbb{N}_o = \mathbb{N}$ 그러므로 $A \cup B \sim \mathbb{N}$. 따라서 $A \cup B$는 가부번집합이다.

(2) $A \cap B \neq \phi$

$C = B - A$라고 하면 $A \cup C = A \cup B$, $A \cap C = \phi$. 여기서 집합 $C \subseteq B$는 유한집합 또는 가부번집합이다(정리 8의 따름정리). 또한 집합 C가 유한집합이면 $A \cup C$는 가부번집합이다[연습문제 4.2 문제 7(223쪽)].

한편 C가 가부번집합이면 (1)에 따라 $A \cup C$ 곧 $A \cup B$도 가부번집합이다.

따름정리 n개의 가부번집합 $A_1, A_2, \cdots, A_k, \cdots, A_n$에 대한 $\bigcup_{k=1}^{n} A_k$도 가부번집합이다.

[증명]* 독자의 연습문제로 돌린다.

* 수학적 귀납법을 활용하여라.

We ask the reader to verify the next example.

Example 4 The set $\mathbb{Z}$ of all integers is denumerable.

Theorem 10. The set $N \times N$ is denumerable.

[Proof] Consider the function $f : N \times N \longrightarrow N$ given by

$$f(j, k) = 2^j 3^k \quad \text{for all } (j, k) \in N \times N$$

This function is injective, so that

$$N \times N \sim f(N \times N) \subseteq N.$$

Since $N \times N$ is infinite so is $f(N \times N)$. By Theorem 8, $f(N \times N)$ is denumerable and hence the set $N \times N$ is denumerable.

Corollary. For each $k \in N$ let A_k be a denumerable set satisfying $A_j \cap A_k = \phi$ for all $j \neq k$. Then $\bigcup_{k \in N} A_k$ is denumerable.[5)]

[Proof] For each $k \in N$, let $f_k : N \longrightarrow N \times \{k\}$ be the function given by $f_k(j) = (j, k)$ for all $j \in N$. Clearly, each $f_k : N \longrightarrow N \times \{k\}$ is a one-to-one correspondence. That is, $N \sim N \times \{k\}$. Since $A_k \sim N$ and $N \sim N \times \{k\}$ for each $k \in N$, we have $A_k \sim N \times \{k\}$ for each $k \in N$. It then follows from Problem 5 of Exercise 4.2 that $\bigcup_{k \in N} N \times \{k\}$ equals the denumerable set $N \times N$. Therefore, $\bigcup_{k \in N} A_k$ is denumerable.

Example 5 The set Q of all rational numbers is denumerable.

[Proof] We shall represent each rational number uniquely as p/q, where $p \in \mathbb{Z}$, $q \in N$ and the greatest common divisor of p and q is 1. Let Q_+ be the set of all such $p/q > 0$, and let $Q_- = \{-p/q \mid p/q \in Q_+\}$ Then $Q = Q_+ \cup \{0\} \cup Q_-$. It is evident that $Q_+ \sim Q_-$.

---

5) This result is true without the hypothesis "$A \cap A_k = \phi$ for all $j \neq k$." See Problem 7.

이제 독자는 다음 예제 4마저 쉽게 확인할 수 있을 것이다.

예제 4 정수의 집합 $\mathbb{Z}$는 가부번집합이다.

정리 10* 집합 $\mathbb{N}\times\mathbb{N}$은 가부번집합이다.

[증명] 함수 $f:\mathbb{N}\times\mathbb{N}\longrightarrow\mathbb{N}$을 각 $(j, k)\in\mathbb{N}\times\mathbb{N}$에 대하여 $f(j, k)=2^j3^k$로 정의된 f는 단사이다(이것을 밝혀라!).

그러므로 $\mathbb{N}\times\mathbb{N}\sim f(\mathbb{N}\times\mathbb{N})$ 여기서 $f(\mathbb{N}\times\mathbb{N})\subseteq\mathbb{N}$

한편 $\mathbb{N}\times\mathbb{N}$은 무한집합이므로 $f(\mathbb{N}\times\mathbb{N})$도 무한집합이다. 그런데 $f(\mathbb{N}\times\mathbb{N})$이 가부번집합이므로 정리 8에 따라 $\mathbb{N}\times\mathbb{N}$도 가부번집합이다.

따름정리 각 $k\in\mathbb{N}$에 대한 A_k는 가부번집합으로서 모든 $j, k\in\mathbb{N}$에 대하여 $A_j\cap A_k=\phi\,(j\neq k)$이면 $\bigcup_{k\in\mathbb{N}} A_k$는 가부번집합[5]이다.

[증명] 각 $k\in\mathbb{N}$에 대한 함수 $f_k:\mathbb{N}\longrightarrow\mathbb{N}\times\{k\}$를 $\forall j\in\mathbb{N}\ f_k(j)=(j, k)$로 정의하면 이 함수는 일대일 대응이다(왜?). 즉 $\mathbb{N}\sim\mathbb{N}\times\{k\}$. 여기서 각 $k\in\mathbb{N}$에 대하여 $A_k\sim\mathbb{N}$, $\mathbb{N}\sim\mathbb{N}\times\{k\}$이므로 $A_k\sim\mathbb{N}\times\{k\}$. 따라서 $\bigcup_{k\in\mathbb{N}} A_k\sim\bigcup_{k\in\mathbb{N}}\mathbb{N}\times\{k\}$. 한편 $\bigcup_{k\in\mathbb{N}}\mathbb{N}\times\{k\}$는 가부번집합 $\mathbb{N}\times\mathbb{N}$과 대등하다(223쪽 연습문제 4.2 문제 5). 따라서 $\bigcup_{k\in\mathbb{N}} A_k$는 가부번집합이다.

예제 5 유리수의 집합 $\mathbb{Q}$는 가부번집합이다.

[증명] 각 유리수를 p/q와 같이 한 가지로 나타낸다. 여기서 $p\in\mathbb{Z}$, $q\in\mathbb{N}$이고 p와 q의 최대공약수는 1이다. 지금

$$\mathbb{Q}_+=\{p/q \mid p/q>0\},\qquad \mathbb{Q}_-=\{-p/q \mid p/q\in\mathbb{Q}_+\}$$

로 각각 놓으면 $\mathbb{Q}=\mathbb{Q}_+\cup\{0\}\cup\mathbb{Q}_-$. 이때 물론 $\mathbb{Q}_+\sim\mathbb{Q}$.

---

* 노영순 · 김기준 · 송형수, 《명제와 집합》 교우사, 2005, 정리 2.6 및 그림 5.4(152-153쪽) 참조.

[5] 이 결론은 가정 "모든 $j, k\in\mathbb{N}$에 대하여 $A_j\cap A_k=\phi(j\neq k)$" 없이도 얻을 수 있다[연습문제 4.3 문제 7 (231쪽) 참조].

Hence, to show that Q is denumerable, it is sufficient to show that Q_+ is denumerable. To this end, we consider the function $f : Q_+ \longrightarrow N \times N$ given by $f(p/q)=(p, q)$. Since this function is injective, we have

$$Q_+ \sim f(Q_+) \subseteq N \times N.$$

Since Q_+, as a superset of N, is infinite, $f(Q_+)$ is an infinite subset of the denumerable set $N \times N$. Therefore, $f(Q_+)$ is denumerable and consequently Q_+ is denumerable. The proof is now complete.

The next theorem indicates that the denumerable sets are, in a sense, the smallest in "size" among the infinite sets.

Theorem 11. Every infinite set contains a denumerable subset.

[Proof] Let X be any given infinite set. Then $X \neq \phi$, so that we can pick an element, call it x_1, from the set X. Next, let x_2 be an element in $X-\{x_1\}$. Similarly, pick an element x_3 from the nonempty set $X-\{x_1, x_2\}$. Having so defined x_{k-1} we let x_k be an element of $X-\{x_1, x_2, \cdots, x_{k-1}\}$. Such an x_k exists for every $k \in N$, because X is infinite, which ensures that

$$X-\{x_1, x_2, \cdots, x_{k-1}\} \neq \phi$$

for every $k \in N$.

The set $\{x_k \mid k \in N\}$ is a denumerable subset of X, and the proof is now complete.

Exercise 4.3

1. Prove the assertion of Example 4 : The set Z of all integers is denumerable.

2. Prove the corollary to Theorem 9.

3. Prove that the union of finitely many countable sets is countable.

그러므로 $\mathbb{Q}$가 가부번집합인 것을 밝히려면 $\mathbb{Q}_+$가 가부번집합임을 지적해야 한다. 여기서 함수 $f: \mathbb{Q}_+ \longrightarrow \mathbb{N}\times\mathbb{N}$를 $f(p/q)=(p, q)$로 정의하면 이 함수는 단사이다(과연?). 따라서 $\mathbb{Q}_+ \sim f(\mathbb{Q}_+)$.

한편 $f(\mathbb{Q}_+)\subseteq\mathbb{N}\times\mathbb{N}$. 또한 $\mathbb{Q}_+$는 무한집합 $\mathbb{N}$의 초집합이므로 무한집합이다. 그러므로 $f(\mathbb{Q}_+)$는 가부번집합 $\mathbb{N}\times\mathbb{N}$의 무한부분집합이므로 가부번집합이다. 마침내 $\mathbb{Q}_+$는 가부번집합이다.

이상으로 증명이 끝난다.

다음 정리는 무한집합 중 가부번집합이 "크기"에 있어서 가장 작다는 것을 지적하고 있다.

정리 11 임의의 무한집합에 대하여 그 부분집합으로서 가부번집합인 것이 존재한다.

[증명] 임의의 무한집합에 대하여 마땅히 X라고 놓으면 $X\neq\phi$. 지금 X에서 하나의 원소(이것을 x_1이라고 부른다)를 택한다. 다음에 집합 $X-\{x_1\}$에서 택한 하나의 원소를 x_2라고 한다. 마찬가지로 집합 $X-\{x_1, x_2\}$에서 하나의 원소 x_3을 택한다. 이와 같이 원소 x_{k-1}를 정한 다음 집합 $X-\{x_1, x_2, \cdots, x_{k-1}\}$의 원소 하나를 x_k라면 각 $k\in\mathbb{N}$에 대한 x_k가 존재한다. 왜냐하면 X는 무한집합이므로

$$X-\{x_1, x_2, \cdots, x_{k-1}\}\neq\phi$$

이기 때문이다. 이 때 집합 $\{x_k \mid k\in\mathbb{N}\}$은 X의 부분집합으로서 가부번집합이다.

연습문제 4.3

다음 1~6을 증명하여라.

1. 예제 4(227쪽)

2. 정리 9의 따름정리(227쪽).

3. m개의 가산집합 $A_1, A_2, \cdots, A_k, \cdots, A_m$에 대하여 $\bigcup_{k=1}^{n} A_k$는 가산집합이다.

4. Prove that if A and B are denumerable sets, then so is $A \times B$. In particular $\mathrm{Z} \times \mathrm{N}$, $\mathrm{Z} \times \mathrm{Z}$, and $\mathrm{Q} \times \mathrm{Q}$ are denumerable.

5. Prove that if for each $k \in \mathrm{N}$, B_k is a denumerable set, then $\bigcup_{k \in \mathrm{N}} B_k$ is denumerable.

6. Prove that the set of all circles in the Cartesian plane having rational radii and centers at points with both coordinates rational is denumerable.

7. Find an injection $f : \mathrm{Q} \to \mathrm{Z} \times \mathrm{N}$ and give an alternate proof for Example 5.

4.4 Nondenumerable Sets

All infinite sets that we have seen so far have been denumerable. This may lead the reader to wonder whether all infinite sets are denumerable. It is commonly thought that Georg Cantor tried to prove that every infinite set is denumerable when he first began his development of set theory. However, he surprised himself by proving that there exist nondenumerable sets.

> ***Theorem 12.*** The open unit interval $(0, 1)$ of real numbers is a nondenumerable set.

[Proof] Let us first express each number x, $0 < x < 1$, as a decimal expansion in the form $.x_1x_2x_3\cdots$, where $x_n \in \{0, 1, 2, \cdots, 9\}$ for all $n \in \mathrm{N}$. For example, $1/3 = .333\cdots$, $\sqrt{2}/2 = .707106\cdots$.

In order to have a unique expression, for those numbers with a terminating decimal expansion such as $1/4 = .25$, let us agree to decrease the last digit by one and append 9's so that $1/4 = .24999\cdots$ not $.2500\cdots$. Under this agreement, two numbers in the interval $(0, 1)$ are equal if and only if the corresponding digits in their decimal expansions are identical. Thus, if two such numbers

$$x = .x_1x_2x_3\cdots \quad \text{and} \quad y = .y_1y_2y_3\cdots$$

4. A, B가 각각 가부번집합이면 $A\times B$는 가부번집합이다. 특히 $\mathbb{Z}\times\mathbb{N}$, $\mathbb{Z}\times\mathbb{Z}$, $\mathbb{Q}\times\mathbb{Q}$는 모두 가부번집합이다.

5. 각 $k\in\mathbb{N}$에 대한 B_k가 가부번집합이면 $\bigcup_{k\in\mathbb{N}} B_k$도 가부번집합이다.

6. 좌표평면에서 중심의 x좌표 · y좌표 및 반지름이 모두 유리수인 원의 집합은 가부번집합이다.

7. 먼저 단사 $f:\mathbb{Q}\to\mathbb{Z}\times\mathbb{N}$를 하나 찾아 별도로 예제 5의 증명을 하여라.

4.4 비가부번집합

지금까지 우리가 보아 온 무한집합은 모두 가부번집합이었으므로 독자 중에는 모든 무한집합이 그러할 것으로 속단할지도 모른다. 혹시 게오르크 칸토어마저 집합론을 처음 발전시킬 때 모든 무한집합이 가부번집합임을 증명하려고 노력하였을는지 모른다. 그렇기에 무한집합으로서 비가부번집합인 것이 존재한다는 사실을 깨닫자 그 자신도 크게 놀랐다고 한다.

정리 12 수직선에서 개단위구간 $(0, 1)$*은 무한집합으로서 비가부번집합이다.

[증명] 먼저 각 수 x, $0<x<1$을 $0.x_1x_2x_3\cdots$와 같은 꼴의 소수로 전개한다(그 가능성?). 여기서 각 $n\in\mathbb{N}$에 대하여 $x_n\in\{0, 1, 2, \cdots, 9\}$. 이를테면

$$1/3=0.333\cdots,\qquad \sqrt{2}/2=0.707106\cdots$$

이 경우 유한소수로 전개된 수도 무한소수 한 가지로 나타내기 위하여 유한소수의 마지막 숫자에서 1을 줄이고 숫자 9를 따라 적도록 한다. 이를테면 유한소수 $1/4=0.25$는 $0.25000\cdots$이 아니라 $0.24999\cdots$로 나타낸다. 이러한 약속하에 구간 $(0, 1)$의 두 수는 각각의 소수전개에 있어서 대응하는 숫자들이 같으면 그리고 그때에만 이 두 수는 같다. 그러므로 $x=0.x_1x_2x_3\cdots$, $y=0.y_1y_2y_3\cdots$

* 개단위구간 $(0, 1)$에 대하여 $[0, 1]$를 폐단위구간으로 부를 수 있다. 물론 이 두 구간의 차이는 독자 누구나 알 것이다.

have one decimal place, say the kth decimal place, such that $x_k \neq y_k$, then $x \neq y$. This is a crucial point upon which our proof rests.

Now suppose that the set $(0, 1)$ is denumerable, then there exists a one-to-one correspondence $f : \mathrm{N} \sim (0, 1)$. So we may list all elements of $(0, 1)$ as follows. :

$$
(\#) \quad
\begin{aligned}
f(1) &= .a_{11}a_{12}a_{13} \cdots \\
f(2) &= .a_{21}a_{22}a_{23} \cdots \\
f(3) &= .a_{31}a_{32}a_{33} \cdots \\
&\ \ \vdots \\
f(k) &= .a_{k1}a_{k2}a_{k3} \cdots \\
&\ \ \vdots
\end{aligned}
$$

where each $a_{jk} \in \{0, 1, 2, \cdots, 9\}$.

We shall construct a number $z \in (0, 1)$ which cannot be found in the above listing of $f(k)$'s. This contradiction will imply that our earlier supposition that $(0, 1)$ is denumerable was wrong and that the set $(0, 1)$ is nondenumerable. Let $z = .z_1z_2z_3 \cdots$ be defined by

$z_k = 5$ if $a_{kk} \neq 5$ and $z_k = 1$ if $a_{kk} = 5$, for each $k \in \mathrm{N}$.

The number $z = .z_1z_2z_3 \cdots$ clearly satisfies $0 < z < 1$; but $z \neq f(1)$ since $z_1 \neq a_{11}$, $z \neq f(2)$ since $z_2 \neq a_{22}$, $\cdots$, and in general $z \neq f(k)$ since $z_k \neq a_{kk}$, for all $k \in \mathrm{N}$. Therefore, $z \notin f(\mathrm{N}) = (0, 1)$. We have now the promised contradiction, and the proof is complete.

Corollary. The set $\mathbb{R}$ of all real numbers is nondenumerable.

[Proof] We have proved, in Example 3(b), that $\mathbb{R} \sim (0, 1)$. Now $(0, 1)$ is nondenumerable ; therefore its equipotent set $\mathbb{R}$ must be nondenumerable (see Problem 1 Exercise 4.4).

Example 6 The set of all irrational numbers is nondenumerable.

에 대하여 소수점 아래 k째 자리에서 $x_k \neq y_k$이면 $x \neq y$. 이같은 주장은 위 정리의 증명에서 주요한 구실을 한다.

여기서 우리는 $(0, 1)$이 가부번집합이라고 가정한다. 그러면 일대일대응 $f : \mathbb{N} \sim (0, 1)$이 존재함으로써 $(0, 1)$의 모든 원소를 다음과 같이 나열할 수 있다.

$$
(\#) \qquad \begin{array}{l}
f(1) = 0.a_{11}a_{12}a_{13}\cdots \\
f(2) = 0.a_{21}a_{22}a_{23}\cdots \\
f(3) = 0.a_{31}a_{32}a_{33}\cdots \\
\quad\vdots \\
f(k) = 0.a_{k1}a_{k2}a_{k3}\cdots \\
\quad\vdots
\end{array}
$$

여기서 각 a_{jk}에 대하여 $a_{jk} \in \{0, 1, 2, \cdots, 9\}$.

지금 위 $(\#)$에서와 같이 나열된 $f(k)$들 중 어느 것과도 같지 않은 하나의 수 $z \in (0, 1)$을 다음과 같이 구성하여 모순을 일으킴으로써 $(0, 1)$이 비가부번집합인 것을 밝히려고 한다. 다시 말하면 각 $k \in \mathbb{N}$에 대하여 $a_{kk} \neq 5$이면 $z_k = 5$로, $a_{kk} = 5$이면 $z_k = 1$로 놓는다.

그러면 $z = 0.z_1z_2z_3\cdots$은 $0 < z < 1$인 수이므로 $z \in (0, 1)$. 한편 $z_1 \neq a_{11}$이므로 $z \neq f(1)$, $z_2 \neq a_{22}$이므로 $z \neq f(2)$, $\cdots$. 대체로 모든 $k \in \mathbb{N}$에 대하여 $z_k \neq a_{kk}$이므로 $z \neq f(k)$. 따라서 $z \notin f(\mathbb{N})$. 여기서 가정에 따라 $f(\mathbb{N}) = (0, 1)$. 그러므로 $z \notin (0, 1)$. 이것은 $z \in (0, 1)$에 모순이다.

이로써 증명이 끝난다.

따름정리 실수의 집합 $\mathbb{R}$은 무한집합으로서 비가부번집합이다.

[증명] 217쪽 예제 3(b)에서 $(0, 1) \sim \mathbb{R}$을 증명한 바 있다. 여기서 $(0, 1)$은 무한집합으로서 비가부번집합이므로 이 집합과 대등한 $\mathbb{R}$도 비가부번집합이다 [연습문제 4.4 문제 1(235쪽) 참조].

예제 6 무리수의 집합은 무한집합으로서 비가부번집합이다.

[Proof] We have shown, in Example 5, that the set Q of all rational number is denumerable. The set of all irrational numbers is equal to, by definition, the set $\mathbb{R}-\mathrm{Q}$. It is easy to see that $\mathbb{R}-\mathrm{Q}$ is an infinite set. To show that $\mathbb{R}-\mathrm{Q}$ is nondenumerable, we suppose the contrary, that $\mathbb{R}-\mathrm{Q}$ is denumerable. It then follows that the union $(\mathbb{R}-\mathrm{Q})\cup\mathrm{Q}=\mathbb{R}$ is denumerable(Theorem 9). This contradicts the corollary to Theorem 12. Therefore, the set $\mathbb{R}-\mathrm{Q}$ of all irrational numbers is nondenumerable.

[Remarks] (1) The method of proof used in theorem 12 is called Cantor's diagonal method, because it was originated by Cantor and the construction of the key number $z=.z_1z_2z_3\cdots$ in the proof is based on the digits $a_{11}, a_{22}, a_{33}, \cdots$ on the principal diagonal of the table (*) of digits. This proof, though it might not be easy for the beginner to appreciate, reveals Cantor's ingenuity.

(2) The existence of nondenumerable sets shows that there are classes of infinite sets. In fact, as the reader shall see in the next chapter, there is an abundance of "equipotence classes" of infinite sets.

Exercise 4.4

1. Let A and B be two equipotent sets. Prove that if A is nondenumerable, then B is nondenumerable.

2. Prove that every superset of a nondenumerable set is nondenumerable.

3. Using the result of Problem 2 above, give an alternate proof for the corollary to Theorem 12.

4. Prove that the set of all irrational numbers between 0 and 1 is nondenumerable.

[증명] 예제 5에서 유리수의 집합 $\mathbb{Q}$는 가부번집합임을 보아왔다. 무리수의 집합은 정의에 따라 집합 $\mathbb{R}-\mathbb{Q}$와 같을 뿐만 아니라 $\mathbb{R}-\mathbb{Q}$가 무한집합인 것도 쉽게 알 수 있다. 지금 이 집합이 비가부번집합인 것을 밝히기 위하여 이 사실을 부정하여 $\mathbb{R}-\mathbb{Q}$가 가부번집합이라고 가정해보자. 그러면 합집합 $(\mathbb{R}-\mathbb{Q})\cup\mathbb{Q}$, 즉 $\mathbb{R}$은 가부번집합이다(225쪽 정리 9). 이것은 정리 12의 따름정리에 모순이다.

따라서 무리수의 집합 $\mathbb{R}-\mathbb{Q}$는 비가부번집합이다.

[참고] (1) 정리 12의 증명법을 칸토어의 **대각법***이라고 부르는 그 이유는 다 다음과 같다.

즉 칸토어가 처음 생각한 이 방법은 표 (#)(233쪽)의 주대각선 위의 숫자들 $a_{11}, a_{22}, a_{33}, \cdots$를 근거로 증명의 열쇠가 되는 수 $z=0.z_1z_2z_3\cdots$를 구성할 수 있기 때문이다. 초보자는 쉽게 이해할 수 없을지도 모르는 이 증명은 칸토어의 예리한 착상을 여실히 드러내고 있다.

(2) 비가부번집합이 존재한다는 사실로 말미암아 무한집합의(동치류와 유사한) 류가 있음을 알 수 있다. 사실 독자는 다음 장에서 "대등한 류"의 무한집합들이 풍부히 존재함을 알아 차린 것이다.

연습문제 4.4

다음 1~5를 증명하여라.

1. 대등한 집합 A, B에서 A가 비가부번집합이면 B도 비가부번집합이다.

2. 비가부번집합의 임의의 초집합은 비가부번집합이다.

3. 정리 12 따름정리의 별도증명. 여기서 위의 문제 2의 결과를 이용하여라.

4. 0과 1 사이의 무리수의 집합은 비가부번집합이다.

5.** 주어진 평면 P에서의 원의 집합은 비가산집합이다.

* 사실 옮긴이는 이 대각법을 대할 때마다 더 수학적(논리적)으로 만족할 만한 별도의 방법이 없을까 기대해 본다.

** 문제 5는 옮긴이가 보충한 것이다.

Chapter 5
Cardinal Numbers and Cardinal Arithmetic

제 5 장
기수와 그것의 셈

5

Cardinal Numbers and Cardinal Arithmetic

The concept of cardinal numbers is introduced. Similarities and distinctions between properties of finite and transfinite cardinal numbers are exhibited in the course of exploring cardinal arithmetic — addition, multiplication, and exponentiation. The chapter ends with a historical remark on the (generalized) continuum hypothesis.

5.1 The Concept of Cardinal Numbers

Very naturally, the concept of numbers entered our lives early. We were able to notice, for example, the similarity between three apples and three oranges and the distinction between two fingers and four fingers. Although we had a concept of number, most of us did not have a precise definition of number. We have known, for example, that

$$2+3=5, \quad 3<4, \quad 6\times 7=42$$

etc. This leads us to believe that we don't need to know what a number really is ; what we should know are equality and order between numbers, and how to calculate with numbers — just as chess players are not concerned with what a knight is, but rather with how it performs.

Therefore, we do not define here what a cardinal number is[1], but just introduce it as a primitive concept relating to the "size" of a set. The important rules guiding this new concept are

---

[1] Perhaps the reader should be informed that it is possible to define the cardinal number as the "initial ordinal." See page 360.

5

기수와 그것의 셈

먼저 *기수*의 개념부터 소개하고 기수의 셈—덧셈, 곱셈, 거듭제곱을 알아보는 과정에 있어서 *유한기수*와 *초한기수*의 유사점과 차이점을 규명한다. 마지막으로 (일반)연속체가설에 관한 역사적 단평으로 이 장의 끝을 맺는다.

5.1 기수의 개념

수의 개념은 일찍부터 우리의 생활에 자연스럽게 등장하였다. 하지만 세 개의 사과와 세 개의 감귤 사이의 유사점, 두 개의 손가락과 네 개의 손가락 사이의 차이점 등 따지고 보면 우리는 수의 개념을 어렴풋이 알고 있을 뿐 정확한 정의는 모르고 있다. 다시 말하여

$$2+3=5, \quad 3<4, \quad 6\times 7=42$$

와 같은 것은 너무나 잘 알고 있으므로 사실 수가 무엇인가를 이해할 필요성을 느끼지 못했던 것이다. 마치 장기를 두는 사람이 포 자체가 무엇인지에 관심을 두는 것보다는 그 말을 어떻게 둬야 할지를 생각하듯이 여기서 살펴봐야 할 것은 등식, 두 수 사이의 순서, 수의 셈을 할 수 있는 방법 등을 꼽을 수 있다. 그러므로 기수[1]가 무엇인지의 정의를 내리기에 앞서 다음에서와 같이 집합의 "크기"와 관련된 몇몇 개념과 규칙부터 소개하기로 한다.

[1] 독자는 기수를 "초기순서수"로서도 정의할 수 있음을 알 것이다(361쪽 중간 구절 참조). 한편 기수를 "집합의 크기" 또는 "농도"라고도 한다.

C-1. Each set A is associated with a cardinal number, denoted by **card** $\mathbf{A}$, and for each cardinal number a there is a set A with card $A=a$.

C-2. Card $A=0$ if and only if $A=\phi$.

C-3. If A is a nonempty finite set, i.e., $A\sim\{1,2,3,\cdots,k\}$ for some $k\in\mathrm{N}$, then card $A=k$.

C-4. For any two sets A and B;

$$\text{card } A=\text{card } B \quad \text{if and only if } A\sim B.$$

The rules C-2 and C-3 define the cardinal numbers of the finite sets—the cardinal number of a finite set is the number of elements in that set. In axiomatic treatments of set theory, C-1 and C-4 are usually postulated as an axiom, called the axiom of cardinality. The beginner might find C-1 and C-4 rather difficult to accept, because these rules do not say much about card A when A is an infinite set. This difficulty will be overcome gradually as we proceed—just as the reader did not know what calculus was until he was about halfway through the course. At this stage, we may say, roughly, that the cardinal number of a set is the property that the set has in common with all sets that are equipotent to it.

Exercise 5.1

1. Show that the natural numbers are cardinal numbers.

2. Give three cardinal numbers which are not natural numbers.

C-1. 임의의 집합 A에 대하여 **card A**로 표시된 하나의 기수가 결부되고 각 기수 a에 대하여 card $A=a$인 하나의 집합 A가 존재한다.

C-2. $A=\phi \Longleftrightarrow$ card $A=0$.

C-3.* A가 공집합이 아닌 유한집합 A(즉 적당한 $k\in\mathbb{N}$에) 대하여

$$A\sim\{1, 2, 3, \cdots, k\} \text{이면} \quad \text{card}\, A=k.$$

C-4. 임의의 집합 A, B에 대하여 이면,

$$A\sim B \Longleftrightarrow \text{card}\, A=\text{card}\, B.$$

위의 규칙 중 C-2, C-3은 유한집합의 기수에 대한 정의로써 주어진 집합의 원소의 개수를 나타낸다. 한편 C-1, C-4는 흔히 공리론적 집합론에서 거론되는 이른바 기수의 공리로서 무한집합의 기수에 대한 언급으로 봐야 한다.

이렇듯 위에서 다룬 C_1-C_4 모두가 지닌 뜻으로 하여금 독자에게 바로 와 닿지 않더라도 염려할 필요는 없다. 이 단계에서는 집합 A의 기수란 **그 집합과 대등한 모든 집합들이 공통으로 지닌 성질****이라고 알아두면 된다.

연습문제 5.1

1.*** 다음 집합의 기수를 각각 말하여라.

$$A=\{0\},\ B=\{\phi\},\ C=\{0, 1, 2, 3\},\ D=\{\phi, \{\phi\}, \{\phi, \{\phi\}\}, a, b\}$$

2.*** card $A=73$인 집합 A를 구하여라.

3. 자연수는 기수임을 보여라(원서 문제 1).

4.*** $A=\{a, b, c\}$, $B=\{1,2\}$에 대하여 다음을 각각 구하여라.

card$(A\times B)$[정의 3(259쪽) 참조], card$(B\times A)$

5.**** 기수로서 자연수가 아닌 것 세 개를 지적하여라.

---

* 원소의 개수를 자연수로 나타낼 수 있는 집합을 유한집합이라고 할 것(75쪽 위에서 5행)은 이 공리에 바탕을 두고 있다.

** 이장우, 《현대집합론의 원리》, 경문사, 1998년, 5.1절 기수의 개념 (164–166쪽) 참조.

*** 문제 1, 2, 4는 옮긴이가 보충한 것이다.

**** 원서 문제 2이고, 이 문제는 현 단계에서 지나쳐 보인다. 다음 2절(243쪽) 이후에서나 다룰 법하다.

5.2 Ordering of the Cardinal Numbers — The Cantor Bernstein Theorem

We shall call the cardinal number of a finite set a finite cardinal number, and the cardinal number of an infinite set a transfinite cardinal number. The rules C - 2 and C - 3 of the previous section show that the finite cardinal numbers are precisely the nonnegative integers. Thus, the finite cardinal numbers have an inherited natural order:

$$0 < 1 < 2 < \cdots < k < k+1 < \cdots$$

For any two transfinite cardinal numbers, the rule C - 4 tells us when they are equal and when they are not equal. But we will not be satisfied with just that; when they are unequal we wish to be able to tell which one is "less" than the other.

Definition 1. Let A and B be sets. Then card A is said to be less than card B, denoted by **card $A <$ card B**, provided that the set A is equipotent to a subset of B but the set B is not equipotent to any subset of A.

Although this definition is designed to order transfinite cardinal numbers, it applies to finite cardinal numbers as well, and when it is applied to finite cardinal numbers the result is the same as the traditional natural ordering mentioned above.

Example 1 card $\mathbb{N} <$ card $\mathbb{R}$.

[Proof] Since the set $\mathbb{N}$ is a subset of $\mathbb{R}$, $\mathbb{N}$ is equipotent to a subset of $\mathbb{R}$, $\mathbb{N} \sim \mathbb{N} \subset \mathbb{R}$, but from Section 4, Chapter 4, we know that the infinite set $\mathbb{R}$ is nondenumerable. Hence, $\mathbb{R}$ is not equipotent to any subset of $\mathbb{N}$. By Definition 1, we have card $\mathbb{N} <$ card $\mathbb{R}$.

So far it is not clear to us how two cardinal numbers card A and card B compare when the set A is equipotent to a subset of B and the set B is equipotent to a subset of A.

5.2 기수의 대소, 칸토어-베른슈타인의 정리

유한집합의 기수를 **유한기수**라 하고 무한집합의 기수를 **초한기수**라고 한다. 앞절의 규칙 C-2, C-3은 유한기수는 다름 아닌 0과 양의 정수임을 말해 주고 있다. 따라서 유한기수는 자연수로부터 이어받은 순서

$$0<1<2<\cdots<k<k+1\ \cdots \qquad \cdots①$$

를 고스란히 지닌다. 한편 C-4에 따르면 임의의 두 초한기수가 어느 때 같고 어느 때 같지 않다는 것을 분간할 수 있다. 그러나 우리는 이런 정도로 만족할 것이 아니라 두 초한기수가 같지 않을 때 어느 쪽이 다른 쪽보다 작은가를 구분하기 위한 다음 정의부터 내린다.

정의 1 집합 A, B에 대하여 A가 B의 하나의 부분집합과 대등하지만 B는 A의 어느 부분집합과도 대등하지 않을 때 card A는 card B**보다 작다** 또는 card B는 card A**보다 크다**고 말하고 이것을 **card A<card B** 또는 **card B>card A**로 나타낸다.

초한기수의 순서에 관한 이 정의는 유한기수에 대해서도 그대로 적용되거니와 그 대소관계는 ①에서와 같이 자연수의 순서와 같다고 말할 수 있다.

예제 1 card $\mathbb{N}$<card $\mathbb{R}$임을 밝혀라.

[증명] 집합 $\mathbb{N}$은 집합 $\mathbb{R}$의 부분집합이므로 $\mathbb{N}$은 $\mathbb{R}$의 부분집합 $\mathbb{N}$과 대등하다. 즉 $\mathbb{N}\sim\mathbb{N}$. 그러나 4.4절에서 알아본 것과 같이 무한집합 $\mathbb{R}$은 비가부번집합이므로 $\mathbb{R}$은 $\mathbb{N}$의 어느 부분집합과도 대등하지 않다. 따라서 정의 1에 의하여 card $\mathbb{N}$<card $\mathbb{R}$.

이상에서 알아본 것만으로는 집합 A, B로 하여금 부분집합과 각각 대등할 때 곧바로 두 기수 card A, card B의 대소를 비교할 수 있는 방법이 채 알려지지 않은 가운데 게오르크 칸토어는 card A와 card B는 같을 것이라고 예측

Georg Cantor conjectured that, in this case, card A should be equal to card B. Later, in the 1890's, this conjecture was proved, independently, both by F. Bernstein in Cantor's seminar, and by E. Schröder on the basis of a logical calculus. This celebrated result is now generally known as the Cantor–Bernstein Theorem.

Theorem 1. Cantor–Bernstein Theorem If A and B are sets such that A is equipotent to a subset of B and B is equipotent to a subet of A, then A and B are equipotent.

We shall first prove the following special case of Theorem 1, from which Theorem 1 follows easily.

Lemma. If B is a subset of a set A such that $f : A \longrightarrow B$ is an injection, then there exists a bijection $h : A \sim B$.

[Proof] If $B=A$, then the identity function on A is such an h. Suppose that B is a proper subset of A, and let C denote the set $\bigcup_{n\geqq 0} f^n(A-B)$, where f^0 is the identity function on A and, for each positive integer k and for each $x \in A$, $f^k(x)=f(f^{k-1}(x))$. For each z in A, define $h(z)$ as follows:

$$h(z)=\begin{cases} f(z) & \text{if } z \in C \\ z & \text{if } z \in A-C \end{cases}$$

Observe that $A-B$ is a subset of C, $f(C)\subseteq C$, and that if m and n are two distinct nonnegative integers, say $m<n$, then $f^m(A-B)$ and $f^n(A-B)$ are disjoint. For otherwise, there exist x and x' in $A-B$ such that $f^m(x)=f^n(x')$, which leads to

$$f^{n-m}(x')=x \in B \cap (A-B),$$

a contradiction.

하였다. 그 후 1890년대에 그 예측을 칸토어의 세미나에서 베른슈타인과 슈뢰더가 제각기 따로 증명하였으나 1902년 알윈 코르셀트는 논리적 계산에 따른 슈뢰더의 증명이 틀렸음을 지적하였다.*

이 유명한 결과는 오늘날 칸토어-베른슈타인의 정리로 알려져 있다.

정리 1 **칸토어-베른슈타인의 정리** 집합 A, B에 대하여 A가 B의 부분집합과 대등하고 B가 A의 부분집합과 대등하면 A와 B는 대등하다.

정리 1은 그것의 특별한 경우인 다음 보조정리로부터 쉽게 유도된다. 그러므로 여기서는 보조정리를 먼저 증명한다.

보조정리 집합 A의 부분집합 B에 대한 단사 $f: A \to B$가 존재하면 전단사 말하자면 $h: A \sim B$도 존재한다.

[증명] $B=A$인 경우 전단사 h로서 A 위의 항등함수 I_A를 지적할 수 있다. 한편 B가 A의 진부분집합일 때 집합 $\bigcup_{n\geq 0} f^n(A-B)$를 C로 놓는다.

여기서 f^0은 A 위의 항등함수이고 각 양의 정수 k와 각 $x \in A$에 대해서는 $f^k(x)=f(f^{k-1}(x))$로 정의한다.

한편 각 $z \in A$에 대한 $h(z)$를 다음과 같이 정의한다.

$$h(z)=\begin{cases} f(z), & z \in C \\ z, & z \in A-C \end{cases}$$

이 경우 $A-B$는 C의 부분집합이고 $f(C) \subseteq C$. 또 음이 아닌 서로 다른 정수 m, n에 대하여, 이를테면 $m<n$일 때는 $f^m(A-B)$와 $f^n(A-B)$는 서로 소이다. 만일 그렇지 않으면 $A-B$의 두 원소 x, x'이 존재함으로써

$$f^m(x)=f^n(x') \text{이면 } f^{n-m}(x')=f^{m-m}(x)=f^0(x)=x$$

그런데 $f^0(x) \in B$이므로 $x \in B$.

한편 $x \in A-B$이므로 $x \notin B$. 이것은 모순이다.

* Gregory H. Moore, *Zermelo's Axiom of Choice*, Springer-Verlag, 1982, p. 48 참조.

Finally, by the definition of h and the last observation, we have

$$\begin{aligned} h(A) &= (A-C)\cup f(C) \\ &= [A-\bigcup_{n\geq 0} f^n(A-B)]\cup f(\bigcup_{n\geq 0} f^n(A-B)) \\ &= [A-\bigcup_{n\geq 0} f^n(A-B)]\cup [\bigcup_{n\geq 1} f^n(A-B)] \\ &= A-(A-B)=B \end{aligned}$$

From these observations and the fact that f is injective, it follows that $h: A\longrightarrow B$ is a bijection. This completes the proof of the lemma.

The main idea behind the above proof may be visualized in the following illustrative diagram, where the whole rectangle represents the set A:

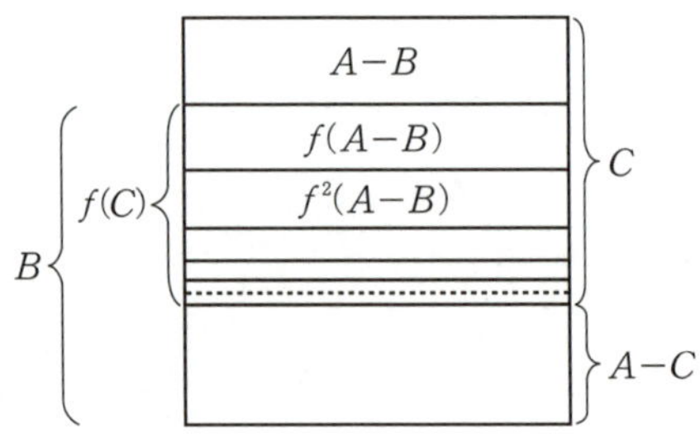

Figure 17

[Proof] (Theorem 1) Let A_0 and B_0 be subsets of A and B, respectively, such that $A\sim B_0$ and $B\sim A_0$, and let $f_0: A\sim B_0$ and $g_0: B\sim A_0$ be two bijections. Let $f: A\longrightarrow A_0$ be given by $f(x)=g_0(f_0(x))$, which is an injection. Hence, by the above lemma, there is a bijection $h: A\sim A_0$. Consequently, the composition $g_0^{-1}\circ h: A\sim B$ of two bijections $h: A\sim A_0$ and $g_0^{-1}: A_0\sim B$ is a bijection[2)].

It is convenient to write $\text{card}\,A\leqq \text{card}\,B$ to mean $\text{card}\,A<\text{card}\,B$ or $\text{card}\,A=\text{card}\,B$. The following corollary is an immediate consequence of the Cantor−Bernstein Theorem.

---

2) This proof and the preceding lemma are adopted from R.H. "A Proof of the Cantor-Bernstein Theorem," American Mathematical Monthly, 75, No. 5(1968), 508.

마침내 함수 h의 정의와 마지막 결론에 따라 다음이 성립한다.

$$\begin{aligned} h(A) &= (A-C)\cup f(C) \\ &= [A-\bigcup_{n\geq 0} f^n(A-B)]\cup f(\bigcup_{n\geq 0} f^n(A-B)) \\ &= [A-\bigcup_{n\geq 0} f^n(A-B)]\cup [\bigcup_{n\geq 1} f^n(A-B)] \\ &= [A-(A-B)]^* = B \end{aligned}$$

그러므로 $h(A)=B$.

즉 h는 전사이다. 한편 가정에 의하여 f는 단사이므로 $h: A\rightarrow B$도 단사이다. 따라서 h는 전단사이다. 이상으로 증명이 끝난다.**

독자는 다음 그림 17을 살펴봄으로써 지금까지의 증명과정을 파악할 수 있을 것이다. 이 그림에서 가장 큰 직사각형은 집합 A를 나타낸다.

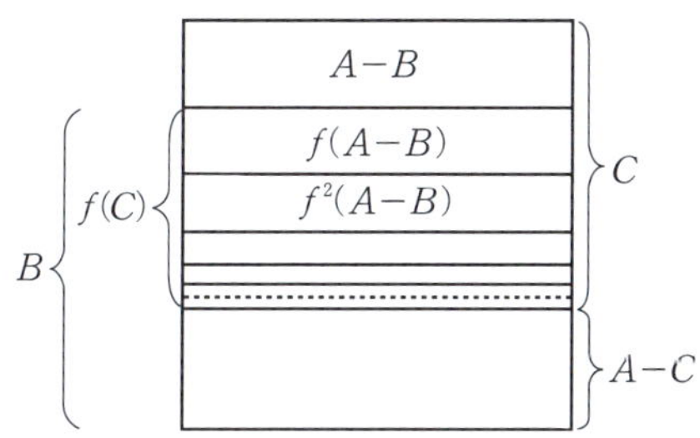

그림 17

[정리 1의 증명] 집합 A, B의 각각의 부분집합 A_0, B_0에 대하여 $A\sim B_0$, $B\sim A_0$라 하고 두 전단사 $f_0: A\sim B_0$, $g_0: B\sim A_0$를 생각한다. 지금 $f(x)=g_0(f_0(x))$로 정의된 함수 $f: A\rightarrow A_0$은 단사(왜?)이므로 앞의 보조정리에 따라 전단사 $h: A\sim A_0$이 존재한다. 따라서 두 전단사

$$h: A\sim A_0,\quad g_0^{-1}: A_0\sim B \text{의 합성 } g_0^{-1}\circ h: A\sim B$$

또한 전단사이다.[2)]

실수의 대소관계에서와 같이 기수에서도 $\text{card}\,A<\text{card}\,B$ 또는 $\text{card}\,A=\text{card}\,B$를 간단히 $\text{card}\,A\leq\text{card}\,B$로 나타내면 편리하다.

이제 우리는 칸토어-베른슈타인의 정리로부터 다음을 직접 얻는다.

---

* 이 등식은 임의의 실수 a, b에 대한 등식 $a-(a-b)=b$와 유사하다.

** 이장우, 《현대집합론의 원리》, 경문사, 1998년, 보제 5.1(168-170쪽) 참조.

2) 이 증명과 그 앞의 보조정리는 미국 수학 월간지, 75, No. 5(1968), 508에서 인용한 콕스(R. H. Cox)의 "칸토어-베른슈타인 정리의 증명"이다.

Corollary. If A and B are sets such that card $A \leqq$ card B and card $B \leqq$ card A, then card $A =$ card B.

So far we know very little about transfinite cardinal numbers, because we have seen only two such cardinal numbers, card $\mathbb{N}$ and card $\mathbb{R}$. Naturally, we would like to know whether there are other transfinite cardinal numbers. The answer to this question is given in the next section—there is indeed an unlimited supply of distinct transfinite cardinal numbers.

Another important question is this: if m and n are two distinct finite cardinal numbers, then either $m<n$ or $n<m$; is this true for transfinite cardinal numbers? The answer is Yes, but the proof depends on a result of the next chapter and is therefore postponed until Theorem 4 of Chapter 6.

Exercise 5.2

Prove the following problems 1 through 6.

1. Let n be any finite cardinal number. Then, $n <$ card $\mathbb{N}$.

2. Let a be any transfinite cardinal number. Then, card $\mathbb{N} \leqq a$. Thus, card $\mathbb{N}$ is the smallest transfinite cardinal number.

3. Let A and B be sets. Then card $A \leqq$ card B if and only if there exists an injection $f: A \to B$.

4. If A, B, and C are sets such that
   (a) card $A \leqq$ card B and card $B \leqq$ card C, then card $A \leqq$ card C.
   (b) card $A <$ care B and card $B <$ card C, then card $A <$ card C.

5. If A and B are sets such that $A \subseteq B$, then card $A \leqq$ card B.

6. If A, B, and C are sets such that $A \subseteq B \subseteq C$, and $A \sim C$, then $A \sim B$.

따름정리* 임의의 집합 A, B에 대하여

$$\text{card}\,A \leqq \text{card}\,B \wedge \text{card}\,B \leqq \text{card}\,A \Longrightarrow \text{card}\,A = \text{card}\,B$$

지금까지 초한기수로서는 단지 $\text{card}\,\mathbb{N}$, $\text{card}\,\mathbb{R}$만을 보아왔다. 이 두 기수 이외의 초한기수가 더 존재하는지에 대해서는 다음 절에서 다룰 것이지만 초한기수는 한없이 많이 존재한다.

한편 임의의 서로 다른 두 유한기수 m, n에 대해서는 $m<n$ 아니면 $n<m$이 성립함은 이미 잘 알고 있다. 독자는 이런 것을 초한기수에 대해서도 주장할 수 있을지 궁금할 것이다. 결과는 긍정적이지만 그 증명은 제6장 정리 4의 따름정리(309쪽)에서 밝히기로 한다.

연습문제 5.2

다음 1~7을 증명하여라.

1. 임의의 유한기수 n에 대하여 $n<\text{card}\,\mathbb{N}$.

2. 임의의 초한기수 a에 대하여 $\text{card}\,\mathbb{N} \leqq a$. 다시 말하면 $\text{card}\,\mathbb{N}$은 초한기수 중 가장 작은 기수이다.

3. 집합 A, B에 대하여 단사 $f : A \to B$가 존재하면 그리고 그때에만

$$\text{card}\,A \leqq \text{card}\,B.$$

4.* 집합 A, B, C에 대하여

(a) $\text{card}\,A \leqq \text{card}\,B \wedge \text{card}\,B \leqq \text{card}\,C \Longrightarrow \text{card}\,A \leqq \text{card}\,C$,

(b) $\text{card}\,A < \text{card}\,B \wedge \text{card}\,B < \text{card}\,C \Longrightarrow \text{card}\,A < \text{card}\,C$.

5. 집합 A, B에 대하여 $A \subseteq B$이면 $\text{card}\,A \leqq \text{card}\,B$.

6. 집합 A, B, C에 대하여 $A \subseteq B \subseteq C \wedge A \sim C$이면 $A \sim B$.

7.** 폐구간 $[a, b]$에 대하여 $(a, b) \subset [a, b]$, $[a, b] \subset (a-1, b+1)$이면

$$\text{card}(a, b) = \text{card}[a, b] = \text{card}(a-1, b+1)$$

* 임의의 실수 a, b, c에 대해서도 따름정리 및 문제 4와 유사한 성질이 있다. 그것들을 적어라.

** 옮긴이가 추가한 것으로서 칸토어-베른슈타인의 정리를 이용하여라.

5.3 Cardinal Number of a Power Set— Cantor's Theorem

Let X be a set. Recall that the power set $\mathscr{P}(X)$ of X is the set of all subsets of X (Section 2, Chapter 2). Georg Cantor himself proved that card $X <$ card $\mathscr{P}(X)$. The significance of this theorem is that it furnishes a way of constructing a far-reaching sequence of new(transfinite) cardinal numbers. For example, we have

$$\text{card } \mathbb{R} < \text{card } \mathscr{P}(\mathbb{R}) < \text{card } \mathscr{P}(\mathscr{P}(\mathbb{R})) < \cdots$$

Theorem 2. Cantor's theorem If X is a set, then

$$\text{card } X < \text{card } \mathscr{P}(X).$$

[Proof] If $X = \phi$, then card $\phi = 0 < 1 =$ card $\mathscr{P}(\phi)$. Hence, it remains to prove the case where $X \neq \phi$. In this case, the function $g : X \longrightarrow \mathscr{P}(X)$ given by $\forall_{x \in X}\ g(x) = \{x\} \in \mathscr{P}(X)$, is injective. Thus, the set X is equipotent to the subset $\{\{x\} \mid x \in X\}$ of $\mathscr{P}(X)$ or, equivalently, card $X \leqq$ card $\mathscr{P}(X)$. From this, to show that card $X <$ card $\mathscr{P}(X)$, it is sufficient to show that X is not equipotent to $\mathscr{P}(X)$.

Assume on the contrary that there is a bijection $f : X \sim \mathscr{P}(X)$; our aim is to prove that this assumption leads to a contradiction. Consider that set $S = \{x \in X \mid x \notin f(x)\}$, which consists of those elements of X that are not contained in their images under f. Since $S \in \mathscr{P}(X)$ and $f : X \sim \mathscr{P}(X)$, there exists an element $e \in X$ such that $f(e) = S$. Either $e \in S$ or $e \notin S$.

Case 1. $e \in S$. It follows, by the definition of S, that $e \notin f(e)$; that is impossible, because $f(e) = S$ and $e \in S$.

Case 2. $e \notin S$. Since $f(e) = S$, we have $e \notin f(e)$. Consequently, by the definition of S, $e \in S$ and hence $e \in f(e)$. This again is impossible.

A contradiction has been obtained and the proof of Cantor's Theorem is complete.

5.3 멱집합의 기수—칸토어의 정리

집합 X의 멱집합 $\mathscr{P}(X)$는 X의 모든 부분집합의 집합(83쪽)임을 알고 있다. 칸토어는 부등식 card $X<$card $\mathscr{P}(X)$를 증명한 바 있다. 사실 이 부등식의 의의는 새로운 (초한)기수를 구성할 수 있는 길을 터 놓았다는 데에 있다. 이를테면 초한기수는 다음과 같이 얼마든지 존재한다.

$$\text{card}\,\mathbb{R}<\text{card}\,\mathscr{P}(\mathbb{R})<\text{card}\,\mathscr{P}(\mathscr{P}(\mathbb{R}),<\cdots$$

정리 2 칸토어의 정리 임의의 집합 X에 대하여 다음이 성립한다.

$$\text{card}\,X<\text{card}\,\mathscr{P}(X)^{*}$$

[증명] $X=\phi$이면 card $\phi=0$, card $\mathscr{P}(\phi)=1$. 그러므로 card $\phi<$card $\mathscr{P}(\phi)$.

$X\neq\phi$일 때 각 $x\in X$에 대하여 $g(x)=\{x\}$, $\{x\}\in\mathscr{P}(X)$로 정의된 함수 $g:X\to\mathscr{P}(X)$는 단사이다. 따라서 X는 $\mathscr{P}(X)$의 부분집합 $\{\{x\}\,|\,x\in X\}$과 대등하다. 다시 말하면 card $X<$card $\mathscr{P}(X)$.

그러므로 정리 2를 증명하려면 X가 $\mathscr{P}(X)$와 대등하지 않다는 것을 보이면 충분하다. 그러기 위하여 전단사 $f:X\sim\mathscr{P}(X)$가 존재한다고 가정함으로써 모순을 유도하기로 한다.

지금 집합 $S=\{x\in X\,|\,x\notin f(x)\}$, 즉 원소 $x\in X$로서 $\mathscr{P}(X)$의 원소, 다름아닌 집합 X의 부분집합인 $f(x)$에 속하지 않는 것의 집합 S를 생각하면 $S\in\mathscr{P}(X)$이고 $f:X\sim\mathscr{P}(X)$이므로 적당한 $e\in X^{**}$가 존재함으로써 $f(e)=S$. 여기서 $e\in S$, $e\notin S$ 두 경우를 살펴보자.

경우 1 $e\in S$ 집합 S의 정의에 따라 $e\notin f(e)$. 한편 $f(e)=S$이고 $e\in S$이므로 $e\in f(e)$. 이것은 $e\notin f(e)$에 모순이다.

경우 2 $e\notin S$ $f(e)=S$이므로 $e\notin f(e)$. 한편 S의 정의에 따라 $e\in S$이므로 $e\in f(e)$. 이것 또한 $e\notin f(e)$에 모순이다.

이상으로 증명이 끝난다.

---

* $X=\{a, b, c\}$일 때 그 부등식이 성립하는지 확인하여 보아라.

** 이러한 원소 e를 악원소라고 한다.

In view of Cantor's Theorem, a very natural question to arise was, Is there a cardinal number x such that

$$\text{card}\,\mathrm{N} < x < \text{card}\,\mathscr{P}(\mathrm{N})$$

This question, called the continuum problem, captured the attention of Cantor and other mathematicians for a long time. More about this problem will be found in Section 8.

Exercise 5.3

1. Show that there is no largest cardinal number.
2. Let A and B be sets. Prove that if $A \sim B$, then card $\mathscr{P}(A) = $ card $\mathscr{P}(B)$.
3. Let A be any denumerable set. Prove that the power set $\mathscr{P}(A)$ of A is nondenumerable.

5.4 Addition of Cardinal Numbers

There is already an arithmetic for finite cardinal numbers. For instance, if k and l are two finite cardinal numbers, the sum $k+l$ and the product kl have their traditional meanings. We now try to generalize these concepts to cover the transfinite cardinal numbers as well; that is, to develop an arithmetic that applies to all cardinal numbers, finite or transfinite, and that will preserve the traditional meanings and properties of the arithmetic of finite cardinal numbers.

Definition 2. Let a and b be cardinal numbers. The cardinal sum of a and b, denoted by $\boldsymbol{a+b}$, is the cardinal number card$(A \cup B)$, where A and B are disjoint sets such that card $A = a$ and card $B = b$.

칸토어의 정리의 관점에서 생각할 때 다음과 같은 지극히 자연스러운 의문이 생길 수 있다. 즉

$$\operatorname{card}\mathbb{N} < x < \operatorname{card}\mathscr{P}(\mathbb{N})$$

를 만족하는 기수 x가 존재하겠는가?

연속체문제로 불리는 이 의문은 오랜 세월 칸토어를 비롯해 여러 수학자들의 주목을 끌어왔다. 우리는 뒤의 8절(279쪽)에서 이 문제에 관한 더 많은 것을 다룰 것이다.

연습문제 5.3

1.* 기수로서 가장 큰 것이 존재하지 않는다. 이 사실을 확인하여라.

2.** 집합 A, B에 대하여 다음을 증명하여라.

$$A \sim B \Longrightarrow \operatorname{card}\mathscr{P}(A) = \operatorname{card}\mathscr{P}(B) \qquad \cdots\cdots ①$$

3. 임의의 가부번집합 A의 멱집합 $\mathscr{P}(A)$는 비가부번집합이다. 이것을 증명하여라.

5.4 기수의 덧셈

유한기수에 대한 셈은 이미 알고 있다. 이를테면 k, l이 두 유한기수일 때 합 $k+l$과 곱 kl이 지닌 뜻은 익히 알고 있다. 이러한 합과 곱의 연산을 초한기수로 확장하고자 다음과 같은 정의부터 내린다.

정의 2 서로 소인 두 집합 A, B의 기수를 각각 a, b라고 할 때 $\operatorname{card}(A \cup B)$를 a, b의 기수의 합이라 하고 이것을 $\boldsymbol{a}+\boldsymbol{b}$로 나타낸다.

---

* 이 물음엔 독자 누구나 곧바로 답할 수 있을 것이다.

** 증명에 앞서 A, B가 유한집합인 경우의 보기부터 들어보아라. 한편 ①의 역은 참일까?

To show that Definition 2 is well-defined, the reader should first observe that for any two cardinal numbers a and b (not necessarily distinct), by the rule C-1 of Section 1, there exist sets X and Y such that card $X=a$ and card $Y=b$, where the sets X and Y need not be disjoint.

But this causes no problem, since we may select $A=X\times\{0\}$ and $B=Y\times\{1\}$; then $A\sim X$, $B\sim Y$, and $A\cap B=\phi$. Thus, $a+b=$card$(A\cup B)$ and this is uniquely defined; for if there are other disjoint sets A_1 and B_1 such that $A_1\sim A$ and $B_1\sim B$, then by theorem 6 of Chapter 4, we have $(A_1\cup B_1)\sim(A\cup B)$ or, equivalently, card$(A_1\cup B_1)=$card$(A\cup B)$.

We have thus proved the following theorem :

Theorem 3. Let a and b be cardinal numbers. Then

(a) There exist disjoint sets A and B such that card $A=a$ and card $B=b$.

(b) If A, B, A_1, and B_1 are sets such that card $A=$card A_1, card $B=$card B_1, $A\cap B=\phi$, and $A_1\cap B_1=\phi$, then card$(A\cup B)=$card$(A_1\cup B_1)$.

The following example shows that Definition 2 agrees with the ordinary sum of two natural numbers when it is applied to two finite cardinal numbers.

Example 2 Find the cardinal sum $4+3$ of the two finite cardinal numbers 4 and 3.

[Solution] Since $N_7=N_4\cup\{5,6,7\}$, card $N_4=4$, card$\{5,6,7\}=3$, and the sets N_4 and $\{5,6,7\}$ are disjoint, we have

$$4+3=\text{card}\,(N_4\cup\{5,6,7\})=\text{card}\,N_7=7$$

which agrees with the ordinary sum of two natural numbers.

Since the union of sets is commutative and associative, we have the following corresponding properties about the cardinal sum.

정의 2로 하여금 잘 정의되었는지를 확인하려면 먼저 다음 사실을 살펴야 한다. 즉 임의의 두 기수 a, b(서로 다르지 않더라도 무방함)에 대하여 규칙 C-1(241쪽)에 따라 집합 X, Y가 존재함으로써 $\text{card}\,X=a$, $\text{card}\,Y=b$. 만일 X, Y가 서로 소가 아닌 경우엔 $A=X\times\{0\}$, $B=Y\times\{1\}$을 각각 택하면 $A\sim X$, $B\sim Y$이고 $A\cap B=\phi$. 그러므로 $a+b=\text{card}(A\cup B)$이고 이런 기수는 유일하게 정의된다. 그 까닭은 $A_1\sim A$, $B_1\sim B$이고 서로 소인 집합 A_1, B_1이 또 존재함으로써 정리 6(219쪽)에 따라 다음이 성립하기 때문이다[정리 6(219쪽) 및 C-4(241쪽)].

$$(A_1\cup B_1)\sim(A\cup B)\Longleftrightarrow \text{card}(A_1\cup B_1)=\text{card}(A\cup B).$$

이상으로 우리는 다음 정리마저 증명한 셈이다.

정리 3 임의의 기수 a, b에 대하여 다음이 성립한다.

(a) 서로 소인 집합, 말하자면 A, B가 존재함으로써

$$\text{card}\,A=a,\quad \text{card}\,B=b$$

(b) 집합 A, B, A_1, B_1에 대하여

$\text{card}\,A_1=\text{card}\,A$, $\text{card}\,B_1=\text{card}\,B$, $A\cap B=\phi$, $A_1\cap B_1=\phi$

이면

$$\text{card}(A_1\cup B_1)=\text{card}(A\cup B)$$

다음 예제는 정의 2를 두 유한기수에 적용할 때 그 결과는 우리가 알고 있는 두 자연수의 합과 일치한다는 것을 보여준다.

예제 2 유한기수 4, 3의 기수의 합 4+3을 구하여라.

[풀이] $\mathbb{N}_7=\mathbb{N}_4\cup\{5,6,7\}$, $\text{card}\,\mathbb{N}_4=4$, $\text{card}\{5,6,7\}=3$, $\mathbb{N}_4\cap\{5,6,7\}=\phi$

그러므로 $4+3=\text{card}(\mathbb{N}_4\cup\{5,6,7\})=\text{card}\,\mathbb{N}_7=7$

따라서 $4+3=7$

이것으로 기수의 합 4+3은 두 자연수 4, 3의 합과 같음을 알 수 있다.

집합에서의 연산 중 합집합에 관해서는 교환법칙, 결합법칙이 각각 성립하므로 우리는 다음 정리를 쉽게 얻는다.

Theorem 4. Let x, y, and z be arbitrary cardinal numbers. Then,

(a) $x+y=y+x$ (Commutativity).

(b) $(x+y)+z=x+(y+z)$ (Associativity).

Following Georg Cantor, the symbols $\aleph_0$ (read aleph-null; $\aleph$ is the first letter of the Hebrew alphabet) and c have been used respectively to denote the cardinal number of a denumerable set and the cardinal number of the continuum, where continuum means the set of real numbers. In other words,

$$\aleph_0 = \text{card}\,\mathbb{N} \quad \text{and} \quad c = \text{card}\,\mathbb{R}.$$

Example 3 Find the cardinal sum $\aleph_0 + \aleph_0$.

[Solution] Let $\mathbb{N}_e$ and $\mathbb{N}_o$ denote, respectively, the set of even natural numbers and the set of odd natural numbers. Then, $\mathbb{N}_e$ and $\mathbb{N}_o$ are disjoint denumerable subsets of the set $\mathbb{N}$, and their union is $\mathbb{N}$. Consequently, by Definition 2,

$$\aleph_0 + \aleph_0 = \text{card}\,\mathbb{N}_e + \text{card}\,\mathbb{N}_o = \text{card}\,(\mathbb{N}_e \cup \mathbb{N}_o) = \text{card}\,\mathbb{N} = \aleph_0$$

The result of Example 3 is distinctive property of the transfinite cardinal numbers; for finite cardinal numbers, $n+m=n$ is true only for $m=0$. The reader should prove, as an exercise, that $c+c=c$.

Example 4 Find the cardinal sum $\aleph_0 + c$.

[Solution] We have learned from Example 3, Section 2, Chapter 4, that the open interval $(0,1)$ and the set $\mathbb{R}$ of real numbers are equipotent. Hence, $\text{card}(0,1) = \text{card}\,\mathbb{R} = c$. Let $S = \mathbb{N} \cup (0,1)$. Then, since $\mathbb{N}$ and $(0,1)$ are disjoint, $\text{card}\,S = \aleph_0 + c$. On the other hand, since

$$\mathbb{R} \sim (0,1) \subset S \quad \text{and} \quad S \sim S \subset \mathbb{R},^{**}$$

by the Cantor–Bernstein Theorem (Theorem 1), we have $S \sim \mathbb{R}$. Therefore, $\aleph_0 + c = c$.

정리 4 임의의 기수 x, y, z에 대하여 다음이 성립한다.

(a) 교환법칙 $x+y=y+x$

(b) 결합법칙 $(x+y)+z=x+(y+z)$

게오르크 칸토어에 따라 두 기호 $\aleph_0$($\aleph$는 히브리어 자모의 첫째 글자로서 $\aleph_0$는 "알레프 널"이라고 읽음), c(독일어 자모의 셋째 글자로서 "체"라고 읽음, 책에 따라서는 $\aleph$)는 각각 가부번집합의 기수와 실수의 집합인 연속체의 기수를 나타내는 데 사용되어 왔다. 다시 말하면

$$\aleph_0 = \aleph_0 = \text{card}\,\mathbb{N}, \qquad c = \aleph = \text{card}\,\mathbb{R}.$$

예제 3 기수의 합 $\aleph_0+\aleph_0$을 구하여라.

[풀이] 짝수인 자연수의 집합 $\mathbb{N}_e$와 홀수인 자연수의 집합 $\mathbb{N}_o$는 각각 집합 $\mathbb{N}$의 가부번부분집합이고 $\mathbb{N}_e \cup \mathbb{N}_o=\mathbb{N}$, $\mathbb{N}_e \cap \mathbb{N}_o=\phi$이므로 정의 2(253쪽)에 따라

$$\aleph_0+\aleph_0=\text{card}\,\mathbb{N}_e+\text{card}\,\mathbb{N}_o=\text{card}\,(\mathbb{N}_e \cup \mathbb{N}_o)=\text{card}\,\mathbb{N}=\aleph_0$$

그러므로 $\aleph_0+\aleph_0=\aleph_0$*.

예제 3의 결과는 초한기수의 독특한 성질 중의 하나이다. 사실 유한기수 m, n에 대하여 $n+m=n$이 참인 것은 $m=0$일 때 뿐인데 반하여 초한기수 C에 대해선 $c+c=c$ [연습문제 5.4 문제 3(c)(259쪽)] 임을 증명하여 봄 직하다.

예제 4 기수의 합 $\aleph_0+c$를 구하여라.

[풀이] 예제 3(217쪽)에 따라 개구간 $(0, 1)$과 실수의 집합 $\mathbb{R}$은 대등하므로 $\text{card}\,(0, 1)=\text{card}\,\mathbb{R}=c$, 즉 $\text{card}\,(0, 1)=c$.

지금 $S=\mathbb{N}\cup(0, 1)$로 놓으면 $\mathbb{N}$, $(0, 1)$은 서로 소이므로 $\text{card}\,S=\aleph_0+c$. 한편 $(0, 1)\subset S$에 대하여 $\mathbb{R}\sim(0, 1)$이고 $S\subset\mathbb{R}$에 대하여 $S\sim S$이므로 정리 1(245쪽)에 따라 $\mathbb{R}\sim S$.

그러므로 $S\sim\mathbb{R}$. 따라서 $\aleph_0+c=c$.

* 앞으로도 초한기수의 덧셈의 경우 유한기수에서는 볼 수 없는 사례에 종종 직면할 것이다.

** 원서 맨 밑에서 셋째 행의 표현도 적당하지 않다.

Exercise 5.4

Prove the following problems 1 through 3.

1. Let x be any cardinal number. Then $x+0=x$.

2. Let x, y and z be cardinal numbers. Then
(a) $x+y=y+x$. (b) $(x+y)+z=x+(y+z)$.

3. Let n be an arbitrary finite cardinal number. Then
(a) $n+\aleph_0=\aleph_0$. (b) $n+c=c$.
(c) $c+c=c$

4. Let x, y, and z be cardinal numbers.
(a) Prove that if $x \leqq y$, then $x+z \leqq y+z$.
(b) Show, by an example, that part (a) above is not true if "$\leqq$" is replaced by "$<$".

5.5 Multiplication of Cardinal Numbers

We now define the multiplication of cardinal numbers in such a way that for finite cardinal numbers, the result agrees with the ordinary multiplication of nonnegative integers.

> ***Definition 3.*** For any cardinal numbers a and b, the cardinal product $\boldsymbol{ab}$ is defined to be the cardinal number of the Cartesian product $A \times B$, where card $A=a$ and card $B=b$.

To see that Definition 3 is independent of the choice of representatives A and B, let X and Y be sets such that $A \sim X$ and $B \sim Y$. Then, by Theorem 7 of Chapter 4, $A \times B \sim X \times Y$ and hence

$$\text{card}\,(A \times B) = \text{card}\,(X \times Y).$$

연습문제 5.4

다음 1~5(a)를 증명하여라.

1. 임의의 기수 x에 대하여 $x+0=x$

2. 임의의 기수 x, y, z에 대하여
(a) $x+y=y+z$ (b) $(x+y)+z=x+(y+z)$

3. 임의의 유한기수 n과 초한기수 $\aleph_0$, c에 대하여
(a) $n+\aleph_0=\aleph_0$ (b) $n+c=c$ (c) $c+c=c$

4.* $\aleph_0+\aleph_0+\cdots+\aleph_0+\cdots=\aleph_0$

5. 임의의 기수 x, y, z에 대하여

(a) $x\leqq y$ 이면 $x+z\leqq y+z$.

(b) (a)에서 "$\leqq$"를 "$<$"로 바꿔 놓은 명제는 참이 아니다. 그러한 보기(반례) 하나를 들어라.

5.5 기수의 곱셈

기수의 연산 중 곱은 유한기수의 곱의 경우를 염두에 두고 다룬다. 다시 말하면 우리에게 익숙한 자연수의 곱과 부합하도록 정의를 내리려는 것이다.

> **정의 3** 집합 A, B의 기수를 각각 a, b라고 할 때 $\mathrm{card}\,(A\times B)$를 a, b의 기수의 곱이라 하고 이것을 $\boldsymbol{ab}$로 나타낸다.

정의 3은 집합 A, B는 서로 독립적이다. 다시 말해서 A, B와는 각각 다른 집합 X, Y가 존재함으로써 $A\sim X$, $B\sim Y$일 때 정리 7(219쪽)에 따라 $A\times B\sim X\times Y$. 그러므로 다음 등식을 얻는다(241쪽 C-4).

$$\mathrm{card}\,(A\times B)=\mathrm{card}\,(X\times Y)$$

* 옮긴이가 보충한 것으로써 이장우, 《현대집합론의 원리》, 경문사, 1998, 예제 5.7 증명(177-178쪽) 참조.

It is also clear that this definition gives the right answer when a and b are finite cardinal numbers. Since we are all familiar with the multiplication of nonnegative integers, our main interest here is the product of transfinite cardinal numbers and the product of a finite and a transfinite cardinal number. First, let us list an easy consequence of Definition 3.

Theorem 5. Let x, y, and z be arbitrary cardinal numbers. Then,
(a) $xy=yx$ (Commutativity).
(b) $(xy)z=x(yz)$ (Associativity).
(c) $x(y+z)=xy+xz$ (Distributivity).

[Proof] Exercise.

Example 5 Let x be an arbitrary cardinal number. Evaluate :
(a) $1x$. (b) $0x$. (c) $\aleph_0 \aleph_0$.

[Solution] Let A be a set such that card $A=x$.
(a) Since the Cartesian product $\{1\}\times A$ is equipotent to the set A, we have $1x=x$.
(b) Since $\phi\times A=\phi$, we have $0x=0$.
(c) Since $\mathrm{N}\times\mathrm{N}\sim\mathrm{N}$(Theorem 10, Chapter 4), we have $\aleph_0 \aleph_0=\aleph_0$.

Example 6 Prove that $cc=c$, where $c=$card $\mathbb{R}$.

[Proof] Since the set $\mathbb{R}$ and the open unit interval $(0, 1)$ of real numbers have the same cardinal number c, to show that $cc\leqq c$, it is sufficient to show that there is an injection from the Cartesian product $(0, 1)\times(0, 1)$ to the interval $(0, 1)$. To this end, let us agree that each $x\in(0, 1)$ is expressed by its infinite decimal expansion so that, for example, the number $\frac{1}{2}$ will be $.4999\cdots$ but not $.5$. Thus we will have a unique expression for each number in $(0, 1)$.

물론 정의 3은 a, b가 유한기수일 때에도 적용된다. 앞으로는 두 초한기수의 곱, 유한기수와 초한기수의 곱을 차례로 알아 보겠거니와 정의 3이 지닌 결과 중 쉬운 것부터 적어보자.

정리 5* 임의의 기수 x, y, z에 대하여 다음이 성립하다.
(a) **교환법칙** $xy=yz$ (b) **결합법칙** $(xy)z=x(yz)$
(c) **분배법칙** $x(y+z)=xy+xz$

[증명] 독자의 연습문제로 돌린다.

예제 5 다음을 계산하여라. 여기서 x는 임의의 기수이다.
(a) $1x$ (b) $0x$ (c) $\aleph_0 \aleph_0$

[풀이] 기수가 x인 집합 하나를 A라고 한다.
(a) 데카르트 곱 $\{1\}\times A$와 A는 대등하므로 (왜?)
$\text{card}(\{1\}\times A)=\text{card}\,A$ 따라서 $1x=x$.
(b) $\phi\times A=\phi$ [예제 2(133쪽)]이므로 $\text{card}(\phi\times A)=\text{card}\,\phi$.
따라서 $0x=0$
(c) $\mathbb{N}\times\mathbb{N}\sim\mathbb{N}$ [정리 10(227쪽)]이므로 $\text{card}(\mathbb{N}\times\mathbb{N})=\text{card}\,\mathbb{N}$
따라서 $\aleph_0 \aleph_0=\aleph_0$

예제 6 $cc=c$임을 증명하여라. 여기서 $c=\text{card}\,\mathbb{R}$

[증명] 집합 $\mathbb{R}$와 실수의 단위개구간 $(0, 1)$의 기수는 모두 c와 같다. 먼저 $c \leqq c$임을 확인하려면 데카르트곱 $(0, 1)\times(0, 1)$로부터 구간 $(0, 1)$로의 단사가 존재한다는 것을 보이면 충분하다. 그러기 위하여 각 $x\in(0, 1)$을 무한소수로 나타내기로 한다. 이를테면 $1/2$은 0.5 대신 $0.4999\cdots$로 나타낸다. 이와같이 구간 $(0, 1)$에 속하는 각 수를 나타냄으로써 함수

$$f:(0, 1)\times(0, 1)\longrightarrow(0, 1)$$

을 다음과 같이 정의한다.

* 임의의 실수 a, b, c에 대해서도 정리 5와 같은 성질이 있음을 주목하자. (c)의 경우 왜 $(x+y)z=xz+yz$에 대해서는 언급이 없을까?

Now we leave it to the reader to verify that the function

$$f:(0,1)\times(0,1)\longrightarrow(0,1)$$

defined by $$f(.x_1x_2x_3\cdots,\ .y_1y_2y_3\cdots)=.x_1y_1x_2y_2\cdots$$

is injective. This completes the proof that $cc\leqq c$. The proof that $cc\geqq c$ is left to the reader.

Exercise 5.5

1. Prove Theorem 5.

2. Let x, y, and z be cardinal numbers such that $x\leqq y$. Prove that $xz\leqq yz$.

3. Prove or disprove the following statement:
   If x, y, and z are cardinal numbers such that $x<y$ and $z\neq 0$, then
   $$xz<yz.$$

4. Let n be a finite cardinal number. Prove that $n\aleph_0=\aleph_0$.

5. Let x and y be cardinal numbers. Prove that
   (a) If $xy=0$, then $x=0$ or $y=0$.
   (b) If $xy=1$, then $x=1$ and $y=1$.

6. Show that the function $f:(0,1)\times(0,1)\longrightarrow(0,1)$ defined by
   $$f(.x_1x_2x_3\cdots,\ .y_1y_2y_3\cdots)=.x_1y_1x_2y_1\cdots$$
   in the proof of Example 6 is bijective.

5.6 Exponentiation of Cardinal Numbers

Let a and b be cardinal numbers, finite or transfinite. In order to give a satisfactory meaning to b^a (*read*: ath power of b), we first examine the finite case: $2^3=2\cdot 2\cdot 2$ and, in general, $n^m=n\cdot n\cdots n$ (m factors).

$$f(0.x_1x_2x_3\cdots, 0.y_1y_2y_3\cdots)=0.x_1y_1x_2y_2x_3y_3\cdots \qquad \cdots①$$

그럴 경우 이 함수는 단사이다(과연?). 이로써 $cc \leqq c$. 한편 $cc \geqq c$(이것의 증명은 독자에게 돌린다). 그러므로 정리 1의 따름정리(249쪽)에 의하여 $cc=c$.

연습문제 5.5

1. 정리 5(261쪽)를 증명하여라.

2. 기수 x, y, z에 대하여 $x \leqq y$이면 $xz \leqq yz$.* 이것을 증명하여라.

3. 다음 명제를 증명하거나 또는 반증(거짓인 경우의 반례를 든다)하여라.
임의의 기수 x, y, z에 대하여

$$x<y \wedge z \neq 0 \Longrightarrow xz<yz$$

4. 임의의 유한기수 n에 대하여 $n\aleph_0=\aleph_0$.** 이것을 증명하여라.

5. 기수 x, y에 대하여 다음을 증명하여라.
(a) $xy=0$이면 $x=0$ 또는 $y=0$ (b) $xy=1$이면 $x=1$ 또는 $y=1$

6. 예제 6(261쪽)의 증명에서 함수 $f:(0, 1)\times(0, 1) \longrightarrow (0, 1)$을 위 ①과 같이 정의한 바 있다. 사실 이 함수는 전단사임을 밝혀라.

7.*** 임의의 자연수 n에 대하여 $\text{card}(\mathbb{R}^n)=c$임을 증명하여라.

5.6 기수의 거듭제곱

지금 기수 a, b에 대하여 b^a(" b의 a 거듭제곱"이라고 읽음)에 관한 정의를 어떻게 내릴 것인가? 우선

$2^3=2\cdot2\cdot2$와 유한기수 n에 대한 $n^m=n\cdot n\cdots n$(m개의 인수 곱)

인 경우를 살펴본다. 그러려면 일반 데카르트곱을 도입함으로써 초한기수의 경우로 넓혀나갈 길도 있겠으나 여기서는 그러한 과정 대신 다음과 같은 방법을 도입하기로 한다.

---

* 여기서도 왜 $zx \leqq zy$에 대해서는 언급이 없는지 생각해보자.

** 연습문제 5.4 문제 4(259쪽) 참조.

*** 옮긴이가 보충한 것으로 기호 $\mathbb{R}^n$에 대해서는 연습문제 5.7 문제 1(277쪽)을 참조하여라.

We could generalize this concept to the transfinite case by introducing "generalized Cartesian products," but there is an approach that works without reference to generalized Cartesian products. Let A be a set with m elements and B a set with n elements. How many functions are there from A to B (see Problem 9, Exercise 3.4)? Since each element of A has n choices for its image, and this selection of image may be made independently m times(once for each element of A), the answer is $n \cdot n \cdots n = n^m$. This concept is generalized as follows :

Definition 4. Let a and b be cardinl numbers with $a \neq 0$. Let A and B be sets such that card $A = a$ and card $B = b$. Denote the set of all functions from A to B by B^A. We define $\boldsymbol{b^a} = \textbf{card}\, \boldsymbol{B^A}$.

Before we can accept Definition 4, we need to verify that this definition is independent of the choice of representatives A and B. The following theorem is what is needed.

Theorem 6. Let A, B, X, and Y be sets such that $A \sim X$, $B \sim Y$. Then
$$B^A \sim Y^X.$$

[Proof] Let $g : A \sim X$ and $h : B \sim Y$ be bijections. Then, we define the function
$$\psi : B^A \longrightarrow Y^X$$
by $\psi(f) : X \longrightarrow Y$, where
$$\psi(f)(x) = h \circ f \circ g^{-1}(x)$$
for all $f \in B^A$.

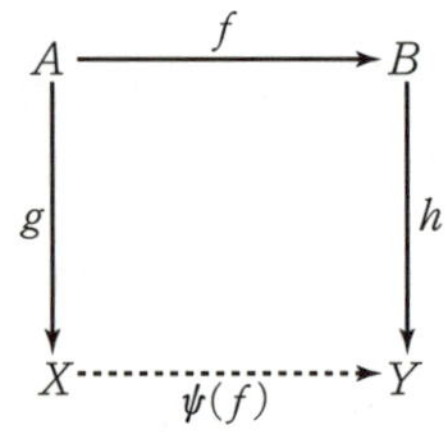

We leave it to the reader to prove that the function $\psi : B^A \longrightarrow Y^X$ is bijective.

두 집합 A, B의 원소가 각각 m, n개일 때 A에서 B로의 함수는 모두 몇 개 있는가?(연습문제 3.4 문제 9 참조). 이 경우 주어진 함수에 따른 A의 각 원소의 상으로서 n가지의 선택이 있으며, 각 선택에 대하여 A의 원소를 하나씩 택할 수 있는 방법은 독립적으로 m가지 있다. 그러므로 구하는 답은 $n \cdot n \cdots n = n^m$. 이 개념을 다음과 같이 확대하기로 한다.

정의 4* 주어진 기수 $a(a \neq 0)$, b에 대하여 $\operatorname{card} A = a(a \neq 0)$, $\operatorname{card} B = b$일 때 A에서 B로의 함수의 집합을 B^A로 나타내고 이것을 다음과 같이 정의한다.

$$\boldsymbol{b^a = \operatorname{card} B^A}$$

다음 정리에서 보듯이 정의 4*에서의 A, B는 대표로서 선택된 특별한 두 집합인 것은 아니다.

정리 6** 집합 A, B, X, Y에 대하여 $A \sim X$, $B \sim Y$이면 $B^A \sim Y^X$.

[증명] 가정에 따른 두 전단사를 $g : A \sim X$, $h : B \sim Y$로 놓고 함수 $\psi : B^A \to Y^X$를 각 $f \in B^A$에 대하여 $\psi(f) = h \circ f \circ g^{-1}$로 정의하자.
그러면

$$\begin{aligned}
\psi(f) = \psi(k) &\Rightarrow h \circ f \circ g^{-1} = h \circ k \circ g^{-1} \\
&\Rightarrow h^{-1} \circ (h \circ f \circ g^{-1}) = h^{-1} \circ (h \circ k \circ g^{-1}) \\
&\Rightarrow (h^{-1} \circ h) \circ (f \circ g^{-1}) = (h^{-1} \circ h) \circ (k \circ g^{-1}) \\
&\Rightarrow I_B \circ f \circ g^{-1} = I_B \circ k \circ g^{-1} \\
&\Rightarrow f \circ g^{-1} = k \circ g^{-1} \\
&\Rightarrow (f \circ g^{-1}) \circ g = (k \circ g^{-1}) \circ g \\
&\Rightarrow f \circ (g^{-1} \circ g) = k \circ (g^{-1} \circ g) \\
&\Rightarrow f \circ I_A = k \circ I_A \\
&\Rightarrow f = k
\end{aligned}$$

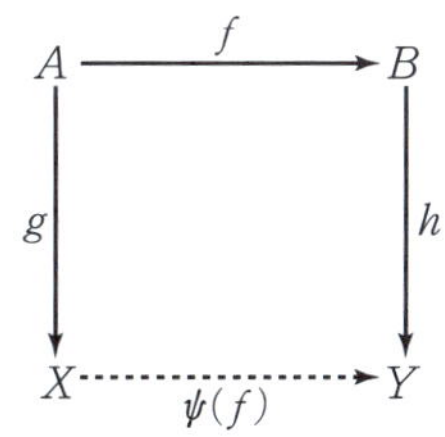

즉 $\psi(f) = \psi(k) \Rightarrow f = k$ 그러므로 ψ는 단사이다.
한편 임의의 $j \in Y^X$에 대하여 $h^{-1} \circ j \circ g \in B^A$이며

$$\psi(h^{-1} \circ j \circ g) = h \circ (h^{-1} \circ j \circ g) \circ g^{-1} = j$$

즉 $\psi(h^{-1} \circ j \circ g) = j$. 그러므로 ψ는 전사이다.
이상으로 함수 $\psi : B^A \to Y^X$는 전단이므로 $B^A \sim Y^X$.

* $A = \{x_1, x_2\}$, $B = \{y_1, y_2, y_3\}$일 때의 B^A, $\operatorname{card} B^A$을 각각 구하여라.

** 이 정리에 대하여 A, B, X, Y가 모두 유한집합인 경우의 보기부터 들어보아라.

Example 7 Let A be a set. Compare the cadinal numbers card $\mathscr{P}(A)$ and $2^{\text{card}\,A}$.

[Solution] Let $B=\{0, 1\}$. We assign to each subset D of A the characteristic function $\chi_D : A \longrightarrow B$ defined in Example 8, Chapter 3. The function from $\mathscr{P}(A)$ to B^A which sends D to χ_D is bijective (Prove it!). Thus, the sets $\mathscr{P}(A)$ and B^A have the same cardinal number; that is,

$$\text{card}\,\mathscr{P}(A)=\text{card}\,B^A=2^{\text{card}\,A}.$$

Theorem 7. Let a, x, and y be cardinal numbers. Then, $a^x a^y = a^{x+y}$.

[Proof] Let A, X, and Y be sets such that card $A=a$, card $X=x$, card $Y=y$, and $X \cap Y=\phi$. Then, by Definition 2, card $(X \cup Y)= x+y$. It is sufficient to show that the sets $A^X \times A^Y$ and $A^{X \cup Y}$ are equipotent. To this end, we assign to each pair (f, g) of functions, $f \in A^X$ and $g \in A^Y$, the function $f \cup g \in A^{X \cup Y}$ (see theorem 8, Chapter 3). We leave it to the reader to verify that this assignment establishes an equipotence between the sets $A^X \times A^Y$ and $A^{X \cup Y}$. Hence,

$$a^x a^y = a^{x+y}.$$

Theorem 8. Let x, y, and z be cardinal numbers. Then $(z^y)^x = z^{yx}$.

[Proof] Let X, Y, and Z be sets with cardinal numbers x, y, and z, respectively. According to Definition 4, the theorem is proved if we establish that $$Z^{Y \times X} \sim (Z^Y)^X$$

Before showing this equipotence, we first need a notational convention: For a given function $f : Y \times X \longrightarrow Z$ and a given element $a \in X$, there exists a function $f^a : Y \longrightarrow Z$ defined by $f^a(b)=f(b, a)$ for all $b \in Y$. We leave it to the reader to show that the function $$\psi : Z^{Y \times X} \longrightarrow (Z^Y)^X$$

which assigns to each $f \in Z^{Y \times X}$ the function $e_f \in (Z^Y)^X$ given by $e_f(a)=f^a$ for all $a \in X$ is a bijection.

예제 7 집합 A에 대한 두 기수 card $\mathscr{P}(A)$, $2^{\text{card}A}$는 같다. 이를 밝혀라.

[풀이] $B=\{0,1\}$로 놓고 임의의 함수 $f\in B^A$에 대하여 $D=\{x \mid f(x)=1\}$이라면 f는 집합 A의 부분집합 D의 특성함수 $\chi_D : A \longrightarrow \{0,1\}$(163쪽 예제 8)과 같다. 여기서 각 $D\in\mathscr{P}(A)$에 χ_D가 대응하는 함수는 일대일 대응(이것을 증명하여라!)이므로 $\mathscr{P}(A)\sim B^A$. 따라서

$$\text{card } \mathscr{P}(A) = \text{card } B^A = 2^{\text{card } A}, \quad \text{즉} \quad \text{card } \mathscr{P}(A) = 2^{\text{card } A}$$

정리 7* 임의의 기수 a, x, y에 대하여 $a^x a^y = a^{x+y}$

[증명] 집합 A, X, Y에 대하여 card $A=a$, card $X=x$, card $Y=y$라고 놓고 $X\cap Y=\phi$라고 가정하면 정의 2(253쪽)에 따라 card$(X\cup Y)=x+y$. 이제 집합 $A^X\times A^Y$, $A^{X\cup Y}$는 대등하다는 것을 보이면 충분하다. 여기서 $f\in A^X$, $g\in A^Y$의 각 순서쌍 (f,g)에 함수 f, g의 합 $f\cup g\in A^{X\cup Y}$를 대응(167쪽 정리 8)시키면 이 대응에 따라 집합 $A^X\times A^Y$, $A^{X\cup Y}$는 대등하다(이것의 증명도 하여라). 따라서 $a^x a^y = a^{x+y}$*.

정리 8* 임의의 기수 x, y, z에 대하여 $(z^y)^x = z^{yx}$

[증명] X, Y, Z를 차례로 기수가 x, y, z인 집합이라고 할 때 $Z^{Y\times X}\sim(Z^Y)^X$인 것을 밝히기만 하면 정의 4(265쪽)에 따라 이 정리는 증명된다. 먼저 $Z^{Y\times X}$와 $(Z^Y)^X$가 대등하다는 것을 보이기에 앞서 다음과 같이 기호에 대한 약속부터 한다.

주어진 함수 $f : Y\times X \longrightarrow Z$와 주어진 원소 $a\in X$에 대한 함수

$f^a : Y \longrightarrow Z$를 $\qquad \forall_{b\in Y}\, f^a(b)=f(b,a)$

로 정의한다. 한편 함수 $\qquad \psi : Z^{Y\times X} \longrightarrow (Z^Y)^X$

로 하여금 각 함수 $f\in Z^{Y\times X}$에 $\forall_{a\in X}\, e_f(a)=f^a$로서 주어진 함수 $e_f\in(Z^Y)^X$가 대응하도록 정의하면 ψ는 전단사이다(과연?). 그러므로 정의 4에 따라 $\qquad (z^y)^x = z^{yx}$.

* a, x, y, z이 모두 유한기수일 때 정리 7, 8에 대한 보기를 각각 하나씩 들어라. 사실 이 두 정리는 초한기수인데에 그 뜻이 있거니와 특히 정리 8에서의 네 함수 f, f^a, e_f, ψ의 역할을 익히는 데 힘이 들 것이다.

Recall that the A-projection $p_A : A\times B \to A$ is a function that assigns a to each ordered pair $(a, b)\in A\times B$; the B-projection $p_B : A\times B \to B$ is similarly defined (see Problem 8, Exercise 3.5).

Theorem 9. Let a, b and x be cardinal numbers. Then

$$(ab)^x = a^x b^x.$$

[Proof] Let A, B and X be sets with cardinal numbers a, b and x, respectively. The function $\psi : (A\times B)^X \to A^X \times B^X$ which pairs each $f : X \to A\times B$ with the function $(p_A \circ f,\ p_B \circ f)$ in $A^X \times B^X$ is bijective (Prove it!). Hence, by Definition 4, $(ab)^x = a^x b^x$.

Recall that the symbols $\aleph_0$ and c denote the cardinal numbers of the sets $\mathbb{N}$ and $\mathbb{R}$, respectively, and that $\mathbb{Q}\sim\mathbb{N}$ (see Example 5, Chapter 4) and $(0, 1)\sim\mathbb{R}$ (see Example 3, Chapter 4). Thus, $\aleph_0$ is the cardinal number of $\mathbb{Q}$ and c is the cardinal number of the interval $(0, 1)$.

Theorem 10. $2^{\aleph_0} = c$.

[Proof] We shall prove this in two steps, showing first that $c \leqq 2^{\aleph_0}$ and then that $2^{\aleph_0} \leqq c$.

Consider the function $f : \mathbb{R} \to \mathscr{P}(\mathbb{Q})$ defined by

$$f(a) = \{x\in\mathbb{Q} \mid x < a\}, \text{ for all } a\in\mathbb{R}$$

This function is injective: If a, b are two distinct real numbers, then there exists a rational number r such that $a < r < b$.[3] Then, $r\in f(b)$ but $r\notin f(a)$, and hence f is injective. This proves, by using the result of Problem 3, Exercise 5.2, and Example 7, that

$$c \leqq \text{card}\, \mathscr{P}(\mathbb{Q}) = 2^{\aleph_0}$$

[3] Because the rational numbers are a dense subset of the set $\mathbb{R}$.

이제 A-사영 $p_A : A\times B \to A$는 각 순서쌍 $(a, b)\in A\times B$에 $a\in A$가 대응하는 함수이고 B-사영 $p_B : A\times B \to B$도 같은 방법으로 정의된 함수임을 상기(177쪽 연습문제 3.5 문제 8)함으로써 다음 정리를 얻는다.

정리 9 임의의 기수 a, b, x에 대하여 $(ab)^x = a^x b^x$.

[증명] A, B, X를 차례로 기수가 a, b, x인 집합이라고 할 때 각 함수 $f : X \to A\times B$에 순서쌍 $(p_A \circ f,\ p_B \circ f)\in A^X \times B^X$가 대응하는 함수

$$\psi : (A\times B)^X \to A^X \times B^X$$

는 전단사이다(이것을 증명하여라). 그러므로 정의 4(265쪽)에 따라

$$(ab)^x = a^x b^x$$

여기서 우리는 다음 몇 가지를 되새겨보기로 한다. 즉

card $\mathbb{N} = \aleph_0$, card $\mathbb{R} = c$ 그런데 $\mathbb{N} \sim \mathbb{Q}$(제 4장 예제 5)이므로 card $\mathbb{Q} = \aleph_0$, 한편 $(0, 1) \sim \mathbb{R}$(제 4장 예제 3)이므로 card$(0, 1) = c$.

정리 10 $2^{\aleph_0} = c$.

[증명] 두 단계 $c \leq 2^{\aleph_0}$, $2^{\aleph_0} \leq c$로 확인한다. 먼저 함수 $f : \mathbb{R} \to \mathscr{P}(\mathbb{Q})$를

$$\forall_{a\in\mathbb{R}}\, f(a) = \{x\in\mathbb{Q} \mid x < a\}$$

로 정의하면 이 함수는 단사이다. 왜냐하면 두 실수 a, b에 대하여 이를테면, $a < b$일 때 적당한 유리수 r이 존재함으로써 $a < r < b$.[3] 그러면 $r\in f(b)$이지만 $r\notin f(a)$, 즉 $a\neq b \Rightarrow f(a)\neq f(b)$*. 그러므로 f는 단사이다. 따라서 249쪽 문제 3과 267쪽 예제 7에 의하여

$$c \leq \mathscr{P}(\mathbb{Q}) = 2^{\aleph_0} \quad \text{곧} \quad c \leq 2^{\aleph_0}.$$

---

[3] 왜냐하면 유리수의 집합 $\mathbb{Q}$는 실수의 집합 $\mathbb{R}$의 진부분집합으로서 조밀하기 때문이다. 여기서 $\mathbb{Q}$, $\mathbb{R}$에 대한 조밀성에 대한 차이점도 생각해볼 만하다.

* 이 함의의 대우를 말하여라. 사실 그것으로 f가 단사임을 정의한 바 있다[정의 10(181쪽) 참조].

To prove the reverse inequality, let $\psi : \{0, 1\}^{\mathrm{N}} \longrightarrow \mathrm{R}$ be the function given by

$$\psi(f) = .f(1)f(2)f(3)\cdots$$

where $f \in \{0,1\}^{\mathrm{N}}$. Note that $\psi(f)$ is a decimal number (consisting of 0's and 1's). If $f, g \in \{0,1\}^{\mathrm{N}}$ and $f \neq g$, then $\psi(f) \neq \psi(g)$ because the decimals which define $\psi(f)$ and $\psi(g)$ are distinct. Therefore, $\psi : \{0,1\}^{\mathrm{N}} \longrightarrow \mathrm{R}$ is injective, and hence $2^{\aleph_0} \leqq c$.

Corollary. $\aleph_0 < c$.

[Proof] By Cantor's Theorem (Theorem 2) and by the result of Example 7, we have

$$\aleph_0 < \mathrm{card}\ \mathscr{P}(\mathrm{N}) = 2^{\mathrm{card N}} = 2^{\aleph_0} = c$$

Exercise 5.6

Prove the following problems 1 through 12.

1. The function $\psi : B^A \longrightarrow Y^X$ in the proof of the Theorem 6 is bijective.
2. Let a be an arbitrary cardinal number. Then $a^0 = 1$, $a^1 = a$, $1^a = 1$, and $0^a = 0$ if $a \neq 0$.
3. $2^a > a$ for any cardinal number a.
4. Let a, b, x and y be cardinal numbers such that $a \leqq b$ and $x \leqq y$. Then,
$$a^x \leqq b^y.$$
5. $n^{\aleph_0} = c = \aleph_0^{\aleph_0}$ for any finite cardinal number $n \geqq 2$.
6. $c^{\aleph_0} = c = c^n$ for any finite cardinal number $n \geqq 1$.
7. Let C denote the set of all complex numbers. Then, card $\mathrm{C} = c$.
8. $\aleph_0 c = c$.
9. That the function from $\mathscr{P}(A)$ to $\{0,1\}^A$ which pairs each D in $\mathscr{P}(A)$ with χ_D is bijective.

한편 $2^{\aleph_0} \leqq c$를 밝히기 위하여 함수 $\psi : \{0, 1\}^{\mathbb{N}} \longrightarrow \mathbb{R}$를

$$\forall_{f \in \{0,1\}^{\mathbb{N}}} \ \psi(f) = 0.f(1)f(2)f(3)\cdots$$

으로 정의한다. 그러면 $\psi(f)$는 0과 1로 이루어진 소수와 같다. 여기서 $\{0, 1\}^{\mathbb{N}}$의 원소 f, g에 대하여 $f \neq g$이면 $\psi(f) \neq \psi(g)$. 즉 $\psi(f)$와 $\psi(g)$로 정의된 소수는 서로 다르므로 함수 $\psi : \{0, 1\}^{\mathbb{N}} \longrightarrow \mathbb{R}$은 단사이다.

따라서 $2^{\aleph_0} \leqq c$. 이상으로 $2^{\aleph_0} = c$

따름정리 $\aleph_0 < c$.

[증명] 칸토어의 정리(251쪽)에 따라 $\mathrm{card}\,\mathbb{N} < \mathrm{card}\,\mathscr{P}(\mathbb{N})$, 즉 $\aleph_0 < \mathrm{card}\,\mathscr{P}(\mathbb{N})$.

한편 예제 7(267쪽)에 따라 $\mathrm{card}\,\mathscr{P}(\mathbb{N}) = 2^{\mathrm{card}\,\mathbb{N}}$, 즉 $\mathrm{card}\,\mathscr{P}(\mathbb{N}) = 2^{\aleph_0}$. 또 정리 10에 따라 $2^{\aleph_0} = c$. 결국 $\aleph_0 < c$

연습문제 5.6

다음 1~12를 증명하여라.

1. 정리 6[이것의 증명은 끝이 났음(265쪽 참조)]의 증명에서 함수 $\psi : B^A \to Y^A$는 전단사이다.

2. 임의의 기수 a에 대하여 $a^0 = 1$, $a^1 = a$, $1^a = 1$, 또 $a \neq 0$이면 0^a* $= 0$

3. 임의의 기수 a에 대하여 $a < 2^a$

4. 기수 a, b, x, y에 대하여 $a \leqq b \wedge x \leqq y$이면 $a^x \leqq b^y$

5. 임의의 유한기수 $2 \leqq n$에 대하여 $n^{\aleph_0} = c = \aleph_0^{\aleph_0}$

6. 임의의 유한기수 $1 \leqq n$에 대하여 $c^{\aleph_0} = c = c^n$

7. 복소수의 집합 $\mathbb{C}$에 대하여 $\mathrm{card}\,\mathbb{C} = c$

8. $\aleph_0 c = c$

9. 집합 $\mathscr{P}(A)$에서 $\{0, 1\}^A$로의 함수를 각 $D \in \mathscr{P}(A)$에 $\chi_D \in \{0, 1\}^A$가 대응하도록 정의된 이 함수는 전단사이다.

* 독자 누구나 a^0은 익숙하겠지만 기호 0^a에 대해서는 낯설지도 모른다.

10. Let A, X and Y be sets such that X and Y are disjoint. Then, the function from $A^X \times A^Y$ to $A^{X \cup Y}$ which pairs each (f, g) in $A^X \times A^Y$ with $f \cup g$ in $A^{X \cup Y}$ is bijective.

11. The function $\psi : Z^{X \times Y} \longrightarrow (Z^Y)^X$ in the proof of Theorem 8 is bijective.

12. The function $\psi : (A \times B)^X \longrightarrow A^X \times B^X$ in the proof of Theorem 9 is bijective.

5.7 Further Examples of Cardinal Arithmetic

In Example 6, it was proved directly that $cc = c$. Now, using Theorem 10, $2^{\aleph_0} = c$, a shorter proof can be given.

Example 8 Prove that $cc = c$ by using Theorem 10 (cf. Example 6).

[Proof] It follows from Theorem 7, 10 and Example 3, $\aleph_0 + \aleph_0 = \aleph_0$, that

$$cc = 2^{\aleph_0} 2^{\aleph_0} = 2^{\aleph_0 + \aleph_0} = 2^{\aleph_0} = c$$

Example 9 Compare the cardinal number of the set $\{f \mid f : \mathbb{R} \longrightarrow \mathbb{R}\}$ of all functions from $\mathbb{R}$ to $\mathbb{R}$ and the cardinal number c of $\mathbb{R}$.

[Solution] We have

$\text{card}\{f \mid f : \mathbb{R} \longrightarrow \mathbb{R}\} = c^c$	Def. 4
$= (2^{\aleph_0})^c$	Th. 10
$= 2^{\aleph_0 c}$	Th. 8
$= 2^c$	Prob. 8, Ex. 5.6
$> c$	Example 7, Th. 2

Therefore, $\text{card}\{f \mid f : \mathbb{R} \longrightarrow \mathbb{R}\} > \text{card}\, \mathbb{R}$.

10. 집합 A, X, Y에 대하여 $X\cap Y=\phi$일 때 $A^X\times A^Y$에서 $A^{X\cup Y}$로의 함수를 각 $(f, g)\in A^X\times A^Y$에 $f\cup g\in A^{X\cup Y}$가 대응하도록 정의하면 이 함수는 전단사이다.

11. 정리 8의 증명에서 함수 $\psi: Z^{Y\times X}\to(Z^Y)^X$는 전단사이다.

12. 정리 9의 증명에서 함수 $\psi:(A\times B)^X\to A^X\times B^X$는 전단사이다.

5.7 또다른 기수의 셈

예제 6에서는 $cc=c$임을 직접 증명한 바 있다. 여기서 정리 10, 즉 $2^{\aleph_0}=c$를 이용하면 그 증명을 더 간단히 할 수 있다.

예제 8 정리 10을 이용하여 $cc=c$임을 증명하여라(예제 6과 비교하여라).

[증명] 정리 7과 10 그리고 예제 3 $(\aleph_0+\aleph_0=\aleph_0)$으로부터 다음을 얻는다.

$$cc=2^{\aleph_0}2^{\aleph_0}=2^{\aleph_0+\aleph_0}=2^{\aleph_0}=c$$

그러므로 $$cc=c$$

예제 9 집합 $\mathbb{R}$에서 그 자신으로의 함수 집합

$$\{f\mid f:\mathbb{R}\to\mathbb{R}\}$$

의 기수를 $\mathbb{R}$의 기수 c와 비교하여라.

[풀이]

$\operatorname{card}\{f\mid f:\mathbb{R}\to\mathbb{R}\}=\operatorname{card}\mathbb{R}^{\mathbb{R}}$	정의 4
$=(\operatorname{card}\mathbb{R})^{\operatorname{card}\mathbb{R}}$	정의 4
$=c^c$	$\operatorname{card}\mathbb{R}=c$
$=(2^{\aleph_0})^c$	정리 10
$=2^{\aleph_0 c}$	정리 8
$=2^c$	연습문제 5.6 문제 8
$>c$	예제 7, 정리 2

그러므로 $$\operatorname{card}\{f\mid f:\mathbb{R}\to\mathbb{R}\}<c^*$$

* 이와 같이 $\operatorname{card}\{f\mid f:\mathbb{R}\to\mathbb{R}\}$은 $\operatorname{card}\mathbb{R}$보다 더 크다. 이 시점에서 독자는 $\operatorname{card}\{f\mid f:\mathbb{R}\to\mathbb{R}\}$보다 더 큰 기수를 한두 개 들 수 있을까?

Example 10 Let $C(\mathbb{R}, \mathbb{R})$ and $C(\mathbb{Q}, \mathbb{R})$ be the sets of continuous real-valued functions with domain $\mathbb{R}$ and domain $\mathbb{Q}$, respectively. Let $K(\mathbb{R}, \mathbb{R})$ be the set of all real-valued constant functions with the domain $\mathbb{R}$. Prove that

$$\text{card } C(\mathbb{R}, \mathbb{R}) = \text{card } C(\mathbb{Q}, \mathbb{R}) = \text{card } K(\mathbb{R}, \mathbb{R}) = c$$

[Proof][4)] To each function $f : \mathbb{R} \to \mathbb{R}$, there corresponds a function $f \mid \mathbb{Q} : \mathbb{Q} \to \mathbb{R}$ defined by $(f \mid \mathbb{Q})(x) = f(x)$ for all $x \in \mathbb{Q}$. The function $f \mid \mathbb{Q}$ is called the restrictions of f to $\mathbb{Q}$. Hence, there is a natural function

$$\psi : C(\mathbb{R}, \mathbb{R}) \longrightarrow C(\mathbb{Q}, \mathbb{R})$$

which takes each $f \in C(\mathbb{R}, \mathbb{R})$ to its restriction $f \mid \mathbb{Q}$ on $\mathbb{Q}$. It is clear that the restriction of a continuous function is continuous. Therefore,

$$\psi : C(\mathbb{R}, \mathbb{R}) \longrightarrow C(\mathbb{Q}, \mathbb{R})$$

is a well-defined function.

It follows from the density property of the rational numbers in the real numbers that for each real number x there is a sequence $\{ x_n \mid n \in \mathbb{N} \}$ of rational numbers such that $\lim_{n \to \infty} x_n = x$

Consequently, if any two continuous functions $f, g : \mathbb{R} \to \mathbb{R}$ have the property that $f(x') = g(x')$ for all $x' \in \mathbb{Q}$, then $f(x) = g(x)$ for all $x \in \mathbb{R}$. In other words, the function $\psi : C(\mathbb{R}, \mathbb{R}) \longrightarrow C(\mathbb{Q}, \mathbb{R})$ is injective. Therefore, we have

$$\begin{aligned} \text{card } C(\mathbb{R}, \mathbb{R}) &\leqq \text{card } C(\mathbb{Q}, \mathbb{R}) \\ &\leqq \text{card } \mathbb{R}^{\mathbb{Q}} \\ &= c^{\aleph_0} \\ &= (2^{\aleph_0})^{\aleph_0} \\ &= 2^{(\aleph_0 \aleph_0)} \\ &= 2^{\aleph_0} \\ &= c \end{aligned}$$

by Theorems 8 and 10.

4) The proof may be omitted at the discretion of the instructor.

예제 10 정의역이 ℝ, ℚ인 연속함수의 집합을 각각 $C(\mathbb{R},\mathbb{R})$, $C(\mathbb{Q},\mathbb{R})$이라 하자. 특히 정의역과 공역이 ℝ인 상수함수의 집합을 $K(\mathbb{R},\mathbb{R})$이라 할 때 다음이 성립한다. 이것을 증명하여라.

$$\operatorname{card} C(\mathbb{R},\mathbb{R})=\operatorname{card} C(\mathbb{Q},\mathbb{R})=\operatorname{card} K(\mathbb{R},\mathbb{R})=c$$

[증명][4] 각 함수 $f:\mathbb{R}\longrightarrow\mathbb{R}$에 $\forall_{x\in\mathbb{Q}}(f\mid\mathbb{Q})(x)=f(x)$로 정의된 함수 $f\mid\mathbb{Q}:\mathbb{Q}\longrightarrow\mathbb{R}$를 대응시킬 수 있다. 이러한 함수 $f\mid\mathbb{Q}$를 f의 ℚ에 대한 **제한**이라고 한다. 이와 같이 각 $f\in C(\mathbb{R},\mathbb{R})$에 그 제한 $f\mid\mathbb{Q}$이 대응되는 함수

$$\psi: C(\mathbb{R},\mathbb{R})\longrightarrow C(\mathbb{Q},\mathbb{R}) \qquad \cdots\cdots①$$

을 쉽게 생각할 수 있다. 여기서 연속함수 f의 제한 $f\mid\mathbb{Q}$ 또한 연속함수(왜?)이므로 ψ는 잘 정의된 함수이다.

유리수의 조밀성에 따라 각 실수 x에 대하여 유리수의 무한수열

$$r_1,\ r_2,\ r_3,\ \cdots,\ r_{n-1},\ r_n,\ \cdots$$

이 존재함으로써 극한 $\lim_{n\to\infty} r_n=x$를 얻는다. 따라서 임의의 두 연속 함수 $f,\ g:\mathbb{R}\longrightarrow\mathbb{R}$에 대하여

$$\forall_{x'\in\mathbb{Q}}\, f(x')=g(x')\text{이면} \quad \forall_{x\in\mathbb{R}}\, f(x)=g(x).$$

다시 말하면 함수 ①은 단사이다. 그러므로 정리 8, 10에 따라 다음을 얻는다.**

$$\begin{aligned}\operatorname{card} C(\mathbb{R},\mathbb{R}) &\leqq \operatorname{card} C(\mathbb{Q},\mathbb{R})\\ &\leqq \operatorname{card} \mathbb{R}^{\mathbb{Q}}\\ &= c^{\aleph_0}\\ &= (2^{\aleph_0})^{\aleph_0}\\ &= 2^{(\aleph_0\aleph_0)}\\ &= 2^{\aleph_0}\\ &= c\end{aligned}$$

따라서 $\qquad \operatorname{card} C(\mathbb{R},\mathbb{R})\leqq c$

[4] 이 증명은 담당 교수의 재량에 따라 생략해도 무방하다.

* 독자는 이런 사실을 미적분학에서의 무리수의 정의로부터 익혔을 것이다.

** 정리 8, 10이 각각 증명과정 어디에 이용되었는지 하나하나 짚어보아라.

Now, consider the set $K(\mathbb{R}, \mathbb{R})$ of all real-valued constant functions with the domain $\mathbb{R}$. Since to each real number a, there exists a constant function $f_a : \mathbb{R} \longrightarrow \mathbb{R}$ defined by $f_a(\mathbb{R})=\{a\}$, we have

$$\text{card } K(\mathbb{R}, \mathbb{R})=c$$

Since each constant function $f_a : \mathbb{R} \longrightarrow \mathbb{R}$ is continuous, we have $K(\mathbb{R}, \mathbb{R}) \subset C(\mathbb{R}, \mathbb{R})$. Therefore,

$$c=\text{card } K(\mathbb{R}, \mathbb{R}) \leqq \text{card } C(\mathbb{R}, \mathbb{R})$$

which combined with the inequality obtained in the last paragraph, gives

$$c=\text{card } K(\mathbb{R}, \mathbb{R}) \leqq \text{card } C(\mathbb{R}, \mathbb{R}) \leqq \text{card } C(\mathbb{Q}, \mathbb{R}) \leqq c$$

This completes the proof that card $C(\mathbb{R}, \mathbb{R})$, card $C(\mathbb{Q}, \mathbb{R})$, and card $K(\mathbb{R}, \mathbb{R})$ are all equal to c.

The result of Example 10 indicates that the constant functions are as "numerous" as the continuous functions. This is another illustration of the curious properties of infinite sets.

Example 11 Find the cardinal number of the set $D(\mathbb{R}, \mathbb{R})$ of all differentiable real-valued functions of a real variable.

[Solution] Since each constant function is differentiable and each differentiable function is continuous, we have

$$K(\mathbb{R}, \mathbb{R}) \subseteq D(\mathbb{R}, \mathbb{R}) \subseteq C(\mathbb{R}, \mathbb{R})$$

By Example 10, we have

$$c=\text{card } K(\mathbb{R}, \mathbb{R}) \leqq \text{card } D(\mathbb{R}, \mathbb{R}) \leqq \text{card } C(\mathbb{R}, \mathbb{R})=c$$

Therefore, card $D(\mathbb{R}, \mathbb{R})=c$.

Exercise 5.7

1. Show that the n-dimensional space $\mathbb{R}^n=\mathbb{R}\times\mathbb{R}\times\cdots\times\mathbb{R}$($n$ factors) contains "just as many" points as the unit open interval $(0, 1)$.

한편 정의역 $\mathbb{R}$에서의 함숫값이 실수인 상수함수의 집합 $K(\mathbb{R}, \mathbb{R})$에 대하여

$$\forall_{a\in\mathbb{R}}\, f_a(\mathbb{R})=\{a\}$$

로 정의된 상수함수 $f_a:\mathbb{R}\longrightarrow\mathbb{R}$이 존재한다.

그러므로 $\text{card}\, K(\mathbb{R}, \mathbb{R})=c$.

여기서 각 상수함수 f_a는 연속이다.

따라서 $$K(\mathbb{R}, \mathbb{R})\subset C(\mathbb{R}, \mathbb{R})$$

그러므로 $$c=\text{card}\, K(\mathbb{R}, \mathbb{R})\leqq\text{card}\, C(\mathbb{R}, \mathbb{R})$$

이 결과와 275쪽 마지막 행의 부등식을 함께 생각함으로써 다음을 얻는다.

$$c=\text{card}\, K(\mathbb{R}, \mathbb{R})\leqq\text{card}\, C(\mathbb{R}, \mathbb{R})\leqq\text{card}\, C(\mathbb{Q}, \mathbb{R})\leqq c$$

따라서 $$\text{card}\, C(\mathbb{R}, \mathbb{R})=\text{card}\, C(\mathbb{Q}, \mathbb{R})=\text{card}\, K(\mathbb{R}, \mathbb{R})=c$$

예제 10의 결과는 상수함수가 연속함수만큼 많다는 것을 뜻한다. 이처럼 무한집합에는 또다른 묘한 성질이 있다.

예제 11 정의역과 공역이 $\mathbb{R}$인 함수로서 미분가능한 것의 집합 $D(\mathbb{R}, \mathbb{R})$의 기수를 구하여라.

[풀이] 상수함수는 미분가능하고 미분가능한 함수는 연속(과연?)이므로

$$K(\mathbb{R}, \mathbb{R})\subseteq D(\mathbb{R}, \mathbb{R})\subseteq C(\mathbb{R}, \mathbb{R}).$$

한편 예제 10에 의하여

$$c=\text{card}\, K(\mathbb{R}, \mathbb{R})\leqq\text{card}\, D(\mathbb{R}, \mathbb{R})\leqq\text{card}\, C(\mathbb{R}, \mathbb{R})=c$$

따라서 $$\text{card}\, D(\mathbb{R}, \mathbb{R})=c$$

연습문제 5.7

1. n차원 공간 $\mathbb{R}^n=\mathbb{R}\times\mathbb{R}\times\mathbb{R}\times\cdots\times\mathbb{R}$의 점 $(x_1, x_2, x_3, \cdots, x_n)$들은 꼭 단위개구간의 점 $x\in(0, 1)$ 만큼 존재한다.* 이것을 확인하여라.

---

* 특히 $n=1, 2, 3$ 즉 수직선 좌표평면, 좌표공간의 점들도 각각 단위개구간 $(0, 1)$의 점들만큼 존재한다. 이렇듯 무한집합의 신비함은 한두 가지가 아니다.

2. The classic Hilbert space consists of all infinite sequences $(x_1, x_2, x_3, \cdots)$ of real numbers, called points, for which the series

$$x_1^2 + x_2^2 + x_3^2 + \cdots$$

converges. Show that the classic Hilbert space contains "just as many" points as the real line $\mathbb{R}$.

3. Let $\mathbb{R}^{\aleph_0}$ be the set of all infinite sequences $(x_1, x_2, x_3, \cdots)$ of real numbers, called points of the space $\mathbb{R}^{\aleph_0}$, A lattice point in $\mathbb{R}^{\aleph_0}$ is a point $(x_1, x_2, x_3, \cdots)$ such that all x_k's are integers. Show that the space $\mathbb{R}^{\aleph_0}$ contains "just as many" points as the set of all lattice points in $\mathbb{R}^{\aleph_0}$.

4. Show that there are "just as many" functions of one real variable which assume only the values 0 and 1 as all real-valued functions of n real variables, where n is any natural number.

5. Let the cardinal number of the set $\{f \mid f : \mathbb{R} \to \mathbb{R}\}$ of all real-valued functions of one variable be denoted by $\mathfrak{f}$. Show that

$$\mathfrak{f}^n = \mathfrak{f}^{\aleph_0} = \mathfrak{f}^c = \mathfrak{f}$$

for all $n \in \mathbb{N}$.

5.8 The Continuum Hypothesis and its Generalization

Since every infinite set contains a denumerable subset (theorem 11, chapter 4), the cardinal number $\aleph_0$ is the smallest transfinite cardinal number. An important question, known as the continuum problem, was raised by Cantor in about 1880: Is there a cardinal number that lies strictly between $\aleph_0$ and $2^{\aleph_0}$ $(=c)$? In set language, are there any nondenumerable infinite subsets of $\mathbb{R}$ with cardinal number less than that of $\mathbb{R}$?

Cantor and many leading mathematicians had tried in vain to solve this problem. Since no such set had been found anywhere in classical mathematics and there seemed to be no way of finding one, it was conjectured by Cantor and others that the answer to the continuum problem must be no. This conjecture is known as the continuum hypothesis.

2. 클래식 힐베르트공간은 실수의 무한수열 $(x_1, x_2, x_3, \cdots)$(이것을 점이라고 부른다)로 이루어진 것이다. 여기서 각 무한수열에 대한 급수

$$x_1^2 + x_2^2 + x_3^2 + \cdots$$

는 수렴한다. 클래식 힐베르트 공간의 점들은 꼭 수직선 $\mathbb{R}$*의 점들만큼 존재한다는 것을 확인하여라.**

3. $\mathbb{R}^{\aleph_0}$을 실수의 무한수열 $(x_1, x_2, x_3, \cdots)$(이것을 공간 $\mathbb{R}^{\aleph_0}$의 점이라고 부른다)의 집합이라 할 때 모든 x_k가 정수인 점 $(x_1, x_2, x_3, \cdots)$을 $\mathbb{R}^{\aleph_0}$에서의 격자점이라고 한다. 공간 $\mathbb{R}^{\aleph_0}$의 점들은 꼭 그 격자점들만큼 존재한다. 이것을 확인하여라.

4. 치역이 $\{0,1\}$인 일변수함수 $\{f \mid f: \mathbb{R} \longrightarrow \mathbb{R}\}$들은 꼭 n변수의 함수(이것을 $\{g \mid g: \mathbb{R}^n \longrightarrow \mathbb{R}\}$로 나타낼 수 있다. 277쪽 문제 1 참조)들만큼 존재한다. 이것을 보여라. 여기서 n은 임의의 자연수이다.

5. 일변수함수인 집합 $\{f \mid f: \mathbb{R} \longrightarrow \mathbb{R}\}$의 기수를 $\mathfrak{f}$로 나타낼 때 임의의 자연수 n과 초한기수 $\aleph_0$, c에 대하여 다음이 성립함을 증명하여라.

$$\mathfrak{f}^n = \mathfrak{f}^{\aleph_0} = \mathfrak{f}^c = \mathfrak{f}$$

5.8 연속체 가설과 그것의 일반화

무한집합은 어느 것이나 가부번집합을 포함한다(227쪽 정리 11). 그러므로 $\aleph_0$은 초한기수 중 가장 작은 기수임을 알 수 있다. 1880년경 칸토어는 이른바 연속체 문제로 알려진 중요한 의문을 제기한 적이 있다. 다름 아니라 엄밀한 뜻에서 두 초한기수 $\aleph_0$, $2^{\aleph_0}(=c)$ 사이에 또 다른 기수가 존재하겠는가에 대한 의문이었다. 이런 의문을 용어 집합을 빌려 다음과 같이 말할 수 있다.

> 집합 $\mathbb{R}$의 비가부번 무한부분집합으로서 그 기수가 $\mathbb{R}$의 기수보다 작은 것이 존재하는가?

이 문제를 놓고 칸토어를 비롯하여 탁월한 수학자들이 많은 노력을 기울였으나 모두 허사로 끝나고 말았다. 고전수학의 어디에도 그러한 집합이 전혀 발견되지 않았으며 발견할 수 있는 방법조차 있어 보이지 않았으므로 칸토어와 여타 수학자들은 이 연속체 문제에 대한 답은 부정적일 것이라고 예측하였다. 오늘날 이 예측은 연속체 가설로 알려져 있다.

---

* 사실 수직선의 점의 집합과 실수의 집합 사이에 일대일 대응이 존재한다는 전제하에 수직선을 $\mathbb{R}$로 나타내고 있다.

** 이 사실을 처음 대하는 독자라면 고등미적분학(급수)에서 확인하여 보아라.

Continuum Hypothesis. There is no cardinal number x satisfying

$$\aleph_0 < x < c(=2^{\aleph_0}).$$

A question closely related to the continuum problem, usually referred to as the generalized continuum problem, is the following:

> Is there any cardinal number that lies strictly between a transfinite cardinal number a and 2^a?

This question too has not been answered. The conjecture that no such cardinal number exists is called the generalized continuum hypothesis.

Generalized Continuum Hypothesis. For any transfinite cardinal number a, there is no cardinal number x such that $a < x < 2^a$.

As early as 1900, at the International Congress of Mathematicians in Paris, the great German mathematician David Hilbert[5](1862-1943) presented a list of 23 important unsolved mathematical problems, the first of which was the continuum problem.

There was no progress made in solving this problem until 1938, when Kurt Gödel[6](1906-1978), the outstanding logician of the century, proved that if the

---

[5] David Hilbert(1862-1943), an outstanding mathematician of all time, was a professor of mathematics at the University of Göttingen, Germany(1895-1943). He influenced the entire world of modern mathematics, ranging from 19th century algebra to modern logic and mathematical physics. The famous Hilbert space is one of his many contributions. Hilbert believed that all mathematical ideas fit together harmoniously.

[6] Kurt Gödel(1906-1978) of the Princeton Institute for Advanced Study in New Jersey was born in Czechoslovakia. He first achieved fame at 25. Famous scholars including Bertrand Russell (1872-1970) and Alfred North Whitehead(1861-1947) had suggested the existence of absolute guides to the truth or falsity of certain mathematical statements. Gödel shocked the world by proving that what Russell and Whitehead sought did not exist. His other major contributions include the proof of the completeness of quantification logic and the proof of the consistency of both the generalized continuum hypothesis and the axiom of choice.

연속체 가설* 두 초한기수 $\aleph_0$, $c(=2^{\aleph_0})$에 대하여 $\aleph_0 < x < c$를 만족하는 기수 x는 존재하지 않는다.

연속체 문제와 밀접한 관련이 있는 의문을 흔히 **일반연속체 문제**라고 한다. 즉 엄밀한 뜻에서

두 초한기수 a, 2^a 사이에 또 하나의 기수가 존재할 것인가?

에 대한 의문을 말한다. 이러한 의문에 대한 답도 찾을 수 없었던 나머지 그런 기수가 존재하지 않을 것이라는 예측을 **일반연속체 가설**이라고 불러왔다.

일반연속체 가설 임의의 초한기수 a에 대하여 $a < x < 2^a$를 만족하는 기수 x는 존재하지 않는다.

일찍이 1900년 독일의 위대한 수학자 다비트 힐베르트[5]는 파리 국제수학자대회에서 23개의 중요한 미해결문제**의 목록을 제기하였다. 그 중 첫 번째 것이 연속체 문제였지만 아무런 진전을 보지못하다가 금세기의 탁월한 논리학자 쿠르트 괴델[6]이 1938년 비로소 그 증명을 마쳤다. 사실 괴델은 이 가설을 다음과 같은 관점에서 증명하였다.

* 수학은 개념상 "존재하는 대상 사이의 관계"를 다루는 학문으로 옮긴이로서 존재하지 않는 것에 대한 어떤 논의에서도 하등 의의를 찾지 못한다.

5) 탁월한 수학자인 다비드 힐베르트는 독일 괴팅겐대학 수학교수였다(1895-1943). 그는 19세기 대수학을 위시하여 논리학, 수리물리학에 이르기까지 현대수학 전분야에 영향을 미쳤다. 그의 많은 공헌 중의 하나로는 유명한 힐베르트 공간이 있다. 힐베르트는 모든 수학적 개념은 서로 조화를 이룬다고 믿었다.

** 다음 책에서 그런 문제들을 살펴보고 과연 그동안 몇 개나 풀렸는지 알아보아라.
Constance Reid Hilbert Springer-Verlag New York, Inc., 1996.
이일해 옮김, 《현대수학의 아버지 힐베르트》, 사이언스북스, 2005년, 10장 수학의 미래(134-147쪽)

6) 미국 뉴저지에 있는 프린스턴 고등연구소의 연구원을 지냈던 쿠르트 괴델(1906~1978)은 체코슬로바키아에서 태어나 25세 때 처음으로 명성을 얻었다. 러셀(1872~1970)과 화이트헤드(1861~1947)를 필두로 유명한 학자들은 수학적 명제의 참, 거짓에 대한 절대적 지침이 존재할 것이라고 암시한 바 있었으나 괴델이 그러한 것이 존재하지 않음을 증명함으로써 수학계에 크나큰 충격을 주었다. 그의 또다른 중요한 공헌으로는 (전칭과 존재에 관한) **작용논리의 완비성**의 증명, 그리고 일반연속체 가설 및 선택공리의 **무모순성**의 증명 등이 있다.

generalized continuum hypothesis is added to the current axioms for set theory, then any contradiction that might be implied by this system of axioms could be formulated as a contradiction implied by the initial axioms (without the generalized continuum hypothesis) alone[7)]. In other words, the generalized continuum hypothesis is relatively consistent with the axioms of set theory.

Finally, in 1963, a significant achievement was made by the young mathematician Paul J. Cohen(1934-) of Stanford University, who declared that the generalized continuum hypothesis is unprovable on the basis of the current axioms for set theory. Therefore, the status of the generalized continuum hypothesis in set theory is analogous to that of Euclid's parallel axiom (the Fifth Postulate) in geometry. We may postulate them or deny them, in either case obtaining a consistent theory of mathematics.

[7)] See K. Gödel, *the Consistency of the Axiom of Choice and of the Generalized Cotinuum Hypothesis with the Axioms of Set Theory*, Princeton University Press, Princeton, N.J., 1940, 66 pp. Rev. ed., 1951, 74 pp.

즉 현행 집합론의 공리계에 일반연속체 가설을 덧붙임으로써 유도되는 임의의 모순은 (일반연속체 가설이 없는) 처음 공리계만으로 유도되는 모순인 것으로 형식화할 수 있다.[7] 다시 말하면 일반연속체 가설은 집합론의 공리들과 상대적으로 모순되지 않는다.

마침내 1963년 미국 스탠퍼드 대학의 젊은 수학자 폴 코헨(1934~2007)에 의하여 뜻깊은 진전을 보기에 이르렀다. 그는 현행 집합론 공리계의 기초 위에서는 일반연속체 가설의 증명이 불가능하다고 주장하였다. 다시 말하면 일반연속체 가가설의 성격은 유클리드기하학에서의 평행선 공리(제 5 공준)*와 유사하다는 것이다. 그러므로 일반연속체 가설과 유클리드기하학에서의 평행선공리를 우리가 받아들이거나 부정하더라도 수학의 이론상 하등의 모순이 일어나지 않는다고 주장할 수 있다.**

비유클리드기하학**(non-Euclidean geometry): 유클리드의 "원본(elements)"에 실린 평행선공리는 당초부터 의문이 있어보여 많은 비판의 대상이 되어왔다. 18세기 초 사케리(G. Saccheri)는 이 공리를 나머지 공리들로부터 유도할 수 있으리라 믿고 증명을 시도한 나머지 여러 거지 결과를 얻었다. 그런 점에서 사케리는 비유클리드기하학의 선구자로 자처할만 하다.

19세기 초에 로바체브스키(N. L. Lobačeveskii), 보여이(J. Bolyai)는 평행선 공리를 대담하게 부정하고 새로운 공리를 세움으로써 하나의 기하학을 구성하였다. 바로 **쌍곡기하학**(hyperbolic geometry) 또는 로바체브스키기하학을 말한다. 가우스(C. F. Gauss) 역시 비슷한 생각을 지니고 있었지만 당시 세상을 풍비한 칸트(Kant)철학의 영향을 벗어나지 못해 발표를 미루기에 이르렀다.

한편 리만(B. Riemann)은 앞선 로바체브스키의 사상을 발전시켜 유클리드기하학, 쌍곡기하학 어느 것과도 다른 이른바 타원기하학(elliptic geometry)를 구성하였다. 이것은 오늘날 리만의 **비유클리드기하학**(Riemann's non-Euclidean geometry)이라고 불리운다. 이렇듯 비유클리드기하학이 성립하는 공간을 **비유클리드공간**이라 하고 재래의 기하학을 **포물기하학**(parabolic geometry)이라고 한다.

아인슈타인(A. Einstein)의 **상대성 이론**(theary of relalivity)에 따르면 우주의 시간·공간의 성격을 설명함에 있어 유클리드기하학보다는 비유클리드기하학쪽이 합리적인 것으로 알려져 있다.

---

[7] 쿠르트 괴델, 《집합론의 공리에서의 선택공리와 일반연속체 가설의 무모순성》 뉴저지, 프린스턴, 프린스턴 대학 간행, 1940년 66쪽. 개정판 1951, 74쪽 참조.

* 독자 누구나 이 공리는 낯익을 것이다.

** 《(일본) 수학사전》, 제2판, 이와나미, 1980년, 211쪽에서 발췌, 우리말로 옮김.

Chapter 6
The Axiom of Choice and Some of Its Equivalent Forms

제 6 장
선택공리, 그와 동치인 형식

6

The Axiom of Choice and Some of Its Equivalent Forms

The axiom of choice and three frequently used equivalent principles — the Hausdroff maximality principle, Zorn's lemma, and the well-ordering principle of Zermelo — are introduced. It is proved, in a circular chain of implications, that the axiom of choice and these three important mathematical principles are all logically equivalent. The principle of transfinite induction is given together with a demonstration of how this induction may be used in mathematical proofs. A short historical comment of the axiom of choice ends the chapter.

6.1 Introduction

Suppose that you enter a fruit market that has a number of (nonempty) baskets of fruit. If you are allowed to choose one (and only one) fruit from each basket as free samples, you know this would not be a difficult task. But the following similar question, which may be seemingly trivial at first glance, is really complex :

> *Given a nonempty family $\mathscr{S}$ whose elements are disjoint nonempty sets S_a, does there exist a set R which has as its elements one element x_a of each S_a?*

6

선택공리, 그와 동치인 형식*

이 장에서는 선택공리 및 그와 동치인 원리로서 흔히 이용되고 있는 *하우스도르프의 극대원리*, *존의 보조정리*, *체르멜로의 정렬원리*를 도입한 다음 이들 세 원리는 선택공리와 논리적으로 동치임을 증명한다.

한편 *초한귀납법의 원리*를 소개하고 이 귀납법이 수학의 여러 증명에서 어떻게 활용되는가를 알아보고 선택공리에 관한 역사적 단평으로 마무리 짓는다.

6.1 도입

과일가게의 과일이 50개의 바구니에 들어 있다고 하자. 지금 각 바구니에서 과일을 하나씩 골라내기란 매우 간단하다.

> 주어진 족 $\mathscr{S}(\mathscr{S}\neq\phi)$의 구성원 S_a들이 둘씩 모두 서로 소인 집합일 때 각 S_a에서 원소 x_a 하나씩을 택한 것을 원소로 하는 집합 R을 믿을 수 있을까?**

---

* 이장우, 《현대집합론의 원리》, 경문사, 1998년, 제7장(231–245쪽) 참조.

** 사실 선택공리에 관한 이 주장에 있어서 족 ψ이 무한 집합이라는 데에 그 뜻이 있다.

The real difficulty lies in the case where $\mathscr{S}$ is infinite. At the beginning of the twentieth century, Ernst Zermelo (1871-1953) and others made unsuccessful attempts to answer this question. Zermelo felt that this question might be unsolvable and that the only way to avoid the difficulty was to postulate an axiom, which has since been known as the axiom of choice. We state this axiom in its general form in terms of functions:

Axiom of Choice. To any nonempty family $\mathscr{S}$ whose elements are nonempty sets, there exists a function called a choice function

$$f : \mathscr{S} \longrightarrow \cup_{A \in \mathscr{S}} A$$

such that $f(A) \in A$ for all $A \in \mathscr{S}$.

The axiom of choice is now indispensable for proving many important results in various areas of contemporary mathematics. In fact, a disguised use of the axiom of choice was made earlier in the proof of theorem 11, Chapter 4.

Mathematicians have since discovered several other principles that often can be used as convenient substitutes for the axiom of choice. These principles, though seemingly bearing no resemblance to the axiom of choice, were soon shown to be equivalent to the axiom of choice. To understand some of these principles and their connection with the axiom of choice, we need a few definitions.

Definition 1. A relation $\leqq$ on a set A is called a partial order relation if and only if the relation $\leqq$ is reflexive and transitive on A and antisymmetric on A, that is, if $a \leqq b$ and $b \leqq a$, then $a = b$. A partially ordered set is a pair $(A, \leqq)$, where A is a set and $\leqq$ is a partial order relation on A.

Example 1 Let X be a nonempty set. Then, the power set $\mathscr{P}(X)$ of X is partially ordered by the inclusion relation $\subseteq$ on $\mathscr{P}(X)$.

이 물음의 어려운 점은 바로 $\mathscr{S}$가 무한족이라는 데 있다. 20세기 초 앞의 질문을 놓고 에른스트 체르멜로(1871~1953)와 그 외 수학자들이 많은 노력을 하였으나 모두 실패하고 말았다. 그럼에도 불구하고 체르멜로는 하나의 공리를 도입함으로써 그 어려움을 피할 수 있을 것으로 믿었다. 우리는 선택공리로 알려진 이 내용을 함수를 써서 다음과 같은 형식으로 나타낸다.

선택공리* 공집합이 아닌 집합을 구성원으로 하는 임의의 족을 $\psi(\psi\neq\phi)$라고 할 때 $\forall_{A_\alpha\in\mathscr{S}} f(A_\alpha)\in A$로 정의된 함수

$$f:\mathscr{S}\to\bigcup_{A_\alpha\in\mathscr{S}}A_\alpha$$

가 존재한다. 여기서 f를 선택함수, $f(A_\alpha)=x_\alpha$를 A_α의 대표원소, $\{x_\alpha \mid f(A_\alpha)=x_\alpha\}$를 ψ에 대한 대표집합이라고 한다.

오늘날 선택공리는 현대수학의 여러 분야에서 빼놓을 수 없게 되었다. 사실 이 공리는 제 4 장 정리 11(229쪽)의 증명에서도 인용한 적이 있다.

선택공리가 알려지자 수학자들은 이 공리를 대치할 만한 몇 가지 원리를 발견하였으나 그것들은 선택공리와 유사점이 있어 보이지 않았다. 그런던 중 원리들 사이에 동치관계가 있음이 밝혀졌다. 앞으로 이들 원리를 이해하고 선택공리와는 어떤 관련이 있는지 파악하기 위해 다음 정의부터 내리자.

정의 1 집합 A 위의 관계 $\leqq$가 반사적, 추이적 그리고 반대칭적**, 즉 A의 임의의 원소 a, b에 대하여

$$a\leqq b \wedge b\leqq a \Longrightarrow a=b$$

이면 그리고 그때에만 이 관계를 반순서 또는 부분순서라고 한다. 이때 집합 A와 그 위의 반순서 $\leqq$를 함께 반순서 집합이라 하고 하나의 쌍 $(A,\leqq)$로 나타낸다.

예제 1*** 집합 X의 멱집합 $\mathscr{P}(X)$는 그 위의 포함관계 $\subseteq$를 일종의 순서로 볼 때 부분순서 집합이다. 이것의 증명은 독자에게 미룬다[293쪽 문제 3(2)].

* 각주 [3] (213쪽) 및 각주 [7](283쪽) 참조. 사실 함수를 인용한 선택공리는 213쪽 및 287쪽에서 언급된 것들보다 수학적으로 다듬어진 표현이라고 본다.

** 반대칭적을 반대칭율이라고도 한다.

*** 반순서집합의 보기를 한두 개 더 들 수 있을까? 이장우, 《현대집합론의 원리》, 경문사, 1998년, 194−196쪽 참조.

Definition 2. A total order relation $\leqq$ on a set A is a partial order relation such that for any pair of elements a and b in A, either $a \leqq b$ or $b \leqq a$. A totally ordered set is a pair $(A, \leqq)$ where A is a set and $\leqq$ is a total order relation.

When the partial (total) order relation $\leqq$ on A is unmistakably clear from the context, we may simply say that A is a partially (totally) ordered set. Total order relations and totally ordered sets are also called linear order relations and linearly ordered sets, respectively. It is evident from the definitions that a totally ordered set is a partially ordered set, but a partially ordered set need not be a totally ordered set(see Example 2 below).

Let B be a subset of a partially ordered set $(A, \mathscr{R})$, and let $\leqq_B$ be the interesection $\mathscr{R} \cap (B \times B)$ of $\mathscr{R}$ with $B \times B$. Then, $(B, \leqq_B)$ is a partially ordered set; it may happen that $\leqq_B$ is a total order relation on B, in which case B is called a totally ordered subset of the partially ordered set $(A, \leqq)$. A totally ordered subset of a partially ordered set is also called a chain.

If $(A, \leqq)$ is a partially ordered (totally ordered) set, then we may say, equivalently, that the set A is partially ordered (totally ordered) by $\leqq$. In either case, we may write

$$a \geqq b \text{ for } b \leqq a, \text{ and } a < b \text{ or } b > a \text{ for } a \leqq b \text{ and } a \neq b.$$

Example 2 Let $\leqq'$ be the relation defined on $\mathbb{R}^2$ by

$$(a_1, a_2) \leqq' (b_1, b_2) \text{ if and only if } a_1 \leqq b_1 \text{ and } a_2 \leqq b_2.$$

The reader should verify that the relation $\leqq'$ is a partial order relation on $\mathbb{R}^2$. Since neither $(1, 2) \leqq' (2, 1)$ nor $(2, 1) \leqq' (1, 2)$, the relation $\leqq'$ is not a total order relation on $\mathbb{R}^2$.

Example 3 In Example 2, the diagonal $\varDelta = \{(x, x) \mid x \in \mathbb{R}\}$ of the plane $\mathbb{R}^2$ is a chain.

정의 2 부분순서집합 $(A, \leqq)$의 임의의 $a, b \in A$에 대하여

$$a \leqq b \text{ 또는 } b \leqq a$$

일 때 ≦를 A 위의 **전순서**, 집합 A와 그 위의 완전순서 ≦의 쌍 $(A, \leqq)$을 **전순서 집합**이라고 한다.

문맥상 집합 A 위의 부분(전)순서 ≦가 분명할 때에는 간단히 A를 부분(전)순서집합이라고 한다. 한편 전순서, 전순서 집합을 각각 **선형순서**, **선형순서집합**이라고도 한다. 전순서는 그 정의에 따라 부분순서이지만 부분순서이더라도 전순서가 아닌 것이 존재한다(다음 보기 2).

지금 부분순서집합 $(A, \mathscr{R})$의 부분순서 부분집합 B에 대한 관계를 A 위의 부분순서 ≦를 이어받은 것으로 볼 때 $\mathscr{R} \cap (B \times B)$을 $\leqq_B$로 나타냄으로써 $(A, \leqq)$의 **반순서부분집합** $(B, \leqq_B)$를 얻는다. 특히 $\leqq_B$가 B 위의 전순서이면 B를 **전순서부분집합** 또는 **사슬***이라고 한다.

예**를 들어 $A=\{a, b, c\}$일 때의 부분순서집합 $(\mathscr{P}(A), \subseteq)$의 부분집합 $B=\{\{a\}, \{a, b\}\}$에 대하여 $\subseteq_B$는 B 위의 전순서이므로 $(B, \subseteq_B)$는 사슬이다. 여기서 $\subseteq_B$는 $\mathscr{P}(A)$ 위의 부분순서 ⊆를 그대로 이어받은 것이다.

부분(전)순서 집합 $(A, \leqq)$를 "A는 ≦에 따른 부분(전)순서집합이다."로 말하기도 하며 어느 경우에도

$$a \leqq b \wedge a \neq b\text{일 때에는 } a < b \text{ 아니면 } b > a$$

로 나타낸다.

예제 2 지금 집합 $\mathbb{R}^2$*** 위의 일종의 순서 하나로 관계 ≦′를

$$a_1 \leqq b_1 \wedge a_2 \leqq b_2 \Longleftrightarrow (a_1, a_2) \leqq' (b_1, b_2)$$

로 정의하면 ≦′은 부분순서이다(과연?). 그러나, 이를테면 $\mathbb{R}^2$의 원소 $(1, 2)$, $(2, 1)$에 대하여

$$(1, 2) \leqq' (2, 1), \quad (2, 1) \leqq' (1, 2)$$

는 모두 참이 아니므로(왜?) ≦′ 전순서가 아니다.

예제 3 예제 2에서 좌표평면 $\mathbb{R}^2$의 대각선 $\Delta_x=\{(x, x) \mid x \in \mathbb{R}\}$(145쪽 참고)은 부분순서집합 $(\mathbb{R}^2, \leqq')$의 하나의 사슬이다(과연?).

---

* '사슬'을 연쇄라고도 한다. 이장우, 《현대집합론의 원리》, 경문사, 1998년, 194−196쪽 및 허걸, 《집합론》, 경문사, 2004년, 169−173쪽 참조. 한편 $\mathbb{R}^2$의 임의의 원소 (x, y)를 좌표평면의 점 하나로 본다.

** 이런 예는 옮긴이가 보충한 것이다.

*** 집합 $\mathbb{R}^2$은 데카르트 곱 $\mathbb{R} \times \mathbb{R}$임에 주목하자.

Exercise 6.1

Prove the following problems 1 through 11.

1. Let $\mathscr{T}$ be a partition of a nonempty set X. Then, there exists a set $R \subseteq X$ such that for every $C \in \mathscr{T}$, $C \cap R$ consists of one and only one element; such a set R is called a set of a representative for $\mathscr{T}$.

2. Let $f : A \to B$ be a surjection. Then, there exists a subset C of A such that C is equipotent to B and hence $\operatorname{card} A \geqq \operatorname{card} B$.

3. The statement of Example 1.

4. (a) The identity relation "=" on the set $\mathbb{R}$ is a partial order relation.
   (b) The ordinary relation "$\leqq$" on the set $\mathbb{R}$ is a total order relation.

5. Let F be the set of all functions $f : \mathbb{R} \to \mathbb{R}$, and let

$$\mathscr{R} = \{(f, g) \in F \times F \mid \forall_{x \in \mathbb{R}}\, f(x) \leqq g(x)\}.$$

   Then, $(F, \mathscr{R})$ is a partially ordered set.

6. The set $\mathbb{N}$ of natural numbers together with the relation "x divides y" is a partially ordered set.

7. Let $(\mathbb{R}^2, \leqq')$ be the partially ordered set of Example 2, and let m be any nonnegative real number. Then, the subset $\{(x, mx) \mid x \in \mathbb{R}\}$ of $(\mathbb{R}^2, \leqq')$ is a chain.

8. Let $\{A_\gamma \mid \gamma \in \Gamma\}$ be a family of nonempty sets; that is, each $A_\gamma \neq \phi$. The generalized Cartesian product $\underset{\gamma \in \Gamma}{\mathrm{P}} A_\gamma$ of the family $\{A_\gamma \mid \gamma \in \Gamma\}$ is defined to be the set of all functions

$$f : \Gamma \to \bigcup_{\gamma \in \Gamma} A_\gamma \text{ such that } f(\gamma) \in A_\gamma, \text{ for all } \gamma \in \Gamma.$$

   Then, $\Gamma \neq \phi \Rightarrow \underset{\gamma \in \Gamma}{\mathrm{P}} A_\gamma \neq \phi$.

9. If $\Gamma = \{1, 2, \cdots, n\}$ is finite, then there is a one-to-one correspondence between $\underset{\gamma \in \Gamma}{\mathrm{P}} A_\gamma$ and the set $\{(x_1, x_2, \cdots, x_n) \mid x_i \in A_i \text{ for all } i \in \Gamma\}$ of n-tuples (Use the same notation as Problem 8).

연습문제 6.1

1. 집합 $X(X=\phi)$의 하나의 분할 $\mathscr{T}$가 있을 때 X의 부분집합 R 중 각 $C\in\mathscr{T}$에 대한 $C\cap R$가 한원소집합인 것이 존재한다. 이것을 증명하여라 (선택공리 인용). 여기서 R를 $\mathscr{T}$에 대한 하나의 **대표집합**이라고 한다.

2. (a) 함수 $f: A\to B$가 전사일 때 A의 하나의 부분집합 C가 존재함으로써 C와 B가 대등하다. 그러므로 $\text{card}\,A\geq\text{card}\,B$. 이것을 증명하여라.

(b) 집합 $X=\{1,2,3,4,5\}$의 부분집합

$$A_1=\{1,2,3\},\ A_2=\{1,5\},\ A_3=\{2,4,5\},\ A_4=\{3,4\}$$

에 대하여 다음 함수는 $\{A_1, A_2, A_3, A_4\}$에서 X로의 선택함수인지 아닌지를 가려라.

- $f_1=\{(A_1,1),(A_2,2),(A_3,3),(A_4,4)\}$
- $f_2=\{(A_1,1),(A_2,1),(A_3,4),(A_4,4)\}$
- $f_3=\{(A_1,2),(A_2,1),(A_3,4),(A_4,3)\}$
- $f_4=\{(A_1,3),(A_2,5),(A_3,1),(A_4,3)\}$

3. (1) 예제 1을 증명하여라.

(2)* $A=\{a,b,c,d\}$의 멱집합 $(\mathscr{T}(A),\subseteq)$의 부분순서부분집합으로써 사슬인 것과 그렇지 않은 것 하나씩을 들어라.

다음 4~11을 증명하여라.

4. (a) 집합 $\mathbb{R}$ 위에서의 상등관계 "$=$"는 부분순서(사실은 전순서)이다.

(b) 집합 $\mathbb{R}$ 위에서의 대소관계 "$\leq$"는 전순서이다.

5. 모든 함수 $f:\mathbb{R}\to\mathbb{R}$의 집합 F에 대하여 $\mathscr{R}=\{(f,g)\in F\times F\mid\forall_{x\in\mathbb{R}}f(x)\leq g(x)\}$로 놓을 때 $(F,\mathscr{R})$은 부분순서집합이다.

6. 집합 $\mathbb{N}$은 "x는 y를 나누어 떨어뜨린다."는 순서에 따른 부분순서집합이다.

7. 보기 2의 부분순서집합 $(\mathbb{R}^2,\leq')$에서 임의의 음수가 아닌 실수 m에 대한 $(\mathbb{R}^2,\leq')$의 부분집합 $\{(x,mx)\mid x\in\mathbb{R}\}$은 하나의 사슬이다.

8. 집합족 $\{A_\gamma\mid\gamma\in\Gamma, A_\gamma\neq\phi\}$에 대한 함수 $f:\Gamma\to\bigcup_{\gamma\in\Gamma}A_\gamma$, $\forall_{\gamma\in\Gamma}f(\gamma)\in A_\gamma$의 집합을 이 족의 **일반데카르트곱**이라 하고 이것을 $\mathbb{P}_{\gamma\in\Gamma}A_\gamma$로 나타낸다. 이때 $\Gamma\neq\phi\Longrightarrow\mathbb{P}_{\gamma\in\Gamma}A_\gamma\neq\phi$.

9.** 유한집합 $\Gamma=\{1,2,\cdots,n\}$에 대한 $\mathbb{P}_{\gamma\in\Gamma}A_\gamma$와 n개 원소의 짝의 집합 $\{(x_1,x_2,\cdots,x_i,\cdots,x_n)\mid\forall_{i\in\Gamma}x_i\in A_i\}$ 사이에 일대일 대응이 존재한다.

---

* 문제 3의 (2)는 옮긴이가 보충한 것이다. 사실 (2)에서와 같이 X가 유한집합인 경우 선택함수는 그 뜻을 잃는다.

** 문제 8의 기호를 이용하여라.

10. Let A and B be nonempty sets and let $\mathscr{R}$ be a relation from A to B with Dom $\mathscr{R}=A$. Then, there exists a function $f: A \longrightarrow B$ such that $f \subseteq \mathscr{R}$.

11. A function $f: A \longrightarrow B$ is surjective if and only if there exists a function $g: B \longrightarrow A$ such that $f \circ g = 1_B$, the identity function on B (cf. Theorem 16, Chapter 3).

6.2 The Hausdorff Maximality Principle

In modern algebra and in topology, we often find it more convenient to use the equivalent Hausdorff maximality principle than to use the axiom of choice. In order to understand this principle, a few more new terms are needed.

Definition 3. Let $(A, \leqq)$ be a partially ordered set.

(a) An element $u \in A$ is said to be an upper bound for a subset B of A if and only if $u \geqq b$ for all $b \in B$.

(b) An upper bound u_0 for B is the least upper bound for B if and only if $u_0 \leqq u$ for every upper bound u for B.

(c) An element $e \in A$ is said to be maximal if and only if $e \leqq a$ implies $e = a$ for all $a \in A$.

Definition 3 has a dual form which we state separately as

Definition 3′ Let $(A, \leqq)$ be a partially ordered set.

(a) An element $v \in A$ is a lower bound for a subset B of A if and only if $v \leqq b$ for all $b \in B$.

(b) A lower bound v_0 for B is the greatest lower bound for B if and only if $v_0 \geqq v$ for every lower bound v for B.

(c) An element $e' \in A$ is minimal if and only if $a \leqq e'$ implies $e' = a$ for all $a \in A$.

10. 집합 $A(\neq\phi)$에서 집합 $B(\neq\phi)$로의 관계 $\mathscr{R}$에 대하여 $\mathrm{Dom}(\mathscr{R})=A$일 때 함수 $f:A\to B$로써 $f\subseteq\mathscr{R}$인 것이 존재한다.

11. 함수 $f:A\to B$에 대하여 함수 $g:B\to A$가 존재함으로써 $f\circ g=1_B$이면, 그리고 그때에만 f는 전단사이다. 여기서 1_B는 B에서의 항등함수이다(선택함수를 인용하여라).

6.2 하우스도르프의 극대원리

현대대수학과 위상수학에서는 선택공리보다 그와 동치인 하우스도르프의 극대원리를 이용하는 것이 더 편리할 때가 많다. 이 원리를 이용하려면 몇 가지 용어부터 알아두어야 한다.

정의 3* 부분순서집합 $(A,\leqq)$에 있어서 A의 부분순서부분집합 B에 대하여 다음과 같이 정의한다.

(a) 모든 $b\in B$에 대하여 적당한 $u\in A$가 존재함으로써 $u\geqq b$이면 그리고 그때에만 u를 B의 **상계**라고 한다.

(b) B의 각 상계 u에 대하여 상계 u_0이 존재함으로써 $u_0\leqq u$이면 그리고 그때에만 u_0을 B의 **최소상계** 또는 **상한**이라 하고 이것을 **sup** B**로 나타낸다.

(c) 모든 $a\in A$에 대하여 $e\in A$가 존재함으로써 $e\leqq a\Rightarrow e=a$이면 그리고 그때에만 e를 A의 **극대원소**라고 한다.

다음은 정의 3에 대한 **쌍대적 형식***을 따로 적어 놓은 것이다.

정의 3′* 부분순서집합 $(A,\leqq)$에서의 부분집합 B에 대하여 다음과 같이 정의한다.

(a) 모든 $b\in B$에 대하여 적당한 $v\in A$가 존재함으로써 $v\leqq b$이면 그리고 그때에만 v를 B의 **하계**라고 한다.

(b) B의 각 하계 v에 대하여 하계 v_0이 존재함으로써 $v_0\geqq v$이면 그리고 그때에만 v_0을 B의 **최대하계** 또는 **하한****이라 하고 이것을 **inf** B 나타낸다.

(c) 모든 $a\in A$에 대하여 $e'\in A$가 존재함으로써 $a\leqq e'\Rightarrow e'=a$이면, 그리고 그때에만 e'을 A의 **극소원소**라고 한다.

---

* 정의 3과 정의 3′에 따른 문제에 관해서는 이장우, 《현대집합론의 원리》, 경문사, 1998년, §6.2 순서집합의 내부구조 200−206쪽을, 한편 쌍대적 형식에 관해서는 같은 책의 203−204쪽 참고.

** sup B, inf B는 각각 "supremum for B", "infimum for B"의 약자이다.

Example 4 Let X be a nonempty set and let $\mathscr{B}$ be a subset of the partially ordered set $(\mathscr{P}(X), \subseteq)$(see Example 1). Then, the least upper bound of $\mathscr{B}$ is $\bigcup_{B\in\mathscr{B}} B$, and the greatest lower bound of $\mathscr{B}$ is $\bigcap_{B\in\mathscr{B}} B$.

The following theorem plays an important role in proving the equivalence of the axiom of choice to other principles. The proof of this theorem is tedious, and perhaps discouraging; therefore the authors suggest that beginners take this theorem for granted, and omit the proof in their first reading.

Theorem 1. Let $(A, \leqq)$ be a nonempty partially ordered set such that every totally ordered subset of A has a least upper bound in A. If $f: A \rightarrow A$ is such that $f(a) \geqq a$ for every $a \in A$, then there exists $p \in A$ such that $f(p) = p$.

[Proof] Let a be an arbitrary element of A that remains fixed throughout the proof. A subset B of A is called admissible if and only if it has the following three properties:

(i) $a \in B$.

(ii) $f(B) \subseteq B$.

(iii) Every least upper bound of a totally ordered subset of B belongs to B.

Let $\mathscr{B}$ be the set of all admissible subsets of A. Then, since the set A itself is admissible, $\mathscr{B} \neq \phi$. An intersection of admissible sets is admissible; hence, the partially ordered set $(\mathscr{B}, \subseteq)$ has a unique minimal element $B_0 = \bigcap_{B\in\mathscr{B}} B$. Since the set $\{x \in A \mid x \geqq a\}$ is clearly admissible, we have $B_0 \subseteq \{x \in A \mid x \geqq a\}$. Thus

(iv) $x \geqq a$ for all $x \in B_0$.

예제 4 (a)* 집합 $A=\{a,b,c\}$에 대한 부분순서집합 $(\mathscr{P}(A),\subseteq)$의 부분순서부분집합 $B=\{\{a\},\{a,b\}\}$에 대하여 $\sup B=\{a,b\}$, $\inf B=\{a\}$.

(b) 집합 $X(X\neq\phi)$에 대한 부분순서집합 $(\mathscr{P}(X),\subseteq)$(289쪽 예제 1)의 부분순서부분집합 하나를 $\mathscr{B}$라고 할 때

$$\sup B=\bigcup_{B\in\mathscr{B}}B,\quad \inf B=\bigcap_{B\in\mathscr{B}}B.$$

이것의 증명은 문제 4(301쪽)으로 미룬다.

다음 정리는 선택공리가 그 외의 원리들과 동치인 것을 증명하는 데 중요한 역할을 하지만 그 증명과정이 지루하게 여겨질 독자는 해당 예제로써 드는 것은 족하다.

정리 1 부분순서집합 $(A,\leq)$의 임의의 전순서 부분집합의 최소 상계가 A에 속할 때 $\forall_{a\in A}f(a)\geq a$로 정의된 함수

$$f:A\longrightarrow A\text{에 대하여 }\exists_{p\in A}f(p)=p.$$

[증명]** 집합 A의 임의의 원소 a는 이 증명이 끝날 때까지 일단 고정된 것으로 생각하고 다음과 같은 정의부터 소개한다.

지금 A의 부분집합 B가 다음 세 조건을 만족하면 그리고 그때에만 B는 **허용가능하다**고 말한다.

(i) $a\in B$ (ii) $f(B)\subseteq B$

(iii) B의 각 전순서 부분집합의 최소상계가 B에 속한다.

이럴 때 A의 부분집합으로써 허용가능한 모든 것의 족을 $\mathscr{B}$라면 적어도 A는 그 자신 허용가능하므로 $\mathscr{B}\neq\phi$. 또 허용가능한 임의의 두 집합의 교집합은 허용가능하므로(과연?) 부분순서집합 $(\mathscr{B},\subseteq)$는 극소원소 $B_0=\bigcap_{B\in\mathscr{B}}B$를 지닌다. 한편 집합 $\{x\in A\mid x\geq a\}$는 허용가능하므로(과연?)

$$B_0\subseteq\{x\in A\mid x\geq a\}.\quad \text{따라서}$$

(iv) 모든 $x\in B_0$에 대하여 $x\geq a$

---

* 예제 4 (a)는 옮긴이가 추가한 것이다.

** 증명을 벅차게 여기는 독자는 스스로 해당보기를 들어라. 그러나 그것마저 벅찰지 모른다.

Let $C=\{x\in B_0 \mid y\in B_0$ and $y<x$ imply $f(y)\leqq x\}$. We shall prove that

(v) $x\in C$ and $z\in B_0$ imply either $z\leqq x$ or $z\geqq f(x)$.

Fix $x\in C$, and denote $D=\{z\in B_0 \mid z\leqq x$ or $z\geqq f(x)\}$. Then, the condition (iv) shows that D satisfies (i). The set D satisfies (ii) : For if $z\geqq f(x)$, then $f(z)\geqq z\geqq f(x)$; if $z=x$, then $f(z)=f(x)$; and if $z<x$, then since $x\in C$ we have $f(z)\leqq x$. The set D also satisfies (iii) : If u is the least upper bound for the totally ordered subset E of D, then either $y\leqq x$ for all $y\in E$, and consequently $u\leqq x$, or $y\geqq f(x)$ for some $y\in E$, and consequently $u\geqq f(x)$. Thus, D is admissible and hence $D=B_0$, which proves the statement (v).

We now prove that C is admissible. The set C satisfies (i) vacuously. To show that C satisfies (ii), we show that if $x\in C$, $y\in B_0$, and $y<f(x)$, then $f(y)\leqq f(x)$. By (v), we have either $y\geqq f(x)$ or $y\leqq x$, so that if $y<f(x)$, then $y\leqq x$. Since $x\in C$, $y<x$ implies $f(y)\leqq x(\leqq f(x))$. If $y=x$, then $f(y)=f(x)$. To prove that C satisfies (iii), let w be the least upper bound for the totally ordered set $G\subseteq C$. To verify that $w\in C$, let $y\in B_0$ and $y<w$. We have, by (v), that each $x\in G$ has the property that either $y\leqq x$ or $y\geqq f(x)$. The inequality $y\geqq f(x)\geqq x$ cannot be true for all $x\in G$, for then $y\geqq w$, which contradicts the choice of y. Therefore, there exists an $x\in G$ such that $y\leqq x$. If $y<x$, then by the definition of C, $f(y)\leqq x\leqq w$. If $y=x$, then since $y<w$, there exists a $z\in G$ such that $y<z$. Whence, by the definition of C, $f(y)\leqq z\leqq w$. Thus, $f(y)\leqq w$ and hence $w\in C$. Now, since the set C is an admissible subset of B_0, we must have $C=B_0$. Consequently, by (v), B_0 is totally ordered. Let p be the least upper bound of B_0 ; then $p\in B_0$ and $p\leqq f(p)\leqq p$, so that $f(p)=p$.

이번에는 $$C=\{x\in B_0 \mid y\in B_0 \wedge y<x \Rightarrow f(y)\leqq x\}$$

라고 놓고 다음을 증명한다.

(v) $x\in C \wedge z\in B_0 \Rightarrow z\leqq x \vee z\geqq f(x)$

먼저 $x\in C$도 고정시키고 $D=\{z\in B_0 \mid z\leqq x \vee z\geqq f(x)\}$로 놓아보자. 그러면 조건 (iv)는 D가 (i)를 만족한다는 사실을 보여주고 있다. 한편 D는 (ii)도 만족한다. 왜냐하면 $z\geqq f(x)$이면 $f(z)\geqq z>f(x)$, $z=x$이면 $f(z)=f(x)$, 또 $z<x$이면 $x\in C$이므로 $f(z)\leqq x$이기 때문이다. 한편 집합 D는 (iii)도 만족한다. 즉 D의 전순서부분집합 E의 최소상계를 u라고 하면 모든 $y\in E$에 대하여 $y\leqq x$이므로 $u\leqq x$ 아니면 적당한 $y\in E$에 대하여 $y\geqq f(x)$. 결국 $u\geqq f(x)$. 따라서 D는 허용가능하므로 $D=B_0$. 이것으로 마지막 (v)마저 증명된다.

이제 C가 허용가능함을 증명할 차례이다. 먼저 C는 (i)을 그대로 만족함을 알 수 있고 C가 (ii)를 만족하는지 $x\in C$, $y\in B_0$, $y<f(x)$이면 $f(y)\leqq f(x)$임을 보이면 된다. 그런데 (v)에 따라 $y\geqq f(x)$ 아니면 $y\leqq x$. 그러므로 $y<f(x)$이면 $y\leqq x$. 한편 $x\in C$이므로 $y<x$이면 $f(y)\leqq x(\leqq f(x))$, $y-x$이면 $f(y)=f(x)$.

끝으로 C가 (iii)을 만족함을 밝히기 위하여 C의 전순서부분집합 G의 최소상계를 w라고 할 때 $w\in C$임을 보이자. 그러려고 $y\in B_0$, $y<w$라고 가정하면 (v)에 따라 모든 $x\in G$에 대하여 $y\leqq x$ 아니면 $y\geqq f(x)$.

여기서 $\forall_{x\in G}\,[y\geqq f(x)\geqq x]$는 참이 아니다. 왜냐하면 그것이 참이라면 $y\geqq w$. 이것은 y의 선택에 모순이기 때문이다. 그러므로 적당한 $x\in G$가 존재함으로써 $y\leqq x$. 그런데 $y<x$이면 C의 정의에 따라 $f(y)\leqq x\leqq w$이므로 $f(y)\leqq w$. 따라서 $w\in C$.

한편 $y=x$이면, $y<w$이므로 적당한 $z\in G$가 존재함으로써 $y<z$. 그러므로 또한 C의 정의에 따라 $f(y)\leqq z\leqq w$. 따라서 $f(y)\leqq w$이므로 $w\in C$. 지금 C는 B_0의 허용가능한 부분집합이므로 $C=B_0$. 그러므로 (v)에 따라 B_0은 전순서집합이다. 여기서 p를 B_0의 최소상계라면 $p\in B_0$이고 $p\leqq f(p)\leqq p$. 따라서 $f(p)=p$.

이상으로 증명이 끝난다.

We are now ready to prove the Hausdorff Maximality Principle by use of the axiom of choice.

Theorem 2. Hausdorff Maximality Principle Let the family $\mathscr{T}$ of all totally ordered subsets of a partially ordered set $(A, \leqq)$ be partially ordered by inclusion $\subseteq$. Then, $(\mathscr{T}, \subseteq)$ has a maximal element.

[Proof] Suppose on the contrary that $\mathscr{T}$ has no maximal element. Then, to each $T \in \mathscr{T}$, there is associated a nonempty set

$$T^* = \{T' \in \mathscr{T} \mid T' \supset T\}$$

By the axiom of choice, there is a function g with domain $\{T^* \mid T \in \mathscr{T}\}$ satisfying $g(T^*) \in T^*$. Consequently, there is a function $f : \mathscr{T} \longrightarrow \mathscr{T}$ defined by $f(T) = g(T^*) \supset T$ for all $T \in \mathscr{T}$. Then, using Example 4 we see that $(\mathscr{T}, \subseteq)$ together with the function f satisfies the hypotheses of Theorem 1. But $f(T) \supset T$ for all $T \in \mathscr{T}$, a contradiction to the conclusion of Theorem 1. Thus, the theorem is proved.

Exercise 6.2

1. Let B be a subset of the partially ordered set $(A, \leqq)$. Prove that a least upper bound (greatest lower bound) of B is unique if it exists.
2. Give an example of a subset of a partially ordered set that has neither a least upper bound nor a greatest lower bound.
3. Let $X = \{a, b, c\}$ and let the power set $\mathscr{P}(X)$ be a partially ordered by the inclusion $\subseteq$. Find all upper bounds, all lower bounds, the least upper bound, and the greatest lower bound for the set $\{\{a, b\}, \{c, a\}\}$.
4. Prove the statement of Example 4.
5. In Problem 3 find maximal and minimal elements of $\mathscr{P}(X)$.
6. Find maximal and minimal elements of the partially ordered set $\mathscr{P}(X)$ in Example 4.

마침내 우리는 선택공리를 써서 하우스도르프의 극대원리를 증명할 수 있다.

정리 2 **하우스도르프의 극대원리** 부분순서집합 $(A, \leqq)$의 전순서 부분집합의 족을 $\mathscr{T}$라고 하자. 지금 $\mathscr{T}$ 위의 포함관계 $\subseteq$를 하나의 순서로 볼 때 부분순서집합 $(\mathscr{T}, \subseteq)$에는 극대원소가 존재한다.

[증명] 이 정리의 결론에 반하여 $\mathscr{T}$에는 극대원소가 존재하지 않는다고 가정하면 각 $T \in \mathscr{T}$에 대한 공집합이 아닌 집합 $T^* = \{ T' \in \mathscr{T} \mid T \supset T' \}$를 짝지을 수 있다. 한편 선택공리에 따라 정의역이 $\{ T^* \mid T \in \mathscr{T} \}$인 함수로서 $g(T^*) \in T^*$인 g가 존재한다. 따라서 각 $T \in \mathscr{T}$에 대한 $f(T) = g(T^*) \supset T$로 정의된 함수 $f : \mathscr{T} \longrightarrow \mathscr{T}$가 존재한다. 이제 보기 4(b)의 결론을 이용하면 $(\mathscr{T}, \subseteq)$는 함수 f와 더불어 정리 1의 가정을 모두 만족함을 알 수 있다. 그러나 각 $T \in \mathscr{T}$에 대하여 $T \subset f(T)$. 이것은 정리 1의 결론에 모순이다.

이로써 배리법에 따라 정리 2는 증명된 것이다.

연습문제 6.2**

1. 부분순서집합 $(A, \leqq)$의 부분집합 B에 대하여 B의 최소상계(최대하계)가 존재할 경우 그것은 단 하나뿐이다. 이것을 증명하여라.
2. 부분순서집합의 부분순서부분집합으로써 최소상계, 최대하계가 모두 존재하지 않는 보기를 하나 들어라.
3. 집합 $X = \{a, b, c\}$에 대한 부분순서집합 $(\mathscr{P}(X), \subseteq)$의 부분순서부분집합 $A = \{\{a, b\}, \{a, c\}\}$가 있다. A의 상계, 하계, 최소상계, 최대하계를 모두 구하여라. [107쪽 그림 참고]
4. 예제 4(b)를 증명하여라.
5. 문제 3에서의 $\mathscr{P}(X)$의 극대원소와 극소원소를 각각 구하여라.
6. 예제 4(b)에서의 부분순서집합 $\mathscr{P}(X)$의 극대원소와 극소원소를 구하여라.

** 이장우, 《현대집합론의 원리》, 경문사, 1998년, §6.2 순서집합의 내부구조, 200쪽–206쪽 참조.

7. Give an example of a partially ordered set that has more than one maximal element and more than one minimal element.

8. Prove that if a totally ordered set has a maximal(minimal) element, then it has a unique maximal(minimal) element.

9. Give an example of a totally ordered set that has neither a maximal element nor a minimal element.

10. Let $(\mathscr{T}, \subseteq)$ be as in Theorem 2, and let $\mathscr{T}_0$ be a totally ordered subset of $\mathscr{T}$. Prove that $T_0 = \cap\{T \in \mathscr{T} \mid T \supseteq T' \text{ for all } T' \in \mathscr{T}_0\}$ is the least upper bound of $\mathscr{T}_0$.

11. Prove the following. Let $(A, \leqq)$ be a nonempty partially ordered set such that every totally ordered subset of A has a greatest lower bound. If $f: A \to A$ is such that $f(a) \leqq a$, for every $a \in A$, then there exists $q \in A$ such that $f(q) = q$.(Hint: Use Theorem 1.)

12. Let $(A, \leqq)$ be a partially ordered set and let B be a totally ordered subset of A. Prove that A has a maximal totally ordered subset C such that $B \subseteq C$.(Hint: Use Theorem 2.)

13. Let $(A, \leqq)$ be a partially ordered set. A subset B of A is called an antichain if and only if for any two distinct elements x and y in B, neither $x \leqq y$ or $y \leqq x$. Prove that every antichain is contained in an antichain which is maximal with respect to inclusion $\subseteq$.(Hint: Use Theorem 2.)

6.3 Zorn's Lemma

Perhaps one of the most widely used equivalent forms of the axiom of choice is Zorn's lemma, which first appeared in 1914. The name Zorn's lemma, which has been popularly used, is somewhat misleading; " Zorn's principle " would have been a more suitable name.

In Theorem 2, we actually proved that the axiom of choice implies the Hausdorff maximality principle. In the proof of the next theorem, we shall show that the Hausdorff maximality principle implies Zorn's lemma.

7. 부분순서집합으로써 극대원소와 극소원소를 각각 둘 이상 존재하는 것의 보기를 들어라.

8. 전순서집합이 극대원소(극소원소)를 지니면 어느 것이고 단 하나뿐이다. 이것을 증명하여라.

9.* 전순서집합의 전순서부분집합으로써 극대원소, 극소원소 어느 것도 존재하지 않는 것의 보기를 들어라.

10. 정리 2(301쪽)에서의 부분순서집합 $(\mathscr{T}, \subseteq)$에 있어서 $\mathscr{T}_0$은 $\mathscr{T}$의 전순서부분집합이

$$T_0 = \cap\{T \in \mathscr{T} \mid \text{모든 } T' \in \mathscr{T}_0 \text{에 대하여 } T' \subseteq T\}$$

은 $\mathscr{T}_0$의 최소상계이다. 이를 증명하여라.

다음 11－13을 증명하여라.

11. 공집합이 아닌 부분순서집합 $(A, \leqq)$에 대하여 A의 각 전순서부분집합이 최대하계를 지닐 때 함수 $f: A \to A$로써 각 $a \in A$에 대하여 $f(a) \leqq a$이면 $f(q) = q$인 $q \in A$가 존재한다[정리 1(297쪽)을 이용하여라].

12. 부분순서집합 $(A, \leqq)$의 전순서부분집합 B가 있을 때 A의 극대원소이며 전순서부분집합 이를테면 C가 존재함으로써 $B \subseteq C$[정리 2(301쪽)를 이용하여라].

13. 부분순서집합 $(A, \leqq)$의 부분순서부분집합 B의 임의의 서로 다른 원소 x, y에 대하여 $x \leqq y$, $y \leqq x$ 중 어느 것도 성립하지 않으면 그리고 그때에만 B는 **반사슬**이라고 한다. 여기서 집합 A의 각 반사슬은 포함관계 $\subseteq$에 따른 A의 극대원소인 하나의 반사슬 C에 포함된다. 즉 $B \subseteq C$[정리 2(301쪽)를 이용하여라].

6.3 조른의 보조정리

아마도 선택공리와 동치인 형식 중에서 가장 널리 이용되는 것으로는 1914년에 비로소 나타난 **조른의 보조정리**라고 본다. 평판이 좋은 이 보조정리는 그 이름을 놓고 약간 오해하기 쉽다. 조른의 보조정리라기보다는 **조른의 원리**라고 부르는 것이 옳을 것이다. 실제로 정리 2에서는 선택공리로부터 하우스도르프의 원리를 유도하였지만 다음 정리의 증명에서는 하우스도르프의 원리로부터 조른의 보조정리를 유도하고 있다.

---

* 보기에 대한 주장의 결론을 부정함으로써 모순을 일으켜 보아라. 한편 극대원소, 극소원소 중 어느 하나만 존재하는 경우가 있을까?

Theorem 3. Zorn's Lemma* Let $(A, \leqq)$ be a partially ordered set in which every totally ordered subset has an upper bound. Then A has a maximal element.

[Proof] By the Hausdorff maximality principle, $(A, \leqq)$ has a totally ordered subset B that is maximal with respect to set inclusion $\subseteq$. Let u be an upper bound of B; u exists by hypothesis. We shall prove that the element u is a maximal element of A. If there is an element $x \in A$ such that $x \geqq u$, then $B \cup \{x\}$ forms a totally ordered subset of $(A, \leqq)$ that includes the maximal totally ordered subset B. Consequently, we must have $B \cup \{x\} = B$, whence $x \leqq u$. This proves that u is a maximal element of $(A, \leqq)$.

Theorem 3 is a typical existential (as opposed to constructive) proposition; it merely asserts the existence of a maximal element in a certain partially ordered set. The proof of Theorem 3 gives no constructive method of finding such a maximal element. A similar comment can be made about Theorem 2 and all the results in the remainder of this chapter.

As an application of Zorn's lemma, we shall establish the following theorem, the proof of which was promised at the end of Section 2, Chapter 5.

Theorem 4 Let A and B be nonempty sets. Then either there is an injection from A to B or there is an injection from B to A.

[Proof] We consider the set $\mathscr{X}$ of all pairs (A_α, f_α), where A_α is a subset of A and $f_\alpha : A_\alpha \to B$ is an injection. We define the relation $\leqq$ on $\mathscr{X}$ by writing

$$(A_\alpha, f_\alpha) \leqq (A_\beta, f_\beta) \text{ if and only if } A_\alpha \subseteq A_\beta \text{ and } f_\alpha \subseteq f_\beta$$

정리 3 **조른의 보조정리*** 부분순서 집합 $(A, \leq)$의 각 전순서부분집합 즉 사슬 상계를 지니면 A의 극대원소가 존재한다.

[증명] 하우스도르프의 원리에 따라 $(A, \leq)$는 집합의 포함관계 ⊆로 볼 때 극대원소의 하나인 전순서부분집합 이를테면 B를 지닌다. 지금 B의 상계를 u(가정에 따라 이러한 u는 존재한다)라고 할 때 u는 A의 극대원소임을 증명하면 된다. 그러면 $x \in A$가 존재함으로써 $x \geqq u$이면 $B \cup \{x\}$는 $(A, \leq)$의 완전순서부분집합으로써 B를 포함한다. 그러므로 $B \cup \{x\} = B$. 따라서 $x \leqq u$. 결국 $x = u$이므로 u는 $(A, \leq)$의 극대원소이다.

오른쪽 그림은 집합 $A = \{a, b, c, d, e, f\}$ 위에서의 반순서를 나타낸 것이다. 이 경우 A의 모든 쇄

$$\{a, b, c\}, \ \{a, b, e, f\}, \ \{d, e, f\}$$

의 상계는 차례로 c, f, f이며 그 중 c, f는 각각 A의 극대원이다(또 다른 사슬에 대해서도 알아보아라).

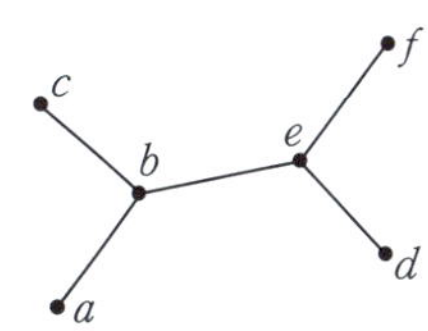

정리 3은 단지 하나의 부분순서집합에서 극대원소의 존재를 주장하는 존재명제라 하지만 정리 3으로는 주어진 부분순서집합의 극대원소를 구체적으로 구할 수 있는 방법을 알 수 없다. 이런 약점은 정리 2(301쪽)와 이 장에서의 나머지 모든 결과에 대해서도 말할 수 있다.

조른의 보조정리의 응용 중 하나로서 다음 정리를 소개한다.

정리 4 공집합이 아닌 집합 A, B에 대하여 A에서 B로의 단사 아니면 B에서 A로의 단사가 존재한다.

[증명]** A의 부분집합 A_α와 단사 $f_\alpha : A_\alpha \to B$에 대한 모든 쌍 (A_α, f_α)의 집합을 $\mathscr{X}$라 하고 그 위의 관계 ≦를 다음과 같이 정의한다.

$$A_\alpha \subseteq A_\beta \text{이고 } f_\alpha \subseteq f_\beta \text{이면 그리고 그때에만 } (A_\alpha, f_\alpha) \leqq (A_\beta, f_\beta)$$

* 원서의 "lemma"는 "보조정리" 또는 "보제"의 뜻으로 사용되는 용어로서 "가정된 것" "당연한 것"의 어원을 지니고 있다.

** 정리 4의 주장은 지극히 단순하나 그 증명이 벅찬 독자라면 해당 보기를 한두 개 들어보아라.

This relation is clearly a partial order relation. In order to apply Zorn's lemma, we need to be sure that any totally ordered subset $\mathscr{T}=\{(A_\gamma, f_\gamma) \mid \gamma\in\Gamma\}$ of $\mathscr{X}$ has an upper bound. One natural candidate for an upper bound for $\mathscr{T}$ is $(\bigcup_{\gamma\in\Gamma} A_\gamma, \bigcup_{\gamma\in\Gamma} f_\gamma)$. Denote $A_1=\bigcup_{\gamma\in\Gamma} A_\gamma$ and $f_1=\bigcup_{\gamma\in\Gamma} f_\gamma$; then

$$f_1 : A_1 \longrightarrow B \text{ is given by } f_1(x)=f_\gamma(x)$$

if $x\in A_\gamma$ and $(A_\gamma, f_\gamma)\in\mathscr{T}$. To prove that $f_1 : A_1 \longrightarrow B$ is well-defined, suppose that x belongs to another subset A_δ, $\delta\in\Gamma$. Then $(A_\gamma, f_\gamma)\leqq(A_\delta, f_\delta)$ or $(A_\delta, f_\delta) \leqq (A_\gamma, f_\gamma)$, and we have $f_\gamma(x)=f_\delta(x)$ in either case. Therefore, $f_1 : A_1 \longrightarrow B$ is a well-defined function. Next, we show that $f_1 : A_1 \longrightarrow B$ is injective. Suppose that $f_1(x)=f_1(y)$ for some x and y in A_1. Then there exist (A_γ, f_γ) and (A_δ, f_δ) in $\mathscr{T}$ such that $x\in A_\gamma$, and $y\in A_\delta$. As before, either $(A_\gamma, f_\gamma) \leqq (A_\delta, f_\delta)$ or $(A_\delta, f_\delta) \leqq (A_\gamma, f_\gamma)$.

We may assume that the first alternative is true; then it follows that $f_\delta(x)=f_\delta(y)$, and hence $x=y$, because f_δ is injective. This proves that $f_1 : A_1 \longrightarrow B$ is injective. Thus, $(A_1, f_1) \geqq (A_\gamma, f_\gamma)$ for all $\gamma\in\Gamma$, so that $(\mathscr{X}, \leqq)$ satisfies the hypothesis of Zorn's lemma. Hence $\mathscr{X}$ has a maximal element, which we denote by $(\tilde{A}, \tilde{f})$. There are two obvious cases:

Case 1. $\tilde{A}=A$.

In this case, $\tilde{f} : A \longrightarrow B$ is an injection and the theorem holds.

Case 2. $\tilde{A}\neq A$.

Let $x_0\in A-\tilde{A}$. In this case, we claim that $\tilde{f} : \tilde{A} \longrightarrow B$ is bijective. For otherwise, there exists an element $y_0\in B-\tilde{f}(\tilde{A})$. The function $\tilde{\tilde{f}} : \tilde{A}\cup\{x_0\} \longrightarrow B$, defined by $\tilde{\tilde{f}}(x_0)=y_0$ and $\tilde{\tilde{f}}(x)=\tilde{f}(x)$ for all $x\in\tilde{A}$, is clearly an injection. Thus $(\tilde{A}\cup\{x_0\}, \tilde{\tilde{f}}) > (\tilde{A}, \tilde{f})$ contradicting the maximality of $(\tilde{A}, \tilde{f})$. This proves that $\tilde{f} : \tilde{A} \longrightarrow B$ is bijective and consequently, $\tilde{f}^{-1}$ defines an injection of B onto the subset $\tilde{A}$ of A. The proof is now complete.

[증명] 이 관계 $\leqq$는 분명히 집합 $\mathscr{X}$ 위의 부분순서이다. 여기서 조른의 보조정리를 적용코자 $\mathscr{X}$의 임의의 전순서부분집합 $\mathscr{T}=\{(A_\gamma, f_\gamma) \mid \gamma\in\Gamma\}$로 하여금 상계를 지니는가를 확인할 필요가 있다. 그런데 $\mathscr{T}$의 상계로서 적절한 후보 중의 하나로 $(\bigcup_{\gamma\in\Gamma}A_\gamma, \bigcup_{\gamma\in\Gamma}f_\gamma)$를 들 수 있으므로

$$A_1=\bigcup_{\gamma\in\Gamma}A_\gamma, \quad f_1=\bigcup_{\gamma\in\Gamma}f_\gamma$$

로 각각 놓고 $x\in A_\gamma$, $(A_\gamma, f_\gamma)\in\mathscr{T}$이면 $f_1(x)=f_\gamma(x)$로써 함수 $f_1 : A_1 \to B$를 정의한다. 지금 $f_1 : A_1 \to B$가 잘 정의되었는가를 확인하기 위하여 x가 다른 집합, 이를테면 A_δ, $\delta\in\Gamma$에 속한다면 다음이 성립한다.

$$(A_\gamma, f_\gamma)\leqq(A_\delta, f_\delta) \text{ 아니면 } (A_\delta, f_\delta)\leqq(A_\gamma, f_\gamma) \qquad \cdots\cdots ①$$

어느 경우에나 $f_\gamma(x)=f_\delta(x)$(과연?). 그러므로 $f_1 : A_1 \to B$는 잘 정의된 함수이다. 이제 $f_1 : A_1 \to B$가 단사인 것을 보이기 위하여 A_1의 적당한 원소 x, y에 대하여 $f_1(x)=f_1(y)$라고 가정하면 $\mathscr{T}$에 (A_γ, f_γ), (A_δ, f_δ)가 각각 존재함으로써 $x\in A_\gamma$, $y\in A_\delta$.

한편 ①에 있어서 $(A_\gamma, f_\gamma)\leqq(A_\delta, f_\delta)$이라면 $f_\delta(x)=f_\delta(y)$. 그런데 함수 f_δ는 단사이므로 $x=y$. 이것은 $f_1 : A_1 \to B$가 단사임을 뜻한다. 그러므로 각 $\gamma\subset\Gamma$에 대하여 $(A_\gamma, f_\gamma)\leqq(A_1, f_1)$.

따라서 $(\mathscr{X}, \leqq)$는 조른의 보조정리의 가정을 만족하므로 그 결과로써 $\mathscr{X}$는 극대원소를 지닌다. 이것을 $(\tilde{A}, \tilde{f})$로 놓다보면 다음과 같은 두 경우가 생긴다.

경우 (1) $\tilde{A}=A$

이런 경우 $\tilde{f} : A \to B$는 단사이므로 정리 4는 성립하다.

경우 (2) $\tilde{A}\neq A$

지금 $x_0\in A-\tilde{A}$라 놓고 $\tilde{f} : \tilde{A} \to B$가 전단사임을 밝히려 한다. 만일 그렇지 않다면 원소 $y_0\in B-\tilde{f}(\tilde{A})$가 존재한다. 여기서 함수 $\tilde{\tilde{f}} : \tilde{A}\cup\{x_0\} \to B$를

$$\tilde{\tilde{f}}(x_0)=y_0 \text{ 이고 모든 } x\in\tilde{A} \text{ 에 대하여 } \tilde{\tilde{f}}(x)=\tilde{f}(x)$$

로 정의하면 이 함수는 분명히 단사이므로 $(\tilde{A}\cup\{x_0\}, \tilde{\tilde{f}})>(\tilde{A}, \tilde{f})$. 이것은 $(\tilde{A}, \tilde{f})$가 $\mathscr{X}$의 극대원소라는 데 모순이므로 $\tilde{f} : \tilde{A} \to B$는 전단사이다.

따라서 $\tilde{f}^{-1}$는 B에서 A의 부분집합 $\tilde{A}$ 위로의 단사이다.

이상으로 증명이 끝난다.

The following is an immediate consequence of Theorem 4.

Corollary. Let A and B be sets. Then, either

$$\text{card}\, A \leqq \text{card}\, B \quad \text{or} \quad \text{card}\, B \leqq \text{card}\, A.$$

Thus, if m and n are two distinct cardinal numbers, then either $m<n$ or $n<m$.

Exercise 6.3

Prove the following problems 1 through 9.

1. Let $(A, \leqq)$ be a partially ordered set in which every totally ordered subset has a lower bound. Then, A has a minimal element.
2. Zorn's lemma implies the Hausdorff maximality principle.
3. Every ring with identity has a proper maximal ideal.
4. Every vector space has a basis.
5. The corollary to Theorem 4.
6. A partially ordered set $(L, \leqq)$ is called a lattice if and only if every subset $\{x, y\}$ with any two elements of L has a (unique) least upper bound, denoted by $x \vee y$, and a (unique) greatest lower bound, denoted by $x \wedge y$. A lattice $(L, \leqq)$ in which every chain has an upper bound has a unique maximal element.
7. Let $(A, \leqq)$ be a partially ordered set in which every totally ordered subset has an upper bound, and let $a \in A$. Then, A has a maximal element u such that $u \geqq a$.
8. Let B be a set. A family $\mathscr{F}$ of subsets of B is said to have finite character provided that $A \in \mathscr{F}$ if and only if every finite subset of A belongs to $\mathscr{F}$. If $\mathscr{F}$ has finite character, then $(\mathscr{F}, \subseteq)$ has a maximal element.
9. Let A be an arbitrary set with more than one element. Then there exists a bijection $f: A \to A$ such that $f(x) \neq x$, for all $x \in A$.

정리 4로부터 직접 얻은 결과로서 다음과 같은 따름정리가 있다.

따름정리 임의의 집합 A, B에 대하여 다음이 성립하다.

$$\text{card } A \leqq \text{card } B \quad \text{아니면} \quad \text{card } B \leqq \text{card } A$$

특히 서로 다른 기수 m, n에 대해서는 $m<n$ 아니면 $n<m$.

연습문제 6.3

다음 1(b)~9를 증명하여라.

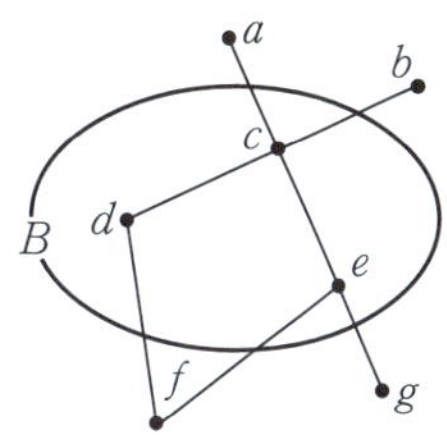

1. (a)* 오른쪽 그림은 부분순서집합 $A=\{a,b,c,d,e,f\}$의 부분순서부분집합 $B=\{c,d,e\}$를 나타낸다(x가 y로 올라가면 $x\leqq y$). 다음을 구하여라.

B의 상계, B의 하계, $\sup B$, $\inf B$

(b) 부분순서집합 $(A,\leqq)$의 각 전순서부분집합이 하계를 지니면 A의 극소원소가 존재한다(305쪽 정리 3의 증명을 참고하여라).

2. 조른의 보조정리는 하우스도르프의 극대원리를 함의한다.

3. **항등원**을 지닌 **환**엔 **진극대아이디얼**이 존재한다(추상대수학의 환이론 참조).

4. 임의의 **벡터공간**에는 **기저**가 존재한다[선형대수학(Linear Algebra)의 벡터공간 참조].

5. 정리 4의 따름정리

6. 부분순서집합 $(L,\leqq)$에 있어서 L의 임의의 두 원소로 구성된 부분순서부분집합 $\{x,y\}$의 최소상계, 최대하계가 각각 L에 꼭 하나씩 존재하면 그리고 그때에만 $(L,\leqq)$은 **격자****라고 하며 $\{x,y\}$의 최소상계, 최대하계를 각각 $x\vee y$, $x\wedge y$로 나타낸다. 이를테면 집합 $A(A\neq\phi)$의 멱집합 $\wp(A)$는 포함관계를 순서로 하는 격자이다. 한편 지금 격자 $(L,\leqq)$의 각 사슬이 상계를 지니면 그 격자에는 극대원소가 단 하나 존재한다.

7. 부분순서집합 $(A,\leqq)$에 있어서 A의 각 전순서부분집합이 상계를 지니고 $a\in A$일 때 A에는 $a\leqq u$인 극대원소 u가 존재한다.

8. 임의의 집합 B의 부분집합이 원소인 집합족 $\mathscr{F}$로 하여금

$$A\in\mathscr{F}\Longleftrightarrow A\text{의 각 유한부분집합이 }\mathscr{F}\text{에 속한다.}$$

를 만족할 때 $\mathscr{F}$는 **유한지표**를 지닌다고 한다. 실제로 $\mathscr{F}$가 유한지표를 지니면 $(\mathscr{F},\subseteq)$에는 극대원소가 존재한다.

9. 원소가 두 개 이상인 임의의 집합 A에 대하여 $\forall x\in A\ f(x)\neq x$로 정의된 전단사 $f: A\rightarrow A$가 존재한다.

---

* (a)는 옮긴이가 보충한 것이다.

** 이장우, 《현대집합론의 원리》, 경문사, 1998년, 6.3 속 격자(206−215쪽). 특히 그림 6.8 및 6.9 참조. 이석종, 《집합과 논리》, 교우사, 2005년, 6.4 격자(199−210쪽) 참조.

6.4 The Well-Ordering Principle

We now make another application of Zorn's lemma to prove an astonishing principle in set theory, the Well-Ordering Principle of Ernst Zermelo (1871-1953).

Definition 4. A totally ordered set $(A, \leqq)$ is said to be well-ordered if and only if every nonempty subset B of A contains (unique) minimal element; that is, if there exists an element $b \in B$ such that $b \leqq x$ for every $x \in B$. Such an element b is called the least element of B. If $(A, \leqq)$ is a well-ordered set, then the relation $\leqq$ is called a well-order relation.

Example 5 (a) The set N of natural numbers is well ordered under the ordinary "less than or equal" relation.

(b) The set of rational numbers under the ordinary "less than or equal" relation is not well-ordered.

Let us say that a set A can be well-ordered if there exists a well-order relation on A.

Theorem 5. Well-Ordering Principle Any nonempty set can be well-ordered.

[Proof] Let A be an arbitrarily given set which is to be well-ordered. Consider the family A^* of all well-ordered sets $(A_0, \leqq_0)$, where $A_0 \subseteq A$. We partially order A^* by writing $(A_0, \leqq_0) \leqq^* (A_1, \leqq_1)$ if and only if

(i) $A_0 \subseteq A_1$

(ii) $x, y \in A_0$ and $x \leqq_0 y$ imply $x \leqq_1 y$ and

(iii) $x \in A_1 - A_0$ implies $y \leqq_1$ for all $y \in A_0$.

6.4 정렬원리

이제 또다시 조른의 보조정리를 응용하여 에른스트 체르멜로(1871~1953)의 정렬원리를 알아보기 전에 다음 정의부터 내리자.

정의 4* 전순서집합 $(A, \leqq)$의 공집합이 아닌 각 부분집합 B가 (단 하나의) 극소원소를 지니면, 그리고 그때에만 $(A, \leqq)$를 **정렬집합**이라 한다. 즉 각 $x \in B$에 대하여 $b \leqq x$를 만족하는 B의 원소 b가 존재함을 뜻하며 이러한 b를 B의 **최소원소**라고 한다. 그리고 정렬집합 $(A, \leqq)$에서의 관계 $\leqq$를 A 위의 **정렬순서**라고 한다.

보기 5 (a) 집합 $(\mathbb{N}, \leqq)$은 대소 관계 $\leqq$에 따른 정렬집합이다.
(b) 집합 $\mathbb{Q}$는 대소 관계 $\leqq$에 따른 정렬집합이 아니다(왜?).

앞으로 정렬순서가 존재하는 집합 A를 두고 **정렬이 가능하다**고 말한다.

정리 5 **정렬원리** 공집합이 아닌 임의의 집합은 정렬이 가능하다.

[증명] 임의의 주어진 집합 A에 대한 정렬이 가능하도록 모든 정렬순서집합 $(A_0, \leqq_0)$(여기서 $A_0 \subseteq A$)의 족 A^*를 생각하고 그 위의 부분순서 $\leqq^*$를 다음 세 조건

(i) $A_0 \subseteq A_1$
(ii) $x, y \in A_0$에 대하여 $x \leqq_0 y \Longrightarrow x \leqq_1 y$
(iii) $x \in A_1 - A_0$일 때 $\forall_{y \in A_0}\, y \leqq_1 x \Longleftrightarrow (A_0, \leqq_0) \leqq^* (A_1, \leqq_1)$.

* (1) 여기서는 전순서이지만 책에 따라서는 부분순서를 놓고 다룬 경우가 많다.
(2) "최대원소 : the greatest element"의 정의를 내리고 최소원소, 극소원소 사이의 차이점에 주목하여라.

The reader should verify that this relation $\leqq^*$ is really a partial order relation on A^*. In order to apply Zorn's lemma, we show that any totally ordered subset $\mathscr{B}$ of $(A^*, \leqq^*)$ has an upper bound. The natural candidate for this upper bound is $(\bigcup_{A\in\mathscr{B}} A, \leqq')$[1] where $x\leqq' y$ if both x and y belong to some A_0 such that $(A_0, \leqq_0)\in\mathscr{B}$, and $x\leqq_0 y$. Obviously, $(\bigcup_{A\in\mathscr{B}} A, \leqq')$ is an upper bound for $\mathscr{B}$ if $(\bigcup_{A\in\mathscr{B}} A, \leqq')$ belongs to A^*. We shall prove that $(\bigcup_{A\in\mathscr{B}} A, \subseteq')$ is well-ordered and hence belongs to A^*. Part of this proof includes the routine verification, which we leave to the reader, that $\leqq'$ is a total order relation on $\bigcup_{A\in\mathscr{B}} A$. Let S be a nonempty subset of $\bigcup_{A\in\mathscr{B}} A$. Then, there exists $(A_0, \leqq_0)\in\mathscr{B}$ such that A_0 intersects S. Since $(A_0, \leqq_0)$ is well-ordered, $S\cap A_0$ contains a unique least element, say x_0, (of $S\cap A_0$). It follows that for any $y\in S$, there exists

$$(A_1, \leqq_1)\in\mathscr{B} \text{ such that } A_0, \leqq_0)\leqq^*(A_1, \leqq_1) \text{ and } x_0, y\in A_1;$$

whence, $x_0\leqq_1 y$ and hence $x_0\leqq' y$. Thus, x_0 is the least element for S under $\leqq'$, so that $(\bigcup_{A\in\mathscr{B}} A, \leqq')$ is well-ordered.

By Zorn's lemma, $(A^*, \leqq^*)$ has a maximal element $(A_1, \leqq_1)$. We claim that $A_1=A$ and hence $(A, \leqq_1)$ is well-ordered. For, if $A_1\neq A$, take any $x_1\in A-A_1$ and extend $\leqq_1$ to $A_1\cup\{x_1\}$ by defining $x\leqq_1 x_1$ for all $x\in A_1$; then $(A_1\cup\{x_1\}, \leqq_1)$ is strictly greater than $(A_1, \leqq_1)$ under $\leqq^*$, which contradicts the maximality of $(A_1, \leqq_1)$. The proof of the well-ordering principle is now complete.

The well-ordering principle is another outstanding example of a nonconstructive proposition. The proof of Theorem 5 gives no indication of how a "well-ordering" of the elements of A is to be accomplished; it merely asserts the existence of a well-order relation. In fact, it is not even known how the set of real numbers can be well-ordered.

---

[1] For simplicity of notation, hereafter the subscript $A\in\mathscr{B}$ will mean $(A, \leqq)\in\mathscr{B}$.

독자는 이렇게 정의된 $\leqq^*$가 과연 A^* 위의 부분순서인가를 확인하여 봐야한다. 먼저 조른의 보조정리를 적용하기 위하여 $(A^*, \leqq^*)$의 임의의 전순서부분집합 $\mathscr{B}$는 상계를 지닌다는 것을 보이도록 한다. 다행히 그것의 후보의 하나는 $(\bigcup_{A\in\mathscr{B}} A, \leqq')$[1]이다. 여기서 x와 y가 적당한 A_0에 속함으로써 $(A_0, \leqq_0)\in\mathscr{B}$이고 $x\leqq_0 y$이면 $x\leqq' y$. 여기서 $(\bigcup_{A\in\mathscr{B}} A, \leqq')$이 A^*에 속하면 그것은 분명히 $\mathscr{B}$의 상계이다.

이제 $(\bigcup_{A\in\mathscr{B}} A, \leqq')$이 정렬집합으로써 A^*에 속한다는 것을 증명하려면 관계 $\leqq'$은 $\bigcup_{A\in\mathscr{B}} A$ 위의 전순서인 것을 먼저 밝혀야 하지만 그 증명의 일부는 독자에게 맡기기로 한다.

우리는 $\bigcup_{A\in\mathscr{B}} A$의 부분집합으로써 공집합이 아닌 것의 하나를 S라면 $\mathscr{B}$에는 $(A_0, \leqq_0)$인 것이 존재하므로써 S와 A_0의 교집합을 얻는다. 그런데 $(A_0, \leqq_0)$는 정렬집합이므로 $S\cap A_0$에는 단 하나의 최소원소, x_0이 존재한다. 한편 임의의 $y\in S$에 대하여 존재한 $(A_1, \leqq_1)\in\mathscr{B}$이 존재함으로써 $(A_0, \leqq_0)\leqq^*(A_1, \leqq_1)$이고 x_0, $y\in A$. 그러므로 $x_0\leqq_1 y$ 따라서 $x_0\leqq' y$. 결국 x_0은 $\leqq'$에 관한 S의 최소원소이므로 $(\bigcup_{A\in\mathscr{B}} A, \leqq')$은 정렬집합이다.

여기서 $(A^*, \leqq^*)$은 조른의 보조정리에 따라 극대원소 $(A_1, \leqq_1)$을 지닌다. 이제 남은 것으로는 $A_1=A$이면 $(A_1, \leqq_1)$이 정렬집합임을 보이면 충분한다. 이에 반하여 $A_1\neq A$라고 가정하는 경우 임의의 $x_1\in A-A_1$를 택하여 $\forall x\in A_1 x\leqq_1 x_1$와 같이 정의함으로써 $\leqq_1$을 $A_1\cup\{x_1\}$로 확장한다면 $(A_1\cup\{x_1\}, \leqq_1)$은 엄밀한 뜻에서 $\leqq^*$에 관하여 $(A_1, \leqq_1)$보다 크다. 이런 사실은 $(A_1, \leqq_1)$이 $(A^*, \leqq^*)$의 극대원소라는 데에 모순이다. 이상으로 증명이 끝난다.

정렬원리는 다음 세 명제(315쪽 위에서 2, 3행)와 함께 또하나의 좋은 사례가 된다. 그러나 정리 5는 집합 A 위의 정렬이 존재한다는 것을 주장할 뿐 그 정렬에 따라 A의 원소를 배열할 수 있는 방법은 커녕 지금껏 실수의 집합에 대한 정렬가능성 조차도 알려져 있지 않다.**

[1] 앞으로는 $(A, \leqq)\in\mathscr{B}$를 간단히 $A\in\mathscr{B}$로 나타내기로 한다.

** 유한집합 $A=\{a, b, c\}$는

$$a<b<c,\quad a<c<b,\quad b<c<a,\quad c<b<a$$

와 같이 대소관계 $<$에 따른 순서 정렬이 가능하다. 그러나 정렬원리는 역시 무한집합이라야 원리로써 뜻을 지닌다.

We have now shown the following chain of implications:

$$\begin{aligned} \text{axiom of choice} &\xRightarrow{\text{Th. 2}} \text{Hausdorff maximality principle} \\ &\xRightarrow{\text{Th. 3}} \text{Zorn's lemma} \\ &\xRightarrow{\text{Th. 5}} \text{well-ordering principle.} \end{aligned}$$

To complete the proof that all these four principles are equivalent, it is sufficient to show that the well-ordering principle implies the axiom of choice.

Theorem 6. The well-ordering principle implies the axiom of choice.

[Proof] Let $\mathscr{S}$ be any nonempty family whose elements are nonempty sets. By the well-ordering principle, there exists a total order relation $\leqq$ such that $(\bigcup_{A\in\mathscr{S}} A, \leqq)$ is well-ordered. Consequently, each set $A\in\mathscr{S}$ contains a least element. Therefore, $f:\mathscr{S}\rightarrow\bigcup_{A\in\mathscr{S}} A$, defined by $f(A)$=the least element of A, for all $A\in\mathscr{S}$, is a well-defined choice function. This proves the axiom of choice.

We have thus established the equivalence of the axiom of choice, the Hausdorff maximality principle, Zorn's lemma, and the well-ordering principle. In the remainder of this book we will accept the axiom of choice (and the three other equivalent principles) and use it freely.

Exercise 6.4

Prove the following problems 1 through 6.

1. The set of rational numbers Q can be well-ordered(Prove directly, without using Theorem 5).

2. Every subset of a well-ordered set is well-ordered under the inherited ordering.

지금까지 우리는 다음과 같은 함의 사이의 관련성에 대하여 살펴 보았다.

$$\text{선택공리} \overset{\text{정리 2}}{\Longrightarrow} \text{하우스도르프의 극대원리}$$
$$\overset{\text{정리 3}}{\Longrightarrow} \text{조른의 보조정리}$$
$$\overset{\text{정리 5}}{\Longrightarrow} \text{정렬원리}$$

이 네 원리 중 임의의 둘 씩이 동치인 것을 밝히려면 다음과 같이 정렬원리로 하여금 선택공리를 함의한다는 것을 보이면 충분하다.

정리 6 정렬원리는 선택공리를 함의한다.

[증명] 주어진 집합(ϕ가 아님)들이 원소인 족을 $\mathscr{S}$라고 하자. 그러면 정렬원리에 따라 $\bigcup_{A\in\mathscr{S}} A$ 위의 전순서 $\leq$가 존재함으로써 정렬집합 $(\bigcup_{A\in\mathscr{S}} A, \leq)$을 얻는다. 그러므로 각 집합 $A\in\mathscr{S}$에는 최소원소가 존재한다. 따라서

$$\forall_{A\in\mathscr{S}} f(A) = (A\text{의 최소원소})$$

로 정의된 함수 $f: \mathscr{S} \longrightarrow \bigcup_{A\in\mathscr{S}} A$는 바로 선택함수이다. 이로써 선택공리를 얻는다.

이상으로 선택공리를 비롯하여 그것과 동치인 하우스도르프의 극대원리, 조른의 보조정리, 정렬원리 차례로 살펴보았다.

앞으로 이 책에서는 선택공리는 물론 그것과 동치인 어느 것이나 자유로이 활용할 것이다.

연습문제 6.4

다음 1~7을 증명하여라.

1.* 유리수의 집합 Q는 정렬이 가능하다(정리 5를 피하고 직접 밝혀라).

2. 정렬집합 $(A, \leq)$의 임의의 부분집합 $B(B\neq\phi)$는 이어받은 순서 $\leq$에 따른 정렬집합이다.

* 여기서도 집합 Q에 대한 정렬가능성만 주장하고 있을 뿐이다.

3. The set of real numbers $\mathbb{R}$ under the usual "less than or equal" relation of real numbers is not well-ordered.

4. Let $(A, \leqq)$ be a partially ordered set such that every nonempty subset B contains a lower bound; that is, there exists $b \in B$ such that $b \leqq x$, for all $x \in B$. Then the partially ordered set $(A, \leqq)$ is totally ordered and hence well-ordered.

5. The relation $\leqq^*$ defined in the proof of Theorem 5 is a partial order relation on A^*.

6. The relation $\leqq'$ defined in the proof of Theorem 5 is a total order relation on $\bigcup_{A \in \mathscr{B}} A$.

7. Let $(A, \leqq)$ be a totally ordered set. A sequence of elements $a_1, a_2, a_3, \cdots$ in A is said to be strictly decreasing if $a_1 > a_2 > a_3 > \cdots$. Prove that a totally ordered set is well-ordered if and only if the totally ordered set contains no infinite strictly decreasing sequence.

6.5 The Principle of Transfinite Induction

For convenience in expressing the principle of transfinite induction and for the development of ordinal numbers in the next chapter, we now introduce the notion of a segment.

Definition 5. Let $(A, \leqq)$ be a totally ordered set. A segment of A is a subset S of A such that if $y \in S$, $x \in A$, and $x \leqq y$, then $x \in S$. A proper segment of A is a segment which is a proper subset of A.

Example 6 Let $(A, \leqq)$ be a well-ordered set, and let x be an arbitrary element of A. Then, the empty set ϕ, the set A, and the set $A_x = \{a \in A \mid a < x\}$ are segments of A.

3. 실수의 집합 $\mathbb{R}$은 그 위에서의 낯익은 대소 관계 $\leqq$에 따른 정렬집합이 아니다(먼저 반례 하나를 들어라. 이 경우도 정리 5에 따라 $\mathbb{R}$에 대한 정렬가능성만은 인정한다).

4. 부분순서 집합 $(A, \leqq)$에 대한 A의 각 부분집합 $B(B \neq \phi)$가 하계를 지닐때 즉 $b \in B$가 존재함으로써 모든 $x \in B$에 대하여 $b \leqq x$ 이면 $(A, \leqq)$는 전순서집합이므로 정렬집합이다.

5. 정리 5의 증명에서 관계 $\leqq^*$는 족 A^* 위의 부분순서이다.

6. 정리 5의 증명에서 관계 $\leqq'$은 $\bigcup_{A \in \mathscr{B}} A$ 위의 전순서이다.

7. 전순서집합 $(A, \leqq)$에 있어서 A의 원소의 수열 $a_1, a_2, a_3, \cdots$에 대하여

$$a_1 > a_2 > a_3 > \cdots$$

이면 이 수열은 엄밀한 뜻에서 감소한다고 말한다. 지금 전순서집합으로써 엄밀한 뜻에서 감소하는 무한순열을 지니지 않으면, 그리고 그때에만 주어진 집합은 정렬집합이다.

6.5 초한귀납법의 원리**

초한귀납법의 원리를 다룰 때나 다음 장에서 다룰 순서수(335쪽)에 대비하여 편의상 절편의 정의로부터 먼저 도입한다.

정의 5 전순서집합 $(A, \leqq)$의 임의의 부분집합 S에 대하여

$$x \in A,\ y \in S \text{ 이고 } x \leqq y \text{ 일 때 } x \in S$$

이면 그리고 그때에만 S는 A의 절편이라고 한다. 특히 S가 A의 진부분집합인 경우 이것을 A의 진절편***이라고 한다.

예제 6 (a) 정렬집합 $(A, \leqq)$의 임의의 원소 $x \in A$에 대하여 공집합 ϕ과 A 자신은 A의 절편이다(하지만 ϕ는 별로 뜻이 없다).

(b) $a \in A$에 대한 집합 $A_a = \{x \in A \mid x < a\}$는 A의 절편이다(과연?).

(c) 305쪽의 그림이 나타내는 집합 A에 대하여 $Se = \{a, b, d\}$(책에 따라 부분순서집합에서의 절편을 정의하기로 한다).

** 이장우, 《현대집합론의 원리》, 경문사, 1998년, 6.6절 초한귀납법(225–230쪽) 참조.

*** 정렬집합 $(\mathbb{N}, \leqq)$의 진절편의 하나로 $\mathbb{N}_{10} = \{a \in \mathbb{N} \mid a < 10\}$(325쪽 연습문제 6.5 문제 1)이 있다. 전체집합 $\mathbb{N}$에 대한 $\mathbb{N}_{10}^c$에 대해서도 알아보아라.

Theorem 7. Let $(A, \leqq)$ be a well-ordered set. Then,

(a) Any union or intersection of segments of A and all segments of a segment of A are again segment of A.

(b) For each segment S of A, except A itself, there exists an element $x \in A$ such that $S = A_x$, where $A_x = \{a \in A \mid a < x\}$.

[Proof] (a) Let $\mathscr{F}$ be a family of segments of A, and let $y \in \bigcup_{S \in \mathscr{F}} S$. If $x \in A$ and $x \leqq y$, then since y belongs to some segment $S_0 \in \mathscr{F}$ of A, we have $x \in S_0$ and hence $x \in \bigcup_{S \in \mathscr{F}} S$. Therefore, $\bigcup_{S \in \mathscr{F}} S$ is a segment of A.

Similarly, the intersection of the family $\mathscr{F}$ of segments of A is a segment of A (see Problem 3).

Let S be a segment of A and let T be a segment of S. Let $y \in T$, $x \in A$, and $x \leqq y$; we are to show that $x \in T$. First, since y belongs to the segment S, we have $x \in S$; then from $y \in T$, $x \in S$, $x \leqq y$, and the hypothesis that T is a segment of S, we have $x \in T$. This completes the proof that T is a segment of A.

(b) Let S be a segment of A such that $S \neq A$. Then, the nonempty set $A - S$ has the least element, say x. It is an easy exercise for the reader to verify that $S = A_x$.

The following theorem is a prelude to the principle of transfinite induction.

Theorem 8. Let $(A, \leqq)$ be a well-ordered set, and let $\mathscr{S}$ be a family of segments of A such that

(1) any union of members of $\mathscr{S}$ belongs to $\mathscr{S}$.

(2) if $A_x \in \mathscr{S}$, then $A_x \cup \{x\} \in \mathscr{S}$.

Then $\mathscr{S}$ contains all segments of A.

[Proof] Suppose that there exists an element $x \in A$ such that the segment $A_x \notin \mathscr{S}$. Then, since $(A, \leqq)$ is well-ordered, the nonempty subset $B = \{x \in A \mid A_x \notin \mathscr{S}\}$ has a least element a. Since $a \in B$, $A_a \notin \mathscr{S}$. If $y \in A$ and $y < a$, then $y \notin B$ and hence $A_y \in \mathscr{S}$.

정리 7 정렬집합 $(A, \leq)$에 대하여 다음이 성립한다.

(a) A의 절편들의 합집합, 교집합은 각각 A의 절편이다.

(b) A의 절편의 임의의 절편은 A의 절편이다.

(c) A의 각 절편 $S(S \neq A)$에 대하여 적당한 $x \in A$가 존재함으로써 $S = A_x$. 여기서 $A_x = \{a \in A \mid a < x\}$.

[증명] (a) A의 절편의 족을 $\mathscr{F}$, $y \in \bigcup_{S \in \mathscr{F}} S$라고 각각 놓고 $x \in A \wedge x \leqq y$이라면 y는 A의 절편 S_0에 속하므로 $x \in S_0$.

따라서 $x \in \bigcup_{S \in \mathscr{F}} S$. 그러므로 $\bigcup_{S \in \mathscr{F}} S$는 A의 절편이다. 마찬가지로 $\mathscr{S}$에서의 교집합도 A의 절편이다[325쪽 문제 3(a) 참조].

(b) A의 절편 S의 임의의 절편 T에 대하여 $y \in T$, $x \in A$, $x \leqq y$일 때, $x \in T$인 것을 보이고자 한다. 먼저 y는 절편 S에 속하므로 $x \in S$. 따라서 $y \in T$, $x \in S$, $x \leqq y$와 더불어 T는 S의 절편이란 가정으로부터 $x \in T$. 이로써 T는 A의 절편임을 알 수 있다.

(c) A의 절편을 $S(S \neq A)$라고 하면 공집합이 아닌 집합 $A - S$는 최소원소, 말하자면 x가 존재함으로써 $S = A_x$.

이것의 증명은 연습문제로 돌린다.

초한귀납법의 원리에 앞서 다음 정리부터 도입한다.

정리 8* 정렬집합 $(A, \leq)$의 절편의 족 $\mathscr{S}$에 대하여 다음 두 가정하에 아래 결론을 얻는다.

(1) $\mathscr{S}$에서의 임의의 합집합이 $\mathscr{S}$에 속한다.

(2) $A_x \in \mathscr{S} \Longrightarrow A_x \cup \{x\} \in \mathscr{S}$

결론 A의 각 절편은 $\mathscr{S}$에 속한다.

[증명] 적당한 $x \in A$에 대하여 $\mathscr{S}$에 속하지 않는 A의 절편 A_x가 존재한다고 가정하면 $(A, \leq)$는 정렬집합이므로 A의 부분집합 $B = \{x \in A \mid A_x \notin \mathscr{S}\}$ $(B \neq \phi)$은 최소원소, 말하자면 $a(a \in B)$가 존재함으로써 $A_a \notin \mathscr{S}$. 여기서 $y \in A$이고 $y < a$이면 $y \notin B$이므로 $A_y \in \mathscr{S}$.

* 우선 이 정리에 해당하는 보기부터 들어보고 증명은 피해가도 무방하다.

By the hypothesis (1), $\bigcup_{y<a} A_y \in \mathscr{S}$. By Theorem 7, there exists an element $b \in A$ such that $\bigcup_{y<a} A_y = A_b$. Consequently, $b<a$ (see Problem 9). Now using the hypothesis (2) and Theorem 7, we have

$$A_b \cup \{b\} = A_c \in \mathscr{S} \quad \text{for some } c \in A$$

Hence, we have $b<c<a$. Consequently,

$$b \in A_c \subseteq \bigcup_{y<a} A_y = A_b$$

which contradicts the fact that $b \notin A_b$. Therefore, $A_x \in \mathscr{S}$ for all $x \in A$.

It remains to be proved that $A \in \mathscr{S}$. Now by the hypothesis (1) we have $\bigcup_{x \in A} A_x \in \mathscr{S}$. If $\bigcup_{x \in A} A_x = A$, then there is nothing left to be proved. Suppose that $\bigcup_{x \in A} A_x \neq A$; then there exists an element $d \in A$ such that $\bigcup_{x \in A} A_x = A_d$. By the hypothesis (2), $A_d \cup \{d\} \in \mathscr{S}$. It follows that $x \leqq d$ for all $x \in A$, and hence $A = A_d \cup \{d\} \in \mathscr{S}$, completing the proof.

Remark. In Theorem 8, the hypothesis (1) implies $\phi \in \mathscr{S}$. For "$\phi \in \mathscr{S}$" may be deduced from the hypothesis (1), "any union of members of $\mathscr{S}$ belongs to $\mathscr{S}$", by taking the "empty union" of the members of $\mathscr{S}$ [see Theorem 7(a), Chapter 2]. Therefore, $\mathscr{S}$ is nonempty.

Theorem 9. Principle of Transfinite Induction

Let $(A, \leqq)$ be a well-ordered set. For each $x \in A$, let $p(x)$ be a statement about x. If for each $x \in A$, the hypothesis "$p(y)$ is true for every $y<x$" implies that "$p(x)$ is true," then $p(x)$ is true for every $x \in A$.

[Proof] Suppose there is some x such that $p(x)$ is false. Then $B = \{x \in A \mid p(x)$ is false$\}$ is a nonempty subset of A, and hence has the least element x_0. Since $x_0 \in B$, $p(x_0)$ is false. If $y \in A$ and $y < x_0$, then $y \notin B$, and hence $p(y)$ is true. Thus $p(y)$ is true for every $y < x_0$. Thus $p(x_0)$ is true, a contradiction. Therefore, $p(x)$ is true for all x in A.

따라서 가정 (1)에 의하여 $\bigcup_{y<a} A_y \in \mathscr{S}$. 한편 정리 7에 따라 적당한 $b \in A$가 존재함으로써 $\bigcup_{y<a} A_y = A_b$. 그러므로 $b<a$(325쪽 문제 9 참조). 이제 가정(2)와 정리 7을 이용하여 다음을 얻는다.

적당한 $c \in A$에 대하여 $A_b \cup \{b\} = A_c$, $A_c \in \mathscr{S}$ 즉 $A_b \cup \{b\} \in \mathscr{S}$. 그러므로 $b<c<a$. 따라서 $b \in A_c \subseteq \bigcup_{y<a} A_y = A_b$. 곧 $b \in A_b$.

이것은 $b \notin A_b$에 모순이다. 그러므로 각 $x \in A$에 대하여 $A_x \in \mathscr{S}$.

다음에는 $A \in \mathscr{S}$임을 증명하자. 가정 (1)에 따라 $\bigcup_{x \in A} A_x \in \mathscr{S}$. 지금 $\bigcup_{x \in A} A_x = A$일 때에는 더 증명할 것도 없다.

한편 $\bigcup_{x \in A} A_x \neq A$라고 가정하면 적당한 $d \in A$가 존재함으로써 $\bigcup_{x \in A} A_x = A_d$. 이 경우 가정 (2)에 따라 $A_d \cup \{d\} \in \mathscr{S}$. 여기서 각 $x \in A$에 대하여 $x \leqq d$이므로 $A = A_d \cup \{d\}$. 그런데 $A_d \cup \{d\} \in \mathscr{S}$이므로 $A \in \mathscr{S}$. 이 사실과 정리 7(a)의 두 번째 주장에 따라 주어진 결론을 얻는다.

주의 정리 8의 가정 (1)은 암암리에 $\phi \in \mathscr{S}$를 전제하고 있다. 왜냐하면 가정 (1) "$\mathscr{S}$에서의 임의의 합집합은 $\mathscr{S}$에 속한다."에 따라 원소로써 $\mathscr{S}$의 구성원으로 "공합집합[115쪽 정리 7(a)]"을 택하다보면 뜻밖에 "$\phi \in \mathscr{S}$"로 이어질 수 있기 때문이다. 그런 까닭에 $\mathscr{S} \neq \phi$이어야 한다.

정리 9 초한귀납법의 원리* 정렬집합 $(A, \leqq)$에서의 명제함수 $p(x)$에 대하여 다음 가정 하에 아래 결론을 얻는다.

가정 $\forall_{x \in A}\, y<x$인 모든 y에 대하여 $p(y)$가 참이면 $p(x)$도 참이다.
결론 $\forall_{x \in A}\, p(x)$는 참이다.

[증명] 이 정리의 결론에 반하여 $p(x)$로 하여금 거짓인 $x \in A$가 존재한다면 집합 $B = \{x \in A \mid p(x)\text{는 거짓이다}\}$는 A의 부분집합으로서 공집합이 아니다. 따라서 B에는 최소원소, 말하자면 $x_0(x_0 \in B)$이 존재하므로 $p(x_0)$은 거짓이다. 지금 $y \in A$, $y<x_0$이라면 $y \notin B$. 그러므로 $p(y)$는 참이다. 이때 모든 $y<x_0$에 관한 $p(y)$가 참이므로 가정에 따라 $p(x_0)$도 참이다. 이것은 $p(x_0)$가 거짓이라는 데에 모순이다.

따라서 주어진 결론을 얻는다. 이 원리에 바탕을 둔 증명법을 **초한귀납법**이라고 한다.

* 325쪽 문제 5 (a), (b) 참조. 한편 이 초한귀납법의 원리와 수학적 귀납법의 원리(61쪽 아래 구절) 사이의 차이를 말하여 보아라. 성급한 질문일지 모르지만 이 원리를 인용한 보기를 지금 들 수 있을까?

The next theorem is a typical example of how transfinite induction is used in proving mathematical results. This theorem will be used in the development of the ordinal numbers in Chapter 7. First we need another definition.

Let $(A, \leqq)$ and $(B, \leqq')$ be well-ordered sets. A function $f : A \rightarrow B$ is said to be increasing if and only if

$$a \leqq a' \text{ in } (A, \leqq) \text{ implies } f(a) \leqq' f(a') \text{ in } (B, \leqq'),$$

and to be strictly increasing if and only if

$$a < a' \text{ in } (A, \leqq) \text{ implies } f(a) <' f(a') \text{ in } (B, \leqq').$$

Theorem 10. Let $(A, \leqq)$, $(B, \leqq')$ be well-ordered sets. If $f : A \rightarrow B$ is increasing, $f(A)$ is a segment of B, and $g : A \rightarrow B$ is strictly increasing, then $f(x) \leqq' g(x)$ for all $x \in A$.

[Proof] In order to use transfinite induction, for each $x \in A$, let $p(x)$ be the statement "$f(x) \leqq' g(x)$." To show that the hypothesis of Theorem 9 is satisfied, we suppose on the contrary that there exists an element $a \in A$ such that $p(x)$ is true for all $x < a$, but $p(a)$ is false. That is,

$$f(x) \leqq' g(x) \text{ for all } x < a \quad \text{and} \quad g(a) <' f(a).$$

Since g is strictly increasing and f is increasing, we have

$$f(x) \leqq' g(x) <' g(a) <' f(a) \leqq' f(y)$$

for all $x < a$ and for all $a \leqq y$. It follows that

$$g(a) <' f(a) \text{ and } g(a) \notin f(A)$$

contradicting the fact that $f(A)$ is a segment of B. Therefore, for each $x \in A$,

$$\text{if } f(y) \leqq' g(y) \text{ for all } y < x, \text{ then } f(x) \leqq' g(x).$$

By the principle of transfinite induction, $f(x) \leqq' g(x)$ for all $x \in A$.

수학적 결과(다소 막연한 표현)를 얻기 위하여 어떻게 초한귀납법을 이용할 것인가에 대한 좋은 보기로서 다음과 같은 정리가 있다. 이 정리는 제 7 장에서 순서수를 다룰 때도 이용되겠지만 우선 다음 정의부터 내리자.

정렬집합 $(A, \leqq)$, $(B, \leqq')$에 있어서 함수 $f: A \to B$로 하여금 $a \leqq a'$인 a, $a' \in A$에 대하여 $(B, \leqq')$에서는 $f(a) \leqq f(a')$이면, 그리고 그때에만 f는 **증가한다**고 말한다. 특히 $a < a'$인 a, $a' \in A$에 대하여 $f(a) <' f(a')$이면 f는 **엄밀한 뜻에서 증가한다***고 말한다.

정리 10** 두 정렬집합 $(A, \leqq)$, $(B, \leqq')$에 대하여 다음과 같은 세 가정 하에 아래 결론을 얻는다.

가정 (1) 함수 말하자면 $f: A \to B$가 존재함으로써 증가한다.
(2) $f(A)$는 B의 절편이다.
(3) 함수 말하자면 $g: A \to B$가 존재함으로써 엄밀한 뜻에서 증가한다.

결론 $\forall_{x \in A} f(x) \leqq' g(x)$

[증명] 초한귀납법을 적용하고자 각 $x \in A$에 관한 명제함수 "$f(x) \leqq' g(x)$"를 $p(x)$로 놓는다. 여기서 $p(x)$가 정리 9의 가정을 만족하는가를 보고자 이에 반하여 적당한 $a \in A$가 존재함으로써 $x < a$인 모든 x에 대하여 $p(x)$가 참이지만 $p(a)$는 거짓, 즉 $x < a$인 모든 x에 대하여

$$f(x) \leqq' g(x) \wedge g(a) <' f(a)$$

라고 가정한다. 한편 이 정리의 가정 (1), (3)에 따라 $x < a$인 모든 x, $a \leqq y$인 모든 y에 대하여 다음을 얻는다.

$$f(x) \leqq' g(x) <' g(a) <' f(a) \leqq' f(y)$$

그러므로 $g(a) <' f'(a)$이고 $g(a) \notin f(A)$(이를 확인하여라).

이것은 가정 (2)에 모순이다. 따라서 각 $x \in A$에 대하여, $y < x$인 모든 y에 대하여 $f(y) \leqq' g(y)$이면 $f(x) \leqq' g(x)$.

그러므로 초한귀납법의 원리에 따라 모든 $x \in A$에 대하여

$$f(x) \leqq' g(x).$$

---

* 그러한 일차함수 하나를 들어라. 용어 "엄밀한 뜻에서 감소한다."(317쪽 문제 7)와 비교하여 봄직하다.

** 당장 안 떠오를지 모르지만 이 정리에 해당하는 보기를 한두 개 들어 보아라. 그런 훈련으로 차차 실력이 쌓인다. "기하학에 왕도(王道)는 없다."하지 않던가?

Exercise 6.5

1. Let N be the set of all natural numbers, and for each $k \in \mathrm{N}$, let $\mathrm{N}_k = \{1, 2, 3, \cdots, k\}$. Find all the segments of $(\mathrm{N}, \leqq)$, where $\leqq$ denotes the usual "less than or equal" relation for natural numbers.

2. Prove the statement of Example 6.

3. Let $(A, \leqq)$ be a well-ordered set. Prove that
   (a) The intersection of an arbitrary family of segments of A is a segment of A.
   (b) For each $x \in A$, $A_x \cup \{x\}$ is a segment of A.

4. Complete the proof of Theorem 7(b).

5. Give a direct proof of Theorem 9. (Hint : Use Theorem 8.)

6. Let $(A, \leqq)$ and $(B, \leqq')$ be well-ordered sets, and let $f : A \rightarrow B$ be increasing such that $f(A)$ is a segment of B. Prove that f takes every segment of A to a segment of B.

7. Prove that in a well-ordered set every subset that is bounded above has the (unique) least upper bound.

8. Let $(A, \leqq)$ be well-ordered. Prove that for each $x \in A$, x is the least upper bound of the segment A_x of A.

9. Prove that $b < a$ in the proof of Theorem 8. [Hint : Consider (i) $b = a$ and (ii) $a < b$.]

6.6 Historical Remarks

It may be worth knowing a brief history of the axiom of choice ; in many respects it is similar to that of Euclid's parallel axiom and of the continuum hypothesis (see Section 8, Chapter 5).

연습문제 6.5

1. 집합 N의 각 $k\in\mathbb{N}$에 대하여 $\mathbb{N}_k=\{1,2,3,\cdots,k\}$일 때 $(\mathbb{N},\leqq)$의 모든 절편을 구하여라. 물론 ≦은 누구에게나 잘 알려진 두 자연수 사이의 대소 관계를 나타낸다.

2. 317쪽 예제 6(a), (b)를 증명하여라.

3. $(A,\leqq)$가 정렬집합일 때 다음을 증명하여라.
(a) A의 절편의 임의의 족 $\mathscr{T}$에서 교집합은 A의 절편이다.
(b) 각 $x\in A$에 대한 $A_x\cup\{x\}$는 A의 절편이다.

4. 정리 7(c)의 증명을 완성하여라.

5. 집합 $A=\mathbb{N}\cup\{-10^{100},-10^{100}+1,\cdots,-2,-1\}$에 대하여
(a)* $(A,\leqq)$는 대소 관계 ≦에 따른 정렬집합임을 보여라.
(b) 정리 8을 이용하여 정리 9를 직접 증명하여라.

6. 정렬집합 $(A,\leqq)$에서 정렬집합 $(B,\leqq')$로의 함수 $f:A\to B$는 증가하고 $f(A)$가 B의 절편이면 A의 임의의 절편 X에 대한 $f(X)$는 B의 절편이다. 이것을 증명하여라.

7. 정렬집합 $(A,\leqq)$에 있어서 A의 각 부분집합 B가 위로 유계이면 A에는 단 하나의 최소상계가 존재한다. 이것을 증명하여라.

8. 정렬집합 $(A,\leqq)$에 대해서 A의 각 원소 x는 A의 절편 A_x의 최소상계이다. 이것을 증명하여라.

9. 정리 8의 증명에 있어서 $b<a$. 이것을 밝혀라[((i) $b=a$ (ii) $a<b$인 경우를 각각 따져 보아라].

6.6 역사적 단평**

이제 여러 점에서 유클리드의 평행선 공리나 연속체 가설(279~283쪽)과 유사하다는 선택공리에 관한 역사를 간단히 살펴보자.

---

* 문제 5 (a)는 옮긴이가 보충한 것이고 (b)는 원서 문제 5인 것이다.

** 이장우, 《현대집합론의 원리》, 경문사, 1998년, 7.6절 간단한 역사절 고찰(251–254쪽) 참조.

As early as the 1880's, Georg Cantor had already implicitly used, in the proof of some theorems, reasoning which was essentially equivalent to the axiom of choice; yet he was not aware of using a new powerful axiom. In 1904, Ernst Zermelo (1871-1953) after careful study stated the axiom of choice explicitly and used it to prove the earth-shaking well-ordering theorem (we also called it the well-ordering principle). Because there is no known way of well-ordering even the familiar set of real numbers, despite the assertion of the well-ordering theorem, for a period of at least six years after the appearance of that theorem, many papers appeared that were critical of Zermelo's proof. Most rejected the axiom of choice. Most critics had to admit, however, that if the axiom of choice were accepted they could find no mistake in Zermelo's proof of the well-ordering theorem. Therefore, the rejection of the well-ordering theorem would amount to the rejection of the axiom of choice. There seemed to be only two alternatives:

(a) Accept in principle only constructive but not purely existential results, and consequently, restrict the methods and domains of mathematics to such an extent that, outside of arithmetic, only narrow areas could be investigated.

(b) Accept constructive as well as purely existential results including the axiom of choice, and consequently, solve more problems and expand mathematics.

Before anyone could intelligently decide which alternative to follow, two difficult questions had to be considered:

(1) Is the axiom of choice independent of the other axioms, or can it be proved by means of other existing axioms of mathematics?

(2) Is the axiom of choice consistent with the classical axioms of mathematics, or can the addition of the axiom of choice to the classical axioms of mathematics result in a contradiction?

As in the case of the continuum hypothesis, many mathematicians spent a great deal of energy trying to answer these two questions. Many years later, in 1938, Kurt Gödel (1906-1978) answered the second question by proving that addition of the axiom of choice to the existing axioms of mathematics would

일찍이 1880년대 칸토어는 어떤 정리들의 증명에 선택공리와 본질적으로 동치인 추론을 이미 은연 중에 쓰고 있었으나 새롭고 강력한 공리의 필요성은 깨닫지 못하였다. 1904년 체르멜로는 많은 연구 끝에 선택공리를 옹호한 나머지 이것을 이용하여 정렬정리(정렬원리라고도 한다)를 증명함으로써 수학계에 크나큰 파문을 일으켰다.

이렇듯 정렬원리의 주장에도 불구하고 낯익은 실수의 집합에 대한 정렬 방법조차 발견되지 못했던 까닭에 이 원리가 발견된지 적어도 6년 동안 체르멜로의 증명을 비판하는 논문 중에는 선택공리를 부인한 것들이 적지 않았다. 그러나 대부분의 비평가들은 일단 선택공리를 받아들이면 체르멜로가 시도한 정렬정리의 증명에는 전혀 오류가 없음을 인정하지 않을 수 없었다. 따라서 정렬정리의 부인은 선택공리의 부인과 마찬가지이다.

이제 남은 대안으로서 다음과 같은 두 길밖에 없을 것이다.

(a) 원칙적으로 구성적 결과만을 수용하고 순전히 존재성에 관한 결과는 수용하지 않는다. 결국 수학의 방법과 영역을 제한함으로써 산수 밖의 영역에 대해서는 단지 좁은 분야만을 연구할 수 있게 한다.

(b) 선택공리와 함께 순전히 존재성에 관한 결과뿐만 아니라 구성적 결과도 수용함으로써 더 많은 문제를 풀어 수학의 범위를 넓혀나간다.

어느 대안에 따를 것인가를 재치있게 결정하기에 앞서 누구나 다음 두 가지 어려운 의문을 생각하지 않을 수 없다.

(1) 선택공리는 여타공리에 대하여 독립적인가 또는 그 공리는 수학의 다른 공리로 증명될 수 있는가?

(2) 선택공리는 수학의 전통적 공리에 모순되지 않는가 또는 수학의 전통적 공리들에 선택공리를 덧붙이면 모순이 생기는가?

연속체가설의 경우처럼 이 두 의문에 대한 답을 얻기 위하여 여러 수학자들이 많은 노력을 기울인지 몇 해가 지나 1938년 쿠르트 괴델은 기존의 수학의 공리들에 선택공리를 덧붙여도 어떤 모순도 야기되지 않는다는 사실을 증명함으로써 두 번째 의문에 대한 답을 얻었다. 괴델의 공헌은 수학계, 특히 선택공리를 쓰는 이들에게 큰 확신과 위안을 주었다는 데 있다고 말할 수 있다. 한편 첫 번째 의문에 대한 답의 추구도 계속되어오다가 마침내 1963년 폴 코헨에 의하여 완전히 풀렸다. 그는 선택공리는 기존의 다른 공리와는 실제로 *독립적*(그 뜻을 알아 보아라)이란 것을 증명하였다.

not produce any contradiction.[2] Gödel's contribution gave the mathematical community and especially the users of the axiom of choice great confidence and comfort. But the search for the answer to the first question went on and on. Finally, in 1963, Paul Cohen solved the question completely. He proved that the axiom of choice was indeed independent of the other existing axioms. In other words, the axiom of choice could not be proved as a theorem using the classical axioms of mathematics.

Today, the axiom of choice has been widely accepted as a new axiom, and it has proved indispensable for modern real analysis, the theory of transfinite cardinal and ordinal numbers, modern algebra, and for wide areas of topology.

학생시절의 괴델(1930)

괴델과 아인슈타인(1949)

[2] K. Gödel, *The Consistency of the Axiom of Choice and of the Generalized Continuum Hypothesis with the Axioms of Set Theory,* Princeton University Press, Princeton, N.J., 1940, 66 pp. Rev. ed., 1951, 74 pp.

다시 말하면 선택공리는 종래 활용해 온 공리를 써서 하나의 정리로서 증명될 수 없음을 밝혀냈다.* 오늘날 새로운 공리로서 널리 수용되고 있는 선택공리는 현대실해석학, 초한기수와 순서수의 이론, 현대대수학 및 위상수학 등 수학의 넓은 분야에서 빼 놓을 수 없는 공리들 중의 하나로 입증되고 있다.**

괴델의 정리를 연구하는 수학자 중 한 사람 대게우치 가이시(竹內外史 : 일리노이대학 교수)의 정서**에는 "수학과 완전성"이란 항목으로 다음과 같은 요지의 글이 실려 있다.

현대수학의 모든 개념은 집합론의 범위에서 재구성할 수 있다. 더구나 집합론에서 쓰이는 개념은 논리적개념과 함께 집합 A가 집합 B에 속한다고 언급하듯이 "속한다"란 단 하나의 기본개념으로 구성되어진다.

이렇듯 집합론의 체계를 공리와 논리로 나누어 생각하면 다음 두 문제가 대두된다.

(a) 집합론의 공리는 현재 인용되고 있는 것이 모두인가. 다시 말하면 현재의 집합론의 공리계는 완성된 것인가.

(b) 논리의 법칙은 현재까지 알려진 것으로서 전부라 할 수 있는가.

이 두 문제에 대한 해답은 모두 괴델에 의해 풀렸거니와 문제 (b)는 "완전성정리"로서 긍정적으로 해결을 보았다(당시 나이 24세). 이 완전성정리를 한마디로 말해본다면 "(현재의) 논리는 완전하다."로 요약된다. 어느 것하나 타당한 논리의 규칙은 현재의 논리의 기본규칙을 재구성하여 얻을 수 있음을 보여주고 있다.

한편 문제 (a)는 "불완전성 정리"란 이름으로 부정적인 결론을 내렸다(당시 나이 25세). 이 불완전성 정리를 살펴본다면

> 산술의 공리계를 포함한 현재의 공리계는 모순이 없지만 완전하지 못하다. 즉 현재의 공리계의 범위에서는 긍정 부정도 할 수 없는 산술적 명제가 존재한다.

이렇듯 괴델의 정리는 우리의 이성의 한계를 드러낸 반면 수학발전의 가능성도 열어놓았다고 볼 수 있다.

괴델은 프린스턴대학 고등연구소에서 수십 년간 아인슈타인과 연구생활을 지냈다. 사진의 왼쪽이 괴델 오른쪽이 아인슈타인이다. 두뇌해부가 유언이던 아인슈타인과 음식을 끊어 굶어 죽은 괴델. 당시 그의 키 168cm 몸무게 29.5kg 나갔다고 한다.

괴델과 아인슈타인(1949)

* 이장우, 《현대집합론의 원리》, 경문사, 1998년, 229쪽 참고 및 7.6절 간단한 역사적 고찰(251–254쪽) 참조.

** 林楨垈, 《數學基礎論의 이해》, 淸文閣, 1995년 5. 치환공리와 선택공리(86-91쪽) 참조.

Chapter 7
Ordinal Numbers and Ordinal Arithmetic

제 7 장
순서수와 그것의 덧셈, 곱셈

7
Ordinal Numbers and Ordinal Arithmetic

The concept of ordinal numbers is introduced, and the curiosities of ordinal arithmetic are pointed out as the ordinal sum and product are explored. The cardinal numbers are revisited as the initial ordinals, and the Burali-Forti paradox is exhibited.

7.1 The Concept of Ordinal Numbers

Loosely speaking, in finite arithmetic, the cardinal numbers are the "counting" numbers : $1, 2, 3, \cdots$, and the "ordinal numbers" are the "ranking" numbers : first, second, third, and so on. The distinction between the finite cardinal numbers and the finite ordinal numbers is so trivial that the natural numbers may be used to serve in both capacities. But what exactly is an "infinite ordinal numbers"? Just as a transfinite cardinal number arises from an infinite set, an infinite (transfinite) ordinl number arises from an infinite well-ordered set. The following definition for well-ordered sets resembles equipotence relation for general sets.

Definition 1. Two well-ordered sets $(A, \leqq)$ and $(B, \leqq')$ are said to be order-isomorphic if there exists a bijection $f : A \rightarrow B$ such that if $a_1, a_2 \in A$ and $a_1 \leqq a_2$, then $f(a_1) \leqq' f(a_2)$. Such a function $f : A \rightarrow B$ is called an order-isomorphism.

7

순서수와 그것의 덧셈, 곱셈*

*순서수*의 개념을 설명하고 그것의 합과 곱을 통하여 순서수가 지닌 신기한 성질을 지적한다. 한편 제 5장에서 알아봤던 기수를 *초기순서수*의 관점에서 다시 돌아보고 *부랄리-포르티의 역리*도 소개한다.

7.1 순서수의 개념

우리가 간단한 계산을 할 때 기수라 하면 "셈수" 1, 2, 3…이고 "순서수"는 첫째, 둘째, 셋째 등과 같이 차례를 나타내는 데 쓰이는 수로 알 만큼 유한기수와 유한순서수 사이의 차이는 분명하므로 자연수는 양쪽 모두의 경우 활용되고 있다. 그렇지만

"무엇이 무한순서수란 말인가?"

초한기수가 무한집합으로부터 생겨나듯 무한(초한)순서수는 무한 정렬집합으로부터 얻어진다. 정렬집합에 관한 다음 정의는 지금까지 우리가 다루어온 두 집합 사이의 대등관계와 비슷하다.

정의 1** 임의의 두 정렬집합 $(A, \leqq)$, $(B, \leqq')$에 대하여 전단사 $f: A \to B$가 존재함으로써 A의 원소 a_1, a_2에 대하여 말하자면

$$a_1 \leqq a_2 \Leftrightarrow f(a_1) \leqq' f(a_2)$$

일 때 A, B **순서동형**이라 하고 함수 f를 **순서동형사상**이라고 한다.

---

* 이장우, 《현대집합론의 원리》, 경문사, 1998년, 정의 6, 7(197–198쪽) 및 제8장 순서수와 그 연산 (255–281쪽) 참조.

** 정의 1에 해당하는 보기로서 짝수의 집합 N_e, 홀수의 집합 N_o를 놓고 N_e에서 N_o으로의 순서 동형 사상 f를 어떻게 정의할 것인가 생각하여 보아라.

It follows that if $f: A \to B$ is an order-isomorphism, then so is $f^{-1}: B \to A$, and that if furthermore $g: B \to C$ is an order-isomorphism, then the composition $g \circ f: A \to C$ is an order-isomorphism. If $(A, \leqq)$ and $(B, \leqq')$ are order-isomorphic, we write, $(A, \leqq) \approx (B, \leqq')$ or simply $A \approx B$. Like the equipotence relation $\sim$, the order-isomorphic relation "$\approx$" is reflexive, symmetric, and transitive.[1)]

In general, we shall consider the ordinal numbers, finite or transfinite, as a primitive concept subject to the following rules:

O-1. Each well-ordered set $(A, \leqq)$ is assigned an ordinal number, denoted by $\operatorname{ord}(A, \leqq)$, and if α is an ordinal number, then there is a well-ordered set $(A, \leqq)$ such that $\operatorname{ord}(A, \leqq) = \alpha$.

O-2. Let $(A, \leqq)$ and $(B, \leqq')$ be well-ordered sets. Then $\operatorname{ord}(A, \leqq) = \operatorname{ord}(B, \leqq')$ if and only if $(A, \leqq) \approx (B, \leqq')$. Since any two finite well-ordered sets having the same number of elements are order-isomorphic (see Problem 1), we shall adopt the following convenient notations:

O-3. $\operatorname{ord}(A, \leqq) = 0$ if and only if $A = \phi$.

O-4. If $(A, \leqq)$ is well-ordered set such that $A \sim \{1, 2, \cdots, k\}$, for some $k \in \mathrm{N}$, then $\operatorname{ord}(A, \leqq) = k$.

The ordinal number of the set N of natural numbers, with the usual less than or equal relation, is customarily denoted by the Greek omega, ω. Thus,

$$\omega = \operatorname{ord}\{1, 2, 3, \cdots\}.^{2)}$$

A given set has only one cardinal number, but a set may have distinct ordinal numbers under different well-orderings. For example, the set of natural numbers may be well-ordered as $(\mathrm{N}, \leqq) = \{1, 2, 3, \cdots\}$ and as

$$(\mathrm{N}, \leqq') = \{1, 3, 5, \cdots, 2, 4, 6, \cdots\}$$

We leave it to the reader to verify that these two well-ordered sets are not order-isomorphic. Consequently, $\operatorname{ord}(\mathrm{N}, \leqq') \neq \omega$.

[1)] Here we call $\approx$ a relation in the sense that it is a relation on any given set of well-ordered sets. Cf. Theorem 5 of Chapter 4.

[2)] Throughout the remainder of this chapter, the order in which the elements of a set are listed will indicate the ordering of the set. Thus, $\{1, 2, 3, \cdots\}$ denotes the usual well-ordered set of natural numbers, and $\{1, 3, 5, \cdots, 2, 4, 6, \cdots\}$ denotes another ordering of the set N.

함수 $f: A \to B$가 순서동형사상이면 $f^{-1}: B \to A$도 순서동형사상이다(과연?). 또한 $g: B \to C$가 순서동형사상이면 합성함수 $g \circ f: A \to C$ 역시 순서동형사상이다. 지금 $(A, \leqq)$와 $(B, \leqq)$가 순서동형일 때 이것을 $(A, \leqq) \approx (B, \leqq')$ 또는 간단히 $A \approx B$로 나타낸다. 여기서 순서동형을 나타내는 관계 "$\approx$"는 대등관계 "$\sim$"와 같이 반사적, 대칭적, 추이적이다.[1] 말하자면 하나의 동치관계이다.

기수의 정의(239쪽)에서와 같이 다음 네 가지 규칙에 따른 것(말하자면 동시관계 $\approx$에 따른 동치율의 공통적 성질)을 **순서수**라고 한다.

O-1. 각 정렬집합 $(A, \leqq)$에 대하여 $\mathrm{ord}(A, \leqq)$로 표시된 하나의 순서수가 결부되고, 각 순서수 α에 대하여 $\mathrm{ord}(A, \leqq) = \alpha$인 하나의 정렬집합 $(A, \leqq)$가 결부된다.

O-2. 두 정렬집합 $(A, \leqq)$, $(B, \leqq')$에 대하여

$$(A, \leqq) \approx (B, \leqq') \quad \mathrm{ord}(A, \leqq) = \mathrm{ord}(B, \leqq')$$

O-3. $A = \phi \Leftrightarrow \mathrm{ord}(A, \leqq) = 0$

O-4. 정렬집합 $(A, \leqq)$에 대하여 적당한 $k \in \mathbb{N}$가 존재함으로써

$A \sim \{1, 2, 3, \cdots, k\}$이면 $\mathrm{ord}(A, \leqq) = k$

특히 우리에게 낯익은 대소 관계에 따른 자연수의 집합 $\mathbb{N}$의 순서수는 그리스어 **ω**(오메가라고 읽는다)로 나타낸다. 즉 $\omega = \mathrm{ord}\{1, 2, 3, \cdots\}$[2]이다.

다음에서 보듯이 유한집합의 기수는 단 하나이지만 무한집합의 경우는 순서를 어떻게 매기느냐에 따라 순서수는 다를 수 있다. 이를테면 집합 $\mathbb{N}$ 위의 순서가 다음과 같이 두 가지일 수도 있다[연습문제 7.1 문제 3(337쪽) 참조].

$$(\mathbb{N}, \leqq) = \{1, 2, 3, \cdots\}, \quad (\mathbb{N}, \leqq') = \{1, 3, 5, \cdots, 2, 4, 6, \cdots\}$$

여기서 $\mathrm{ord}(\mathbb{N}, \leqq) = \omega$. 그러나 $\mathrm{ord}(\mathbb{N}, \leqq') \neq \omega$.

---

[1] 여기서 "$\approx$"을 주어진 정렬집합 위의 하나의 관계로 본다(제4장 정리 5와 비교).

[2] 앞으로 이 장에서는 집합의 원소가 나열된 순서로 하여금 그 집합의 순서를 나타내는 것으로 한다. 이를테면 $\{1, 2, 3, \cdots\}$은 자연수의 집합 위에서의 대소 관계에 따른 정렬집합을, $\{1, 3, 5, \cdots, 2, 4, 6, \cdots\}$은 집합 $\mathbb{N}$의 또다른 순서를 나타낸다[연습문제 7.1 문제 3(337쪽) 참조].

Exercise 7.1

Prove the following problems 1 through 4.

1. Let A and B be two equipotent finite sets, and let $(A, \leqq)$ and $(B, \leqq')$ be well ordered. Then $(A, \leqq)$ and $(B, \leqq')$ are order-isomorphic.
2. Let $(A, \leqq)$ be a totally ordered set which is the union of two subsets B and C such that both B and C are well-ordered under the ordering inherited from the ordering $\leqq$ of A. Then $(A, \leqq)$ is well-ordered.
3. Let the set N of natural numbers be well-ordered as

   $$(\mathrm{N}, \leqq)=\{1, 2, 3, \cdots\} \text{ and as } (\mathrm{N}, \leqq')=\{1, 3, 5, \cdots, 2, 4, 6, \cdots\}$$

   Then $(\mathrm{N}, \leqq')$ and $(\mathrm{N}, \leqq)$ are not order-isomorphic.
4. The set $(\mathrm{N}, \leqq')$ in Problem 3 is well ordered.

7.2 Ordering of the Ordinal Numbers

The rule O-2, given in Section 1, tell us when two ordinal numbers are equal, and when they are not equal. If two ordinal numbers are not equal, we wish to be able to say one is " less " than the other. The following definition is designed just to serve that purpose.

Definition 2. Let α and β be ordinal numbers and let $(A, \leqq)$ and $(B, \leqq')$ be well-ordered sets such that $\alpha = \mathrm{ord}(A, \leqq)$ and $\beta = \mathrm{ord}(B, \leqq')$. Then we say that α is less than or equal to β, in symbols $\boldsymbol{\alpha} \leqslant \boldsymbol{\beta}$, or $\boldsymbol{\beta} \geqslant \boldsymbol{\alpha}$, if and only if $(A, \leqq)$ is order-isomorphic to a segment of $(B, \leqq')$. When $\alpha \leqslant \beta$ and $\alpha \neq \beta$, we write $\alpha < \beta$ or $\beta > \alpha$.

It is clear that Definition 2 is independent of the choice of representing well-ordered sets $(A, \leqq)$ and $(B, \leqq')$, and that the relation "$\leqslant$" is reflexive and transitive. The antisymmetry of $\leqslant$ is proved in Theorem 2, so that $\leqslant$ is a partial order "relation." Our eventual goal is to prove that it is a total order " relation."[3)]

---

3) Strictly speaking, "$\leqslant$" is not a relation, because its "domain" is not a set But, we still call it a relation as defined on any given set of ordinal numbers. Cf. Theorem 5 of Chapter 4.

연습문제 7.1

1. 유한정렬집합 $(A, \leqq)$, $(B, \leqq)$에 대하여 A와 B가 대등하면 이 두 정렬집합은 순서동형이다. 이것을 증명하여라.

2. 전순서집합 $(A, \leqq)$은 그 부분집합 B, C의 합집합으로서 각각은 A 위의 전순서 $\leqq$로부터 이어받은 순서에 따른 정렬집합이라고 한다. 이때 $(A, \leqq)$는 정렬집합임을 밝혀라.

3.* 집합 $\mathbb{N}$으로부터 얻은 다음 두 정렬집합은 순서동형이 아님을 밝혀라.

$$(\mathbb{N}, \leqq) = \{1, 2, 3, \cdots\}, \qquad (\mathbb{N}, \leqq') = \{1, 3, 5, \cdots, 2, 4, 6, \cdots\}$$

4. 문제 3의 $(\mathbb{N}, \leqq')$은 정렬집합이다. 이를 증명하여라.

5.* $\mathbb{N}$, $\mathbb{Z}$, $\mathbb{Q}$, $\mathbb{R}$에 대하여 어느 둘로 순서동형이 아님을 밝혀라.

7.2 순서수의 순서매김

7.1절의 규칙 O-2는 어느 경우에 순서수가 같고 어느 경우에 같지 않은가를 말해주고 있다. 여기서 두 순서수가 같지 않을 때 바람직한 것은 그 중의 하나가 "다른 순서수보다 작다."라고 말할 수 있는 일이다. 다음은 바로 그 같은 목적을 위하여 내려진 정의이다.

정의 2 순서수 α, β와 정렬집합 $(A, \leqq)$, $(B, \leqq')$에 대하여 $\alpha = \operatorname{ord}(A, \leqq)$, $\beta = \operatorname{ord}(B, \leqq')$일 때 $(A, \leqq)$가 $(B, \leqq')$의 하나의 절편과 순서동형이면, 그리고 그때에만 α는 β보다 작다 또는 같다고 말하고 이것을 $\boldsymbol{\alpha \leqslant \beta}$ 또는 $\boldsymbol{\beta \geqslant \alpha}$로 나타낸다. 한편 $\alpha \leqslant \beta$이고 $\alpha \neq \beta$일 때 $\boldsymbol{\alpha < \beta}$ 또는 $\boldsymbol{\beta > \alpha}$로 나타낸다.

정의 2는 정렬집합 $(A, \leqq)$, $(B, \leqq')$의 선택과는 무관할 뿐만 아니라 관계 "$\leqslant$"는 반사적이고 추이적인 것이 분명하다(과연?). 한편 $\leqslant$의 반대칭적인 것은 정리 2에서 증명되고 있으므로 $\leqslant$는 부분순서로서의 "관계"를 나타낸다. 우리의 궁극적 목표는 이런 관계 $\leqslant$[3]가 전순서임을 밝히려는 데 있다.

* 문제 5는 옮긴이가 보충한 것이다.

[3] 엄밀히 말하여 "$\leqslant$"은 관계가 아니다. 왜냐하면 그것의 정의역은 집합이 아니기 때문이다. 하지만 $\leqslant$은 순서수의 집합 위에 정의된 것으로서 여전히 관계라고 부른다[정리 5(217쪽)와 비교].

Theorem 1. The only order-isomorphism of a well-ordered set $(A, \leqq)$ onto a segment of $(A, \leqq)$ is the identity function of A onto A.

[Proof] Suppose on the contrary that there exists an order-isomorphism $f: A \longrightarrow A_a$, for some segment A_a of A. Then, $f(a) < a$ and consequently the set $B = \{x \in A \mid f(x) < x\}$ is not empty. Let b denote the least element of B. Then $f(b) < b$, and since an order-isomorphism is strictly increasing, we have $f(f(b)) < f(b)$. This proves that B contains the element $f(b)$, which is less than the least element b of B, a contradiction. The contradiction shows that a well-ordered set cannot be order-isomorphic to any of its proper segments. To complete the proof of the theorem, it remains to be shown that the identity function $1_A: A \longrightarrow A$ is the only order-isomorphism of A onto A.

Let $g: A \longrightarrow A$ be an order-isomorphism. Then, since both $g: A \longrightarrow A$ and the identity function $1_A: A \longrightarrow A$ are strictly increasing, by using Theorem 10 of Chapter 6 twice, we have

$$1_A(x) \leqq g(x) \leqq 1_A(x)$$

for all $x \in A$. Therefore, $g = 1_A$.

Theorem 2. If α and β are ordinal numbers such that $\alpha \leqslant \beta$ and $\beta \leqslant \alpha$, then $\alpha = \beta$.

[Proof] Let $(A, \leqq)$ and $(B, \leqq')$ be well-ordered sets such that $\operatorname{ord}(A, \leqq) = \alpha$ and $\operatorname{ord}(B, \leqq') = \beta$. By the hypotheses $\alpha \leqslant \beta$ and $\beta \leqslant \alpha$, we have two order-isomorphisms

$$f: A \longrightarrow D \quad \text{and} \quad g: B \longrightarrow C$$

where C and D are segments of A and B, respectively. Consequently, the function $h: A \longrightarrow C$ given by $h(x) = g(f(x))$, for all $x \in A$, is an order-isomorphism of A to a segment, say E, of C. By Theorem 1, and Theorem 7 of Chapter 6, we must have $E = A$. Consequently, $C = A$ and the order-isomorphism $g: B \longrightarrow A$ shows that $\alpha = \beta$.

정리 1 임의의 정렬집합 $(A, \leqq')$에 대하여 A에서 그 절편 B 위로의 순서동형사상으로서는 오로지 A 위의 항등함수 1_A가 존재할 따름이다.

[증명] 이 정리의 결론에 반하여 A의 적당한 절편 A_a에 대하여 순서동형사상 $f: A \to A_a$가 존재한다고 가정하자. 그러면 $f(a) < a$이므로 집합 $B = \{x \in A \mid f(x) < x\}$는 공집합이 아니다. 여기서 B의 최소원소(과연 그런 원소가 존재할까?)를 b라고 하면 $f(b) < b$. 그런데 순서동형사상은 엄밀한 뜻에서 증가하므로 $f(f(b)) < f(b)$. 이것은 B에 그 최소원소 b보다 작은 원소 $f(b)$가 존재한다는 것을 뜻하므로 이것은 모순이다. 따라서 정렬집합에서는 그 어떤 진절편으로의 순서동형사상도 존재하지 않음을 알 수 있다.

이 정리의 증명을 완성하려면 A에서 그 자신 위로의 순서동형 사상은 단지 항등함수 $1_A: A \to A$뿐인 것을 보여야 한다.

지금 $g: A \to A$에 대하여 g는 물론 항등함수 1_A도 엄밀한 뜻에서 증가한다. 그러므로 제6장 정리 10(323쪽)을 두 차례 반복 이용함으로써 다음을 얻는다.

$$\forall_{x \in A}\,[1_A(x) \leqq g(x) \leqq 1_A(x)].$$

따라서 $\forall_{x \in A}\,[1_A(x) = g(x)]$ 곧 $g = 1_A$.

정리 2* 임의의 순서수 α, β에 대하여 $\alpha \leqslant \beta$이고 $\beta \leqslant \alpha$이면 $\alpha = \beta$.

[증명] 정렬집합 $(A, \leqq)$, $(B, \leqq')$에 대하여 $\mathrm{ord}(A, \leqq) = \alpha$, $\mathrm{ord}(B, \leqq') = \beta$라고 놓으면 가정 $\alpha \leqslant \beta$, $\beta \leqslant \alpha$에 따라 두 순서동형 사상

$$f: A \to D, \qquad g: B \to C$$

가 존재한다. 여기서 C, D는 각각 A, B의 절편이다. 그러므로 모든 $x \in A$에 대하여 $h(x) = g(f(x))$로 정의된 함수 $h: A \to D$는 A에서 C의 절편, 말하자면 E로의 순서동형사상이다. 한편 제6장 정리 1과 7에 따라 $E = A$. 그러므로 $C = A$인 사실과 순서동형사상 $g: B \to A$에 따라 $\alpha = \beta$.

* 임의의 두 기수는 물론 실수 a, b에 대해서도 정리 2와 같은 성질이 있다 [따름정리 (249쪽) 참조]. 사실 정리 2와 다음 정리 3은 이론상 차례가 바뀌어야 한다.

We have now completed the proof that the relation $\leqslant$ for the ordinal numbers is a partial order relation. This property will be strengthened, by the following theorem, to total order.

Theorem 3. For any ordinal numbers α and β, either $\alpha \leqslant \beta$ or $\alpha \geqslant \beta$.

[Proof] Let $(A, \leqq)$ and $(B, \leqq')$ be well-ordered sets such that $\text{ord}(A, \leqq) = \alpha$ and $\text{ord}(B, \leqq') = \beta$. We shall prove that either A is order-isomorphic to a segment of B, or B is order-isomorphic to a segment of A. An element a of A will be called admissible if the segment A_a of A is order-isomorphic to some segment B_b, $b \in B$, of B. Let M denote the set of all admissible elements of A. It follows from Theorem 1 that for each admissible element a, there exists a unique element $b \in B$ such that $A_a \approx B_b$ (see Problem 2). Consequently, there is a well-defined function $f : M \longrightarrow B$ given by $f(a) = b$ if $A_a \approx B_b$. It is again a consequence of Theorem 1 that the function f is injective. We leave it to the reader to verify that $f : M \longrightarrow B$ is strictly increasing and that $f(M) = N$ is a segment of B.

Now, suppose that neither A is order-isomorphic to a segment of B nor B is order-isomorphic to a segment of A. Then, $A - M$ and $B - N$ are nonempty; let p and q be the least element of $A - M$ and $B - N$, respectively. Then, we must have $A_p = M \approx N = B_q$, and hence p is admissible, a contradiction. The proof of Theorem 3 is now complete.

The following theorem is another easy consequence of Theorem 1.

Theorem 4. Let $(A, \leqq)$ and $(B, \leqq')$ be two well-ordered sets. Then $\text{ord}(A, \leqq) < \text{ord}(B, \leqq')$ if and only if A is order-isomorphic to a proper segment of B.

[Proof] Exercise.

이상으로 순서수에 대한 관계 $\leqslant$는 부분순서란 것의 증명을 끝냈다. 이같은 성질은 다음 정리에 따라 전순서까지로 강화된다.

정리 3* 임의의 순서수 α, β에 대하여 $\alpha \leqslant \beta$ 이니면 $\beta \leqslant \alpha$.

[증명] 정렬집합 $(A, \leqq)$, $(B, \leqq')$에 대하여 $\text{ord}(A, \leqq) = \alpha$, $\text{ord}(B, \leqq') = \beta$라고 할 때 A는 B의 절편과 순서동형이거나 B가 A의 절편과 순서동형인 것을 증명하면 된다.

A의 원소 a에 대하여 B의 원소 b가 존재함으로써 A의 절편 A_a와 B의 절편 B_b가 순서동형이면, a는 **허용가능하다**(297쪽 중간 구절과 비교하여라)고 말한다. 지금 A의 허용가능한 원소의 집합을 M이라고 하면 정리 1에 따라 허용가능한 각 원소 a에 대하여 B의 원소 b가 단 하나씩 존재함으로써 $A_a \approx B_b$[연습문제 7.2 문제 2(343쪽)]. 결국 $A_a \approx B_b$이면 $f(a) = b$로서 잘 정의된 함수 $f: M \to B$가 존재한다. 또한 정리 1의 결과에 따라 함수 f는 단사이다. 여기서 $f: M \to B$는 엄밀한 뜻에서 증가하고 $f(M) = N$은 B의 절편이다. 이것의 확인은 독자에게 맡긴다.

지금 A는 B의 절편과 순서동형이 아니고 B도 A의 절편과 순서동형이 아니라고 가정하면 $A - M$과 $B - N$은 모두 공집합이 아니다. 여기서 $A - M$, $B - N$의 최소원소를 각각 p, q라면

$$A_p = M \approx N = B_q.$$

따라서 p는 허용가능하므로 $p \in M$. 이것은 $p \notin M$ (왜?)에 모순이다. 이상으로 정리 3의 증명이 끝난다.

정리 1의 또다른 쉬운 결과로서 다음 정리를 얻는다.

정리 4 정렬집합 $(A, \leqq)$, $(B, \leqq')$에 대하여 A가 B의 진절편과 순서동형이면 그리고 그때에만 $\text{ord}(A, \leqq) < \text{ord}(B, \leqq')$이다.

[증명] 독자의 연습문제로 맡긴다.

* 임의의 기수 a, b에 대해서도 정리 3과 유사한 성질이 있다[제 6장 정리 4 따름정리(309쪽) 참조].

We now summarize the results of this section in the following trichotomy theorem.

Theorem 5. Let α and β be any two ordinal numbers. Then, exactly one of the followings is true :

(a) $\alpha < \beta$ (b) $\alpha = \beta$ (c) $\alpha > \beta$.

Exercise 7.2

1. Show that ord$\{1, 3, 5, \cdots, 2, 4, 6, \cdots\} >$ ord$\{1, 2, 3, \cdots\}$ (cf. Problem 3, Exercise 7.1).

2. Prove the following assertion that appears in the proof of Theorem 3 : For each admissible element a, there exists a unique element $b \in B$ such that $A_a \approx B_b$.

3. Show that the set M of admissible elements, in the proof of Theorem 3, is a segment of A.

4. Prove that the function $f : M \longrightarrow B$, which appeared in the proof of Theorem 3 and is given by $f(a) = b$ if $A_a \approx B_b$, is injective and strictly increasing, and that $f(M)$ is a segment of B.

5. Prove Theorem 4.

6. Let k be any natural number. Prove that

(a) ord$\{k, k+1, k+2, \cdots\} = \omega$

(b) ord$\{k, k+1, k+2, \cdots, 0, 1, 2, 3, \cdots, k-1\} > \omega$.

7. What is the smallest transfinite ordinal number?

여기서 이 절의 결과를 다음과 같은 **삼분법**(또는 **삼자택일률**) 정리로 요약할 수 있다.

정리 5* 임의의 두 순서수 α, β에 대하여 다음 중 한 경우만이 꼭 성립한다.
(a) $\alpha<\beta$ (b) $\alpha=\beta$ (c) $\alpha>\beta$

연습문제 7.2

다음 1~6을 증명하여라.

1. $\mathrm{ord}\{1, 2, 3, \cdots\}<\mathrm{ord}\{1, 3, 5, \cdots, 2, 4, 6, \cdots\}$ [연습문제 7.1 문제 3 풀이(409쪽) 참조].

2.** 정리 3의 증명에 나타난 다음 주장을 증명하여라.
허용가능한 각 원소 a에 대한 원소 $b\in B$가 단 하나씩 존재함으로써
$$A_a\approx B_b$$

3. 정리 3의 증명에서 허용가능한 원소의 집합 M은 A의 절편이다.

4. 정리 3에서 $A_a\approx B_b$일 때 함수 $f: M\to B$를 $f(a)=b$와 같이 정의하면, 이 함수는 단사이고 엄밀한 뜻에서 증가한다. 그리고 $f(M)$은 B의 절편이다.

5. 정리 4

6. 임의의 자연수 k에 대하여
(a) $\mathrm{ord}\{k, k+1, k+2, \cdots\}=\omega$
(b) $\omega<\mathrm{ord}\{k, k+1, k+2, \cdots, 0, 1, 2, 3, \cdots, k-1\}$

7. 가장 작은 초한순서수가 존재하는가? 존재한다면 그것을 지적하여라.

8.*** 임의의 순서수 α, β, γ에 대하여 $\alpha<\beta$, $\beta<\gamma$이면 $\alpha<\gamma$임을 밝혀라.

---

* 임의의 두 기수 a, b에 대해서도 정리 5와 유사한 성질이 있는가?
** 원서에는 문제 2와 똑같은 내용의 문제 8이 실려 있어 후자를 삭제하였다.
*** 이 문제는 옮긴이가 보충하였다.

7.3 Addition of Ordinal Numbers

For any two disjoint well-ordered sets $(A, \leqq)$ and $(B, \leqq')$ we shall define the total order relation $\leqq^*$ on $A \cup B$ as follows:

(1) If a and b are both in A (or in B), then we write $a \leqq b^*$ if and only if $a \leqq b$ (or $a \leqq' b$).

(2) If $a \in A$ and $b \in B$, we agree that $a \leqq^* b$.

It is easy to see that $(A \cup B, \leqq^*)$ is a well-ordered set[4] (see Problem 2, Exercise 7.1). Our natural candidate for the "sum," $\operatorname{ord}(A, \leqq) + \operatorname{ord}(B, \leqq')$, is $\operatorname{ord}(A \cup B, \leqq^*)$. In the case where A and B are not disjoint, a little additional work is needed:

Form the Cartesian products $A \times \{0\}$, $B \times \{1\}$, and define in $A \times \{0\}$ the well-ordered relation $\leqq_0$ by $(a, 0) \leqq_0 (b, 0)$ if and only if $a \leqq b$ in $(A, \leqq)$. Similarly, define $\leqq_1'$ on $B \times \{1\}$ by $(c, 1) \leqq_1' (d, 1)$ if and only if $c \leqq' d$. It is obvious that

$$(A \times \{0\}, \leqq_0) \approx (A, \leqq), \quad (B \times \{1\}, \leqq_1') \approx (B, \leqq');$$

where $A \times \{0\}$ and $B \times \{1\}$ are disjoint. Thus, if A and B are not disjoint, we may use the disjoint sets $(A \times \{0\}, \leqq_0)$ and $(B \times \{1\}, \leqq_1')$ as reasonable substitutes for $(A, \leqq)$ and $(B, \leqq')$.

Definition 3. Let α and β be ordinal numbers. The ordinal sum of α and β, denoted by $\boldsymbol{\alpha+\beta}$, is the ordinal number $\operatorname{ord}(A \cup B, \leqq^*)$, where $(A, \leqq)$ and $(B, \leqq')$ are disjoint well-ordered sets such that $\operatorname{ord}(A, \leqq) = \alpha$ and $\operatorname{ord}(B, \leqq') = \beta$.

In Problem 1 the reader is asked to justify that the definition of the ordinal sum, $\alpha + \beta$, is independent of the choice of representing well-ordered sets.

If α and β are two finite ordinal numbers, the ordinal sum $\alpha + \beta$ agrees with the usual sum of two nonnegative integers. But, for transfinite ordinal numbers, the properties of the ordinal sum may be very different from the finite case; for example, $\alpha + \beta$ need not be the same as $\beta + \alpha$ (see Example 2).

[4] It should be noted that in $(A \cup B, \leqq^*)$ any element $a \in A$ is less than any element $b \in B$, and that in $(B \cup A, \leqq^*)$ any element $b \in B$ is less than any element $a \in A$. Therefore, $(A \cup B, \leqq^*)$ and $(B \cup A, \leqq^*)$ should be considered distinct.

7.3 순서수의 덧셈

서로 소인 임의의 두 정렬집합 $(A, \leqq)$, $(B, \leqq')$에 대하여 합집합 $A \cup B$ 위의 완전순서 $\leqq^*$를 다음과 같이 정의한다.

(1) a, b가 모두 A(또는 모두 B)의 원소로써

$$a \leqq b (\text{또는 } a \leqq' b) \Leftrightarrow a \leqq^* b$$

(2) $a \in A$, $b \in B$이면 $a \leqq^* b$

위의 정의에 따르면 $(A \cup B, \leqq^*)$는 정렬집합[4] 임을 쉽게 알 수 있다[연습문제 7.1 문제 2(336쪽)]. 여기서 두 순서수의 "합" $\mathrm{ord}(A, \leqq) + \mathrm{ord}(B, \leqq')$에 대한 후보로는 자연스럽게 $\mathrm{ord}(A \cup B, \leqq^*)$이 꼽힌다. 실제로 A와 B가 서로 소가 아닌 경우에는 다른 처리가 필요하다. 즉 데카르트 곱 $A \times \{0\}$, $B \times \{1\}$를 각각 구한 후 $A \times \{0\}$ 위의 정렬 $\leqq_0$을, $(A, \leqq)$에 있어서 $a \leqq b$이면 그리고 그때에만 $(a, 0) \leqq_0 (b, 0)$인 것으로 정의한다. 마찬가지로 $B \times \{1\}$ 위의 정렬 $\leqq_1'$도 $c \leqq' d$이면 그리고 그때에만 $(c, 1) \leqq_1' (d, 1)$인 것으로 정의한다. 그러면 $A \times \{0\}$과 $B \times \{1\}$은 서로 소일뿐만 아니라

$$(A \times \{0\}, \leqq_0) \approx (A, \leq), \quad (B \times \{1\}, \leqq_1') \approx (B_1, \leqq')$$

이렇듯 A, B가 서로 소가 아닌 경우엔 서로 소인 정열집합 $(A \times \{0\}, \leqq_0)$, $(B \times \{1\}, \leqq_1')$를 각각 구성하면 된다.

정의 3 서로 소인 임의의 정렬집합 $(A, \leqq)$, $(B, \leqq')$의 순서수가 각각 α, β일 때 $\mathrm{ord}(A \cup B, \leqq^*)$를 α와 β의 **순서수의 합**이라 하고 이것을 $\boldsymbol{\alpha + \beta}$로 나타낸다.

순서수의 합의 정의에 있어서 정렬집합의 대표로서 어떤 것을 택할건가 그 방법과는 무관하다. 이런 사실을 독자는 349쪽 문제 1에서 확인할 수 있다.

유한순서수 α와 β의 합 $\alpha + \beta$는 음이 아닌 두 정수의 합과 일치하지만 초한순서수의 합의 경우 그러하지 않을 수 있다. 이를테면 $\alpha + \beta$가 $\beta + \alpha$와 같지 않을 수도 있다[예제 2 (247쪽) 참조].

[4] $(A \cup B, \leqq^*)$에서는 임의의 원소 $a \in A$가 임의의 원소 $b \in B$보다 작다. 한편 $(B \cup A, \leqq^*)$에서는 임의의 원소 $b \in B$가 임의의 원소 $a \in A$보다 작으므로 $(A \cup B, \leqq^*)$와 $(B \cup A, \leqq^*)$는 서로 다른 정렬집합이다.

Example 1 Find the ordinal sum $5+4$ of the two finite ordinal numbers 5 and 4.

[Solution] Since $5=\text{ord}\{0,1,2,3,4\}$ and $4=\text{ord}\{5,6,7,8\}$, we have

$$5+4=\text{ord}\{0,1,\cdots,4,5,\cdots,8\}=9.$$

Example 2 Let k be any nonzero finite ordinal number. Show that $k+\omega=\omega$ and that $\omega+k\neq\omega$. Thus, the ordinal sum is in general not commutative.

[Proof] Since $k=\text{ord}\{0,1,2,3,\cdots,k-1\}$ and $\text{ord}\{k,k+1,\cdots\}=\omega$, we have

$$\begin{aligned} k+\omega &= \text{ord}\{0,1,2,\cdots,k-1,k+1,\cdots\} \\ &= \omega \end{aligned}$$

and

$$\begin{aligned} \omega+k &= \text{ord}\{k,k+1,\cdots,0,1,2,\cdots,k-1\} \\ &> \omega \end{aligned}$$

Thus, $\omega+k\neq\omega$ (See Problem 6(b), Exercise 7.2).

Theorem 6. Let α, β, and γ be any ordinal numbers. Then

(a) $(\alpha+\beta)+\gamma=\alpha+(\beta+\gamma)$ (Associative Law)

(b) $\beta<\gamma$ implies $\alpha+\beta<\alpha+\gamma$

(c) $\alpha+\beta=\alpha+\gamma$ implies $\beta=\gamma$ (Left Cancellation).

[Proof] (a) Exercise.

(b) Let $(A,\leqq)$, $(B,\leqq')$, and $(C,\leqq'')$ be well-ordered sets such that $\text{ord}(A,\leqq)=\alpha$, $\text{ord}(B,\leqq')=\beta$, $\text{ord}(C,\leqq'')=\gamma$. $A\cap B=\phi$, and $A\cap C=\phi$. Since $\beta<\gamma$, by Theorem 4, B is orderisomorphic to some proper segment C_z of C; let $g: B\rightarrow C_z$ denote this order-isomorphism. Let the disjoint unions $A\cup B$ and $A\cup C$ be endowed with the well-order relations $\leqq^*$ defined in the beginning of this section. Then, under these well-order relations, the function

$$f: A\cup B\rightarrow A\cup C_z$$

defined by

$$f(x)=\begin{cases} x & \text{if } x\in A \\ g(x) & \text{if } x\in B \end{cases}$$

예제 1 두 유한순서수 5, 4의 합 5+4를 구하여라.

[풀이] $5=\text{ord}\{0, 1, 2, 3, 4\}$, $4=\text{ord}\{5, 6, 7, 8\}$
그러므로 $5+4=\text{ord}\{0, 1, 2, \cdots, 8\}=9$, 즉 $5+4=9$.

예제 2 0이 아닌 임의의 유한순서수 k에 대하여 $k+\omega=\omega$. 그러나 $\omega+k\neq\omega$. 이것을 밝혀라. 그러고 보면 순서수의 덧셈에 관한 교환법칙이 성립하지 않는 경우도 있다.

[증명] $k=\text{ord}\{0,1,2,\cdots,k-1\}$, $\text{ord}\{k,k+1,k+2,\cdots\}=\omega$
그러므로 $k+\omega=\text{ord}\{0,1,2,\cdots,k-1,k,k+1,\cdots\}$
$=\omega$
즉 $k+\omega=\omega$*
한편 $\omega+k=\text{ord}\{k,k+1,k+2,\cdots,0,1,2,\cdots,k-1\}>\omega$ (왜?)
따라서 $\omega+k\neq\omega$[연습문제 7.2 문제 5(b) 참조].

정리 6 임의의 순서수 α, β, γ에 대하여 다음이 성립한다.
(a) **결합법칙** $(\alpha+\beta)+\gamma=\alpha+(\beta+\gamma)$
(b) $\beta<\gamma \Rightarrow \alpha+\beta<\alpha+\gamma$
(c) **좌소거법칙** $\alpha+\beta=\alpha+\gamma \Rightarrow \beta=\gamma$

[증명] (a) 독자의 연습문제로 돌린다.
(b) 정렬집합 $(A,\leq)$, $(B,\leq')$, $(C,\leq'')$의 순서수를 차례로 α, β, γ로 놓고 $A\cap B=A\cap C=\phi$라고 하자.
$\beta<\gamma$이면 정리 4에 따라 B는 C의 적당한 진절편 C_z와 순서동형이다. 이 순서동형사상을 $g:B\to C_z$로 놓은 다음 서로 소인 합집합 $A\cup B$, $A\cup C$ 위에 이 절의 서두에서 정의한 정렬순서 $\leq$*를 각각 부여하면 그 순서에 따라

$$f(x)=\begin{cases} x, & x\in A \\ g(x), & x\in B \end{cases}$$

로 정의된 함수 $f:A\cup B\to A\cup C$는 $A\cup B$에서 $A\cup C$의 진절편

* 이런 등식에 또 한 번 놀란 독자가 한둘이 아닐 것이다. 기수나 실수의 경우엔 어떨까?

is an order-isomorphism of $A \cup B$ onto the proper segment $A \cup C_z$ of $A \cup C$. Therefore, by Theorem 4, $\alpha+\beta<\alpha+\gamma$.

(c) Suppose on the contrary that there exist ordinal numbers α, β, and γ such that $\alpha+\beta=\alpha+\gamma$ and $\beta \neq \gamma$. By Theorem 5, we may assume that $\beta<\gamma$ (the case $\gamma<\beta$ may be handled similarly). Consequently, by part (b) above, we have $\alpha+\beta<\alpha+\gamma$, a contradiction.

It is worth mentioning here that although the ordinal sum is left cancellative, it is not right cancellative. That is, $\beta+\alpha=\gamma+\alpha$ does not necessarily imply that $\beta=\gamma$ (see Problem 4). Similarly, in contrast to Theorem 6(b), $\beta<\gamma$ does not imply $\beta+\alpha<\gamma+\alpha$ (see Problems 5 and 6).

Exercise 7.3

1. Let $(A_1, \leqq_1)$, $(A_2, \leqq_2)$, $(B_1, \leqq_1')$, and $(B_2, \leqq_2')$ be well-ordered sets such that $A_1 \approx A_2$, $B_1 \approx B_2$, $A_1 \cap B_1=\phi$, and $A_2 \cap B_2=\phi$. Prove that

$$(A_1 \cup B_1, \leqq_1^*) \approx (A_2 \cup B_2, \leqq_2^*).$$

2. Show that $\alpha+0=\alpha=0+\alpha$ for any ordinal number α.

3. Prove that the ordinal sum is associative: $(\alpha+\beta)+\gamma=\alpha+(\beta+\gamma)$ for any ordinal numbers α, β and γ.

4. Show, by giving a counterexample, that the ordinal sum is not right cancellative: There exist ordinal numbers α, β and γ such that $\beta+\alpha=\gamma+\alpha$, but $\beta \neq \gamma$.

5. Prove the following modification of Theorem 6(b): If α, β and γ are ordinal numbers such that $\beta<\gamma$, then $\beta+\alpha \leqslant \gamma+\alpha$ (see Problem 6).

6. Show that, in Problem 5, $\beta<\gamma$ does not necessarily imply that

$$\beta+\alpha<\gamma+\alpha \text{ [cf. Theorem 6(b)].}$$

7. Prove that $\alpha<\beta$ if and only if $\alpha+1 \leqslant \beta$.

$A \cup C_z$ 위로의 순서동형 사상과 같다. 그러므로 정리 4에 따라

$$\alpha+\beta<\alpha+\gamma$$

(c) 지금 (c)의 결론에 반하여 순서수 α, β, γ가 존재함으로써 $\alpha+\beta=\alpha+\gamma$, $\beta\neq\gamma$라고 가정하면 정리 5에 따라 $\beta<\gamma$($\gamma<\beta$일 경우도 같은 방법으로 다룰 수 있다). 따라서 (b)에 의하여 $\alpha+\beta<\alpha+\gamma$. 이것은 가정 $\alpha+\beta=\alpha+\gamma$에 모순이다. 그러므로 (c)는 성립한다.

순서수의 합에 있어서 좌소거법칙은 성립하지만 우소거법칙은 반드시 성립하지는 않음을 여기서 언급할 만하다. 즉 $\beta+\alpha=\gamma+\alpha\rightarrow\beta=\gamma$는 거짓일 수 있다(문제 4 참조). 마찬가지로 정리 6(b)와는 달리 $\beta<\gamma\rightarrow\beta+\alpha<\gamma+\alpha$도 거짓일 수 있다(아래 연습문제 7.3 문제 5와 6).

연습문제 7.3

1. 정렬집합 $(A_1, \leqq_1)$, $(A_2, \leqq_2)$, $(B, \leqq_1')$, $(B, \leqq_2')$에 대하여

$$A_1\approx A_2,\ B_1\approx B_2,\ A_1\cap B_1=\phi,\ A_2\cap B_2=\phi$$

이면 다음이 성립함을 밝혀라.

$$(A_1\cup B_1,\ \leqq_1^*)\approx(A_2\cup B_2,\ \leqq_2^*).$$

2. 임의의 순서수 α에 대하여 $\alpha+0=\alpha=0+\alpha$임을 보여라.

3. 정리 6(a) 결합법칙을 증명하여라.

4. 순서수의 합에 있어서 우소거법칙은 반드시 성립하지는 않는다. 즉 $\beta+\alpha=\gamma+\alpha$이지만 $\beta\neq\gamma$인 순서수 α, β, γ가 존재한다. 이런 반례를 들어라.

5. 다음은 정리 6(b)를 완화한 것이다. 이것을 증명하여라.

순서수 β, γ에 있어서 $\beta<\gamma$이면 임의의 순서수 α에 대하여

$$\beta+\alpha\leqslant\gamma+\alpha \text{(다음 문제 6 참조).}$$

6. 문제 5에서 $\beta<\gamma$는 반드시 $\beta+\alpha<\gamma+\alpha$를 함의하지는 않음을 보여라 [346쪽 정리 6(b)와 비교하여라].

7. 임의의 순서수 α, β에 대하여

$$\alpha+1\leqslant\beta\Leftrightarrow\alpha<\beta$$

이것을 증명하여라.

8. Prove the converse of Theorem 6(b):

$$\text{If } \alpha+\beta<\alpha+\gamma \text{ then, } \beta<\gamma.$$

9. Let α and γ be ordinal numbers such that $\alpha\leqslant\gamma$. Prove that there exists a unique β such that $\alpha+\beta=\gamma$. [Such a β may be denoted by $(-\alpha)+\gamma$].

10. Prove that $\alpha+1>\alpha$ for any ordinal number.

11. Is there an ordinal number that is greater than all the other ordinal numbers?

7.4 Multiplication of Ordinal Numbers

In cardinal arithmetic, the product of two cardinal numbers card A and card B is defined to be card $(A\times B)$. In any attempt to define the " product " of kind of ordinal numbers ord $(A, \leqq)$ and ord $(B, \leqq')$, we must first decide what a well-ordering to impose on the Cartesian product $A\times B$. Our natural preference is the so-called " lexicographic " ordering defined below.

Definition 4. Let $(A, \leqq)$ and $(B, \leqq')$ be well-ordered sets. Then the lexicographic ordering $\leqq^*$ on $A\times B$ is defined by:
(a) If $a<x$, then $(a, b)\leqq^*(x, y)$ for any b and y in B.
(b) If $a=x$ and $b\leqq' y$ then $(a, b)\leqq^*(x, y)$.

Theorem 7. Let $(A, \leqq)$ and $(B, \leqq')$ be well-ordered sets. Then the lexicographic ordering $\leqq^*$ on $A\times B$ is a well-order relation on it.

[Proof] It is clear that $\leqq^*$ is a total order relation on $A\times B$. To show that the total order relation is a well-ordering, let S be an arbitrary nonempty subset of $A\times B$. We shall prove that S contains the least element. First, we observe that the set

8. 정리 6(b)의 역, $\alpha+\beta<\alpha+\gamma \Longrightarrow \beta<\gamma$를 증명하여라.

9. 순서수 α, γ에 대하여 $\alpha\leqslant\gamma$이면 순서수 β가 단 하나 존재함으로써 $\alpha+\beta=\gamma$[이런 β를 $(-\alpha)+\gamma$로 나타낸다]. 이것을 증명하여라.

10. 임의의 순서수 α에 대하여 $\alpha<\alpha+1$. 이것을 밝혀라.

11. 순서수 중에는 가장 큰 것이 존재하는가? 존재하면 그 사실을 증명하고 그렇지 않으면 반례 하나를 들어라.

12.* $\{1,3,5,\cdots\}\cup\{\cdots,6,4,2\}$, $\{\cdots,5,3,1,2,4,6,\cdots\}$의 순서수를 구하라.

7.4 순서수의 곱셈

우리가 기수를 계산할 때 두 기수 card A, card B의 곱을 card $(A\times B)$로서 정의하였다. 지금 두 순서수 ord $(A, \leq)$, ord $(B, \leq')$의 "곱"을 정의하려는 어떤 시도에서도 먼저 데카르트 곱 $A\times B$ 위에 부여할 정렬순서를 정할 필요가 있거니와 그러자면 바로 다음에서 정의한 "사전식 순서"의 뜻을 익혀두어야 한다.

정의 4 정렬집합 $(A,\leq)$, $(B,\leq')$에 대하여 $A\times B$ 위의 사전식 순서(또는 사전식 나열) $\leq^*$를 다음과 같이 정의한다.

(a) $a<x$이면 B의 임의의 원소 b, y에 대하여 $(a,b)\leq^*(x,y)$

(b) $a=x$이고 $b\leq' y$이면 $(a,b)\leq^*(x,y)$

정리 7 정렬집합 $(A,\leq)$, $(B,\leq')$에 대한 $A\times B$ 위의 사전식 순서는 정렬순서이다.

[증명] $A\times B$ 위의 사전식 순서 $\leq^*$가 전순서인 것은 분명하다(과연?). 이제 그 전순서 $\leq^*$가 정렬순서임을 보이기 위해서는 $A\times B$의 임의의 부분집합으로서 공집합이 아닌 것을 S라고 할 때 S에는 최소원소가 존재함을 밝히면 된다. 먼저 집합

$$p_A(S)=\{x\in A \mid (x, y)\in S \text{ for some } y\in B\}$$

is a nonempty subset of the well-ordered set A, and hence contains the least element, say a. Then consider the set $\{y\in B \mid (a, y)\in S\}$ which is a nonempty subset of the well-ordered set B and hence contains the least element, say b. It is now quite obvious that the element (a, b) in S is the least element of S. Therefore, $(A\times B, \leqq^*)$ is a well-ordered set.

Theorem 8. Let $(A_1, \leqq_1)$, $(A_2, \leqq_2)$, $(B_1, \leqq_1')$ and $(B_2, \leqq_2')$ be well-ordered sets such that $(A_1, \leqq_1)\approx(A_2, \leqq_2)$ and $(B_1, \leqq_1')\approx(B_2, \leqq_2')$. Then

$$(A_1\times B_1, \leqq_1^*)\approx(A_2\times B_2, \leqq_2^*)$$

[Proof] Let $f: A_1\to A_2$ and $g: B_1\to B_2$ be order-isomorphisms. We leave it as an exercise for the reader to show that the function

$$f\times g: A_1\times B_1\to A_2\times B_2$$

defined by $$(f\times g)(x, y)=(f(x), g(y))$$

for all $(x, y)\in A_1\times B_1$ is an order-isomorphism.

Theorems 7 and 8 permit us to define the product of two ordinal numbers.

Definition 5. For any ordinal numbers α and β, the ordinal product $\beta\alpha$ is defined by $\beta\alpha=\text{ord}(A\times B, \leqq^*)$, where $(A, \leqq)$ and $(B, \leqq')$ are well-ordered sets such that $\text{ord}(A, \leqq)=\alpha$, $\text{ord}(B, \leqq')=\beta$, and $\leqq^*$ is the lexicographic ordering on $A\times B$.

Note that, according to Definition 5, $\text{ord}(A\times B, \leqq^*)$ is $\beta\alpha$, not $\alpha\beta$. The ordinal product is not commutative.

$$p_A(S)=\{x\in A\,|\,\text{적당한 } y\in B\text{에 대하여 } (x,y)\in S\}$$

는 정렬집합 A의 부분집합(공집합이 아님)인 것을 알 수 있다. 그러므로 집합 p_A에는 최소원소, 말하자면 a가 존재한다. 이제 집합 $\{y\in B\,|\,(a,y)\in S\}$를 생각하면 이것 또한 정렬집합 B의 공집합이 아닌 부분집합이므로 최소원소, 말 하자면 b를 지닌다. 여기서 $(a,b)\in S$는 S의 최소원소인 것이 자명하다(과연?). 그러므로 $(A\times B,\leqq^*)$는 정렬집합이다.

정리 8 정렬집합 $(A_1,\leqq_1),(A_2,\leqq_2),(B_1,\leqq_1'),(B_2,\leqq_2')$에 있어서

$$(A_1,\leqq_1)\approx(A_2,\leqq_2),\ (B_1,\leqq_1')\approx(B_2,\leqq_2')$$

이면

$$(A_1\times B_1,\leqq_1^*)\approx(A_2\times B_2,\leqq_2^*)$$

[증명] 함수 $f:A_1\to A_2, g:B_1\to B_2$를 각각 순서동형사상이라 할 때 임의의 $(x,y)\in A_1\times B_1$에 대하여 $(f\times g)(x,y)=(f(x),g(y))$로 정의된 함수 $f\times g:A_1\times B_1\to A_2\times B_2$는 순서동형 사상이다. 이것의 증명은 독자의 연습문제로 돌린다.

그러므로 $$(A_1\times B_1,\ \leqq_1^*)\approx(A_2\times B_2,\ \leqq_2^*)$$

이제 정리 7, 8을 통해 두 순서수의 곱을 정의할 수 있다.

정의 5 정렬집합 $(A,\leqq),(B,\leqq')$에 있어서 $\text{ord}(A,\leqq)=\alpha$, $\text{ord}(B,\leqq')=\beta$일 때 $\text{ord}(A\times B,\leqq^*)$를 α,β의 **순서수의 곱**이라 하고 이것을 $\boldsymbol{\beta\alpha}$로 나타낸다. 여기서 $\leqq^*$은 $A\times B$ 위의 사전식 순서이다.

정의 5에서 $\text{ord}(A\times B,\leqq^*)$는 $\alpha\beta$가 아니라 $\beta\alpha$임에 주의해야 한다. 또 다음 예제 3에서 보듯이 순서수의 곱셈에서는 기수의 곱셈과는 달리 교환법칙이 성립하지 않는 경우도 있다.

보기 1 $2=\text{ord}(\{1,2\},\leqq)$, $3=\text{ord}(\{1,2,3\},\leqq')$에 대하여

$$\begin{aligned}2-3&=\text{ord}(\{1,2,3\}\times\{1,2\},\leqq^*)\\&=\text{ord}(\{1,1\},\{1,2\},\{2,1\},\{2,2\},\{3,1\},\{3,2\},\leqq^*)=6\end{aligned}$$

그러므로 $2\cdot3=6$

Example 3 Compare the ordinal products 2ω and $\omega 2$.

[Solution] Let $(A, \leqq)=\{1, 2, 3, \cdots\}$ and $(B, \leqq')=\{0, 1\}$ so that $\text{ord}(A, \leqq)=\omega$ and $\text{ord}(B, \leqq')=2$. Let $A\times B$ be endowed with the lexicographic ordering so that $2\omega=\text{ord}(A\times B, \leqq^*)$. Then, the function $f: A\times B\to A$ given by

$$f(j, k)=\begin{cases} 2j-1 & \text{if } k=0 \\ 2j & \text{if } k=1 \end{cases}$$

is an order-isomorphism. Therefore, $2\omega=\omega$.

Next, impose the lexicographic ordering $\leqq'^*$ on $B\times A$ so that $\omega 2=\text{ord}(B\times A, \leqq'^*)$. We leave it to the reader to verify that $\text{ord}(B\times A, \leqq^*)=\omega+\omega$. Therefore, $2\omega\neq\omega 2$.

Theorem 9. Let α, β, and γ be any ordinal numbers. Then,
(a) $(\gamma\beta)\alpha=\gamma(\beta\alpha)$ (Associative Law)
(b) $\gamma(\alpha+\beta)=\gamma\alpha+\gamma\beta$ (Left Distributive Law).

[Proof] Exercise.

Theorem 10. Let α, β, and γ be any ordinal numbers such that $\gamma>0$. Then,
(a) $\alpha<\beta$ implies $\gamma\alpha<\gamma\beta$
(b) $\gamma\alpha=\gamma\beta$ implies $\alpha=\beta$ (Left Cancellation).

[Proof] (a) Let $(A, \leqq)$, $(B, \leqq')$, and $(C, \leqq'')$ be well-ordered sets such that $\text{ord}(A, \leqq)=\alpha$, $\text{ord}(B, \leqq')=\beta$, and $\text{ord}(C, \leqq'')=\gamma$. If $\alpha<\beta$, then there exists an element $p\in B$ such that $A\approx B_p$. Let q denote the least element of C, and let $A\times C$ and $B\times C$ be endowed with the lexicographic orderings. It follows that $A\times C$ is isomorphic to $B_p\times C$ under the inherited ordering from $B\times C$, and that $B_p\times C$ is the segment

$$(B\times C)_{(p,q)}=\{(x, y)\in B\times C \mid (x, y)<'^*(p, q)\}$$

of $B\times C$. This proves that $\gamma\alpha<\gamma\beta$.

예제 3** 두 순서수의 곱 2ω, $\omega 2$에 대하여 $2\omega \neq \omega 2$임을 밝혀라.

[**풀이**] 지금 $(A, \leqq) = \{1, 2, 3, \cdots\}$, $(B, \leqq') = \{0, 1\}$라고 놓음으로써 $\text{ord}(A, \leqq) = \omega$, $\text{ord}(B, \leqq') = 2$. 한편 $A \times B$ 위에 사전식 순서 $\leqq^*$를 여함으로써 $2\omega = \text{ord}(A \times B, \leqq^*)$. 여기서

$$f(j, k) = \begin{cases} 2j-1, & k=0 \\ 2j, & k=1 \end{cases}$$

로 정의된 함수 $f: A \times B \to A$는 순서동형사상이다. 따라서 $2\omega = \omega$.

마찬가지로 $B \times A$ 위에 사전식 순서 $\leqq'^*$를 부여함으로써 $\omega 2 = \text{ord}(B \times A, \leqq'^*)$를 얻는다 [독자는 $\text{ord}(B \times A, \leqq'^*) = \omega + \omega$임을 확인하여 보아라]. 한편 $\omega < \omega + \omega$(왜?) 그러므로 $2\omega \neq \omega 2$.

정리 9 임의의 순서수 α, β, γ에 대하여 다음이 성립한다.
(a) **결합법칙** $(\gamma\beta)\alpha = \gamma(\beta\alpha)$ (b) **좌분배법칙** $\gamma(\alpha + \beta) = \gamma\alpha + \gamma\beta$

[**증명**] 독자의 연습문제로 돌린다.

정리 10 임의의 순서수 α, β, $\gamma\,(\gamma > 0)$에 대하여 다음이 성립한다.
(a) $\alpha < \beta \Longrightarrow \gamma\alpha < \gamma\beta$ (b) **좌소거법칙** $\gamma\alpha = \gamma\beta \Longrightarrow \alpha = \beta$

[**증명**] (a) 정렬집합 $(A, \leqq)$, $(B, \leqq')$, $(C, \leqq'')$에서 $\text{ord}(A, \leqq) = \alpha$, $\text{ord}(\beta, \leqq') = \beta$, $\text{ord}(C, \leqq'') = \gamma$라고 놓는다. 지금 $\alpha < \beta$이면 적당한 $p \in B$가 존재함으로써 $A \approx B_p$. 집합 C의 최소원소를 q라 하고 $A \times C$, $B \times C$ 위에 각각 사전식 순서 $\leqq'^*$를 부여하면 $B \times C$로부터 이어받은 순서에 따라 $A \times C$, $B_p \times C$는 순서동형이고 $B_p \times C$는 $B \times C$의 절편 $(B \times C)_{(p,q)} = \{(x, y) \in B \times C \mid (x, y) <'^* (p, q)\}$ 따라서 $\gamma\alpha < \gamma\beta$이다.

** 독자는 순서쌍의 집합을 인용하여 $\omega \cdot 2 = \omega + \omega > \omega$, $2\omega = \omega$임을 각각 유도하여 보여라(353쪽 각주 * 참조) 사실은 1 이상의 유한순서수 n에 대하여 $n\omega < \omega n$. 이것은 밝힐 수 있을까?

(b) Suppose on the contrary that there exist ordinal numbers α, β, and $\gamma>0$ such that $\gamma\alpha=\gamma\beta$ and $\alpha\neq\beta$. We assume that $\alpha<\beta$ (the case where $\beta<\alpha$ may be treated similarly). Then, by part (a) of this theorem, we have $\gamma\alpha<\gamma\beta$, a contradiction.

It should be noted that although ordinal multiplication is left distributive and left cancellative, it is neither right distributive nor right cancellative. (See Problems 4 and 9.)

Exercise 7.4

1. Complete the proof of Theorem 8.
2. Show that
   (a) $0\alpha=0=\alpha 0$ and $1\alpha=\alpha=\alpha 1$ for any ordinal number α
   (b) $\alpha\beta=0$ if and only if either $\alpha=0$ or $\beta=0$.
3. Verify that
   (a) $\omega 2=\omega+\omega$ (b) $k\omega=\omega$ for any finite ordinal $k\neq 0$.
4. Show, by giving a counterexample, that ordinal multiplication is not right distributive, that is, $(\beta+\gamma)\alpha=\beta\alpha+\gamma\alpha$ need not be true.
5. Prove that ordinal multiplication is left distributive, that is $\gamma(\alpha+\beta)=\gamma\alpha+\gamma\beta$ for all ordinal numbers α, β, and γ.
6. Prove that ordinal multiplication is associative: $(\gamma\beta)\alpha=\gamma(\beta\alpha)$ for all ordinal numbers α, β, and γ.
7. Prove the following modification of Theorem 10(a): If α, β, and γ are such that $\alpha<\beta$ and $\gamma>0$, then $\alpha\gamma\leqslant\beta\gamma$. (See Problem 8, below).
8. Show that, in Problem 7 above, $\alpha<\beta$ and $\gamma>0$ do not necessarily imply that $\alpha\gamma<\beta\gamma$ (cf. Theorem 10(a)).
9. Show, by giving a counterexample, that ordinal multiplication is not right cancellative: There exists ordinal numbers α, β, and $\gamma>0$ such that $\alpha\gamma=\beta\gamma$, but $\alpha\neq\beta$.
10. Prove the following converse of Theorem 10(a): If α, β, and γ are ordinal numbers such that $\gamma\alpha<\gamma\beta$ and $\gamma>0$, then $\alpha<\beta$.

(b) 결론에 반하여 순서수 α, β, $\gamma\,(0<\gamma)$가 존재함으로써 $\gamma\alpha=\gamma\beta$이고 $\alpha\neq\beta$라고 가정한다. 지금 $\alpha<\beta$($\beta<\alpha$인 경우도 같은 방법으로 다룰 수 있다)라고 가정하면 (a)에 따라 $\gamma\alpha<\gamma\beta$. 이것은 (b)의 가정 $\gamma\alpha=\gamma\beta$에 모순이다. 그러므로 (b)는 성립한다.

순서수의 곱셈에 있어서 좌분배법칙과 좌소거법은 성립하지만 우분배법칙과 우소거법이 성립하지 않는 경우가 있음을 주의해야 한다(다음 연습문제 7.4 문제 4, 9).

연습문제 7.4

1. 정리 8의 증명을 완성하여라.

2. 다음이 성립함을 보여라.
(a) 임의의 순서수 α에 대하여 $0\alpha=0=\alpha 0$, $1\alpha=\alpha=\alpha 1$
(b) $\alpha=0$ 또는 $\beta=0 \Leftrightarrow \alpha\beta=0$

3. 임의의 유한순서수 $k\,(k\neq 0)$에 대하여 $k\omega=\omega$임을 확인하여라.

4. 순서수의 곱셈에 관한 우분배법칙이 반드시 성립하지 않는다. 즉 임의의 순서수 α, β, γ에 대하여 $(\beta+\gamma)\alpha=\beta\alpha+\gamma\alpha$가 거짓인 반례를 들어라.

5. 순서수의 곱셈에 관한 좌분배법칙이 성립한다. 즉 임의의 순서수 α, β, γ에 대하여 $\gamma(\alpha+\beta)=\gamma\alpha+\gamma\beta$. 이것을 증명하여라.

6. 순서수의 곱셈에 관한 결합법칙이 성립하다. 즉 임의의 순서수 α, β, γ에 대하여 $(\gamma\beta)\alpha=\gamma(\beta\alpha)$. 이것을 증명하여라.

7. 다음은 정리 10(a)를 수정한 것이다. 이것을 증명하여라.
순서수 α, β, γ에 대하여 $\alpha<\beta$이고 $0<\gamma$이면 $\alpha\gamma\leqslant\beta\gamma$(문제 8 참조).

8. 문제 7에서 $\alpha<\beta$, $0<\gamma$은 반드시 $\alpha\gamma<\beta\gamma$를 함의하지는 않는다(정리 10 (a)와 비교). 이것을 확인하여라.

9. 순서수의 곱셈에 있어서 우소거법이 반드시 성립하지는 않는다. 즉 순서수 α, β, $\gamma\,(0<\gamma)$에 대하여 $\alpha\gamma=\beta\gamma$이지만 $\alpha\neq\beta$인 반례 하나를 들어라.

10. 다음은 정리 10(a)의 역이다. 이것을 증명하여라.
순서수 α, β, γ에 대하여 $\gamma\alpha<\gamma\beta \wedge 0<\gamma \Rightarrow \alpha<\beta$.

7.5 Conclusion

It may seem natural now to explore the exponentiation of the ordinal numbers. We choose, however, to leave this to more advanced books. Instead, we shall first re-examine the ordering of the ordinal numbers, and then revisit the cardinal numbers from the ordinal-number-theoretical point of view.

Theorem 11. Let α be an arbitrary ordinal number. Then the set of all ordinal numbers β such that $\beta < \alpha$ is well-ordered set whose ordinal number is α.

[Proof] Let $(A, \leqq)$ be a well-ordered set whose ordinal number is α. For any ordnal number β with $\beta < \alpha$ and any well-ordered set $(B, \leqq')$ with $\mathrm{ord}(B, \leqq') = \beta$, B is order-isomorphic to a proper segment A_b, $b \in A$, of A. It follows from Theorem 1 that the element $b \in A$ is uniquely determined by the ordinal number β. Consequently, there is a well-defined function

$$f : \{\beta \mid \beta \text{ is an ordinal number with } \beta < \alpha\} \to A$$

given by $f(\beta) = b$ if $\beta = \mathrm{ord}(B, \leqq')$ and $B \approx A_b$. It is quite routine, and therefore left to the reader, to show that the function f is an order-isomorphism. Therefore, the set $\{\beta \mid \beta < \alpha\}$ is well ordered and $\mathrm{ord}\{\beta \mid \beta < \alpha\} = \alpha$.

In view of Theorem 11, it may be convenient to identify the ordinal number α with the set $\{\beta \mid \beta < \alpha\}$ and thus to regard each ordinal number as a well-ordered set (of ordinal numbers). For example:

$$\begin{array}{rlrl}
0 & \equiv \phi & \omega+1 & \equiv \{0, 1, 2, \cdots, \omega\} \\
1 & \equiv \{0\} & \omega+2 & \equiv \{0, 1, 2, \cdots, \omega, \omega+1\} \\
2 & \equiv \{0, 1\} & & \vdots \\
3 & \equiv \{0, 1, 2\} & \omega 2 & \equiv \{0, 1, 2, \cdots, \omega, \omega+1, \cdots\} \\
& \vdots & \omega 2+1 & \equiv \{0, 1, 2, \cdots, \omega, \omega+1, \cdots, \omega 2\} \\
\omega & \equiv \{0, 1, 2, \cdots\} & & \vdots
\end{array}$$

7.5 결론*

다음으로 순서수의 거듭제곱**을 알아 봄직하지만 그것은 이 책의 수준을 넘어서므로 여기서는 단지 순서수의 순서매김를 재음미한 후 기수를 순서수의 관점에서 다시 돌아보자.

정리 11 임의의 순서수 α에 대하여 $\beta<\alpha$인 모든 순서수 β의 집합은 순서수가 α인 정렬집합이다.

[증명] 순서수가 α인 하나의 정렬순서집합을 $(A, \leqq)$라 하자. 지금 $\beta<\alpha$인 순서수 β에 대하여 $\mathrm{ord}(B, \leqq')=\beta$인 정렬집합을 $(B, \leqq')$이라 하면 B는 $b\in A$에 대한 A의 진절편 A_b와 순서동형이다(그 근거는?). 더구나 정리 1에 의하여 $b\in A$는 순서수 β에 따라 단 하나로 정해진다. 결국 $\beta=\mathrm{ord}(B, \leqq')$, $B\approx A_b$이면 $f(\beta)=b$로서의

$$f:\{\beta \mid \beta\text{는 } \beta<\alpha\text{인 순서수이다}\}\longrightarrow A$$

는 잘 정의된 함수이나. 독자는 함수 f가 순서동형 사상임을 쉽게 확인할 수 있을 것이다(밝혀보아라!).

이상으로 $\{\beta \mid \beta<\alpha\}$는 $\mathrm{ord}\{\beta \mid \beta<\alpha\}=\alpha$인 정렬집합이다.

보기 2 $\mathrm{ord}(\{\beta \mid \beta\text{는 순서수, } \beta<2, \leqq'\})=\mathrm{ord}(\{1,2\}, \leqq)=2$

정리 11의 관점에서 순서수 α를 집합 $\{\beta \mid \beta<\alpha\}$와 동일시함으로써 순서수를 다음과 같이 그것에 수반되는 정렬집합으로 볼 수 있다.

$$\begin{array}{ll}
0\equiv\phi & \omega+1\equiv\{0,1,2,\cdots,\omega\} \\
1\equiv\{0\} & \omega+2\equiv\{0,1,2,\cdots,\omega,\omega+1\} \\
2\equiv\{0,1\} & \qquad\vdots \\
3\equiv\{0,1,2\} & \omega2\equiv\{0,1,2,\cdots,\omega,\omega+1,\cdots\} \\
\quad\vdots & \omega2+1\equiv\{0,1,2,\cdots,\omega,\omega+1,\cdots,\omega2\} \\
\omega\equiv\{0,1,2,\cdots\} & \qquad\vdots
\end{array}$$

* 이장우, 《현대집합론의 원리》, 경문사, 1998년, 275−279쪽 참조.

** 순서수의 거듭제곱에 관심이 많은 독자를 위하여 다음 책을 소개한다.
Charles C. Pinter, *Set Theory*, 경문사, 2002년, 282−289쪽.

Theorem 12. Any set of ordinal numbers is well-ordered.

[Proof] Suppose on the contrary that there exists a set A of ordinal numbers that is not well-ordered. Then there is a subset B of A that does not have a least element. Consequently, the set B contains a strictly decreasing infinite sequence of ordinal numbers $\alpha_1 > \alpha_2 > \alpha_3 > \cdots$. This sequence is contained in $\{\beta \mid \beta < \alpha_1\}$, and hence, contradicting Theorem 11, the set $\{\beta \mid \beta < \alpha_1\}$ is not well-ordered. This completes the proof of the theorem.

Viewing ordinal numbers as sets, let us consider, for example, the set $\mathscr{N}$ of all ordinal numbers α (α's are sets!) with α equipotent to the set N of natural numbers. This set includes :

$$\omega,\ \omega+1,\ \cdots,\ \omega 2,\ \omega 2+1,\ \cdots,\ \omega 3,\ \cdots,\ \omega 4,\ \cdots,\ \omega n,\ \cdots$$

$$\omega^2,\ \cdots,\ \omega^3,\ \cdots,\ \omega^n,\ \cdots,\ \omega^{\omega}\cdots,\ \omega^{\omega^{\omega}},\ \cdots,\ \omega^{\omega^{\omega^{\cdot^{\cdot^{\cdot}}}}},\ \cdots$$

According to Theorem 12, the set $\mathscr{N}$ is well-ordered and hence there is a unique least ordinal number that is equipotent to N. It is called the initial ordinal for the set N, (It is not difficult to see that the initial ordinal for N is equal to ω.) In general, by the well-ordering principle and by Theorem 12, every set X has a (unique) initial ordinal. The initial ordinals satisfy the guiding rules C-1 to C-4 for the cardinal numbers(see Section 1, Chapter 5).

This suggests an alternate approach to the cardinal numbers. As a matter of fact, some authors introduce the ordinal numbers first and then define the cardinal number of a set as the initial ordinal of that set. There is a logical advantage to treating the cardinal numbers as the initial ordinals, but from a pedagogical and practical point of view we chose to explore the cardinal numbers before the ordinal numbers.

In Theorem 12, we have carefully avoided the phrase "the set of all ordinal numbers." As in the case of Russell's paradox, the assumption of the existence of this set leads to a contradiction, known as the Burali-Forti paradox.

정리 12 순서수가 원소인 임의의 집합은 정렬집합이다.

[증명] 이 정리에 반하여 적당한 순서수의 집합 A가 존재함으로써 A는 정렬순서 집합이 아니라고 가정한다. 그러면 A의 부분집합으로서 최소원소를 지니지 않는, 말하자면 B가 존재한다. 따라서 B는 엄밀한 뜻에서 감소하는 무한순서수 순열 $\alpha_1 > \alpha_2 > \alpha_3 > \cdots$를 포함한다[연습문제 6.4 문제 7(317쪽) 참조]. 한편 이 수열은 집합 $\{\beta \mid \beta < \alpha_1\}$에 포함된다. 그러므로 $\{\beta \mid \beta < \alpha_1\}$은 정렬집합이 아니다. 이것은 정리 11에 모순이다. 이로써 증명이 끝난다.

순서수를 집합으로 볼 때 이를테면 자연수의 집합 $\mathbb{N}$과 대등한 모든 순서수 α(α는 집합이다!)의 집합 $\mathscr{N}$에는 다음과 같은 원소들이 존재한다.*

$$\omega,\ \omega+1,\ \cdots,\ \omega 2,\ \omega 2+1,\ \cdots,\ \omega 3,\ \cdots,\ \omega 4,\ \cdots,\ \omega n,\ \cdots$$

$$\omega^2,\ \cdots,\ \omega^3,\ \cdots,\ \omega^\omega,\ \cdots,\ \omega^n,\ \cdots,\ \omega^{\omega^\omega},\ \cdots,\ \omega^{\omega^{\omega^{\cdot^{\cdot^{\cdot}}}}},\ \cdots$$

그런데 정리 12에 따르면 집합 $\mathscr{N}$은 정렬집합이므로 $\mathbb{N}$과 대등한 가장 작은 순서수가 단 하나 존재한다. 이 순서수를 집합 $\mathbb{N}$의 **초기순서수**(또는 **처음순서수**)라고 한다(이 초기순서수는 ω임을 쉽게 알 수 있다).

대체로 정렬원리와 정리 12에 따라 각 집합 X는(단 하나의) 초기순서수를 지닌다. 이런 초기순서수는 기수에 대한 규칙 C-1~C-4(241쪽)를 모두 만족한다. 이같은 사실로 미루어 보아 기수에 앞서 순서수를 먼저 다룰 수 있음을 알 수 있다. 실제로 저자에 따라 순서수를 먼저 도입하고 그 집합의 초기순서수로써 집합의 기수를 정의하기도 한다. 이처럼 기수를 초기순서수로써 다루는 데 논리적 장점이 있다 하더라도 우리는 교육적이고 실질적 관점에서 초기순서수에 앞서 기수를 다루어 왔던 것이다.

사실 정리 12에서 "모든 순서수의 집합"이란 표현을 조심스럽게 피했는데 그 이유는 러셀의 역리의 경우에서와 같이 그러한 집합의 존재를 인정한다면 **부랄리-포르티(Burali-Forti)의 역리**로 알려진 다음 정리 13에서의 모순이 일어나기 때문이다.

---

* 이처럼 무궁무진한 순서수들! 사실 집합론의 핵심은 바로 "무한"에 있다고 말해도 과언이 아니다.

Theorem 13. There does not exist a set of all ordinal numbers.

[Proof] Suppose on the contrary that there is a set S of all ordinal numbers. By Theorem 12, S is well-ordered. The ordinal number of S, denoted by σ, must be a member of S. It follows from Theorem 11 and 4 that

$$\sigma = \text{ord}\{\beta \in S \mid \beta < \sigma\} = \text{ord } S_\sigma < \text{ord } S = \sigma$$

which is a contradiction.

[Alternative Proof] Let S and σ be as defined above. Then, σ is the largest ordinal number, which contradicts the fact that $\sigma < \sigma + 1$(see Problem 10, Exercise 7.3).

Exercise 7.5

Prove the following problems 1 through 5.

1. The function
$$f : \{\beta \mid \beta < \alpha\} \to A,$$
given in the proof of Theorem 11, such that $f(\beta) = b$ if $\beta = \text{ord}\,(B, \leqq')$ for some well-ordered set $(B, \leqq')$ and $B \approx A_b$, is an order-isomorphism.

2. The initial ordinal for the set $\mathbb{N}$ is equal to ω.

3. Every set has a unique initial ordinal.

4. Let X and Y be sets. Then $X \sim Y$ if and only if X and Y have the same initial ordinal.

5. Let X and Y be sets. Then card $X <$ card Y if and only if the initial ordinal of X is less than the initial ordinal of Y.

정리 13* 모든 순서수의 집합은 존재하지 않는다.

[증명] 이 정리의 결론을 부정하여 모든 순서수의 집합 S가 존재한다고 가정하면 정리 12에 따라 S는 정렬집합이다. 지금 S의 순서수를 σ라면, σ는 S의 원소이어야 한다. 이 경우 정리 11과 4에 따라 다음을 얻는다.

$$\sigma = \text{ord}\{\beta \in S \mid \beta < \sigma\} = \text{ord}\, S_\sigma < \text{ord}\, S = \sigma^{**}$$

그러므로 $\sigma < \sigma$. 이것은 $\sigma = \sigma$에 모순이다.

이로써 증명이 끝난다.

[별증]*** 위에서와 같이 S, σ를 각각 정의하면 σ는 가장 큰 순서수이다. 이것은 $\sigma < \sigma + 1$[연습문제 7.3 문제 10(351쪽)]이란 사실에 모순이다.

연습문제 7.5

다음 1~5를 증명하여라.

1. 정리 11의 증명에서 주어진 정렬집합 $(B, \leqq')$에 대하여 $\beta = \text{ord}(B, \leqq')$이고 $B \approx A_b$일 때, $f(\beta) = b$로 정의된 함수 $f : \{\beta \mid \beta < \alpha\} \longrightarrow A$는 순서동형사상이다.

2. 집합 $\mathbb{N}$의 초기순서수는 ω와 같다.

3. 임의의 집합에 대한 초기순서수가 단 하나씩 존재한다.

4. 집합 X, Y에 대하여

$$X, Y \text{ 각각의 초기순서가 같다.} \Leftrightarrow X \sim Y.$$

5. 집합 X, Y에 대한

X의 초기순서수가 Y의 초기순서수보다 작다. $\Leftrightarrow \text{card}\, X < \text{card}\, Y$.

---

* 기수에 대해서도 정리 13과 유사한 성질이 있는가?

** 물론 이런 표현도 피하고 싶다. 어떻게 나타낼 것인가?

*** 이 별증이 한결 간단해 보인다.

Appendix

부록

Appendix

The Peano Axioms for Natural Numbers

The system of natural numbers 1, 2, 3, ⋯ serves as the starting point for constructing the system of integers, the system of rational numbers, the system of residue classes modulo an integer, etc. What really are natural numbers? Giuseppi Peano (1858–1932) answered this question with five axioms, the Peano axioms for natural numbers.[1]

The Peano Axioms for Natural Numbers. There exists a set N of elements, called the natural numbers, satisfying the following axioms :

1. there exists a special element in N which we denote by 1 ;
2. for each element $n \in \mathrm{N}$, there exists a unique element n^+, called the successor of n, in N ;
3. for all $n \in \mathrm{N}$, $n^+ \neq 1$;
4. if $n, m \in \mathrm{N}$ and $n^+ = m^+$, then $n = m$;
5. if P is a subset of N such that $1 \in \mathrm{P}$, and $n \in \mathrm{P}$ implies $n^+ \in \mathrm{P}$, then $\mathrm{P} = \mathrm{N}$.

It may be more convenient to write $n+1$ for the successor n^+ of n, and we write $2=1+1$, $3=2+1$, $4=3+1$, etc.

Axiom 5 is the basis of mathematical induction.

[1] See. G. Peano, Arithmetices Principia, Bocca, Turin, 1889.

부록

자연수에 관한 페아노의 공리*

누구에게나 낯익은 자연수 1, 2, 3, …에 관한 체계는 *정수계*, *유리수계*, *잉여류계* 등을 구성하는 출발점이 되고 있다. 하지만 도대체 무엇이 자연수란 말인가?

페아노(1858~1932)는 이 질문에 자연수에 관한 페아노의 공리[1]로 알려진 다음 다섯 공리로 답하고 있다.

자연수에 관한 페아노의 공리 다음 공리 1~5를 모두 만족하는 집합 N이 존재한다. 이때 그 원소를 자연수**라고 부른다.

1. N에는 1로 나타낸 특별한 원소가 존재한다.
2. 각 $n\in$N에 대하여 그의 후임자인 원소 n^+가 N에 단 하나씩 존재한다.
3. 모든 $n\in$N에 대하여 $n^+\neq 1$
4. N의 원소 m, n에 대하여 $m^+=n^+\Longrightarrow m=n$
5.*** 각 $n\in$N과 N의 부분집합 P에 대하여

$$(1\in \mathrm{P}\wedge n\in \mathrm{P}\Longrightarrow n^+\in \mathrm{P})\text{이면}\quad \mathrm{P}=\mathrm{N}$$

임의의 $n\in$N의 후임자 n^+를 $n+1$로 나타내면 편리하다. 이를테면 앞으로 $2=1+1$, $3=2+1$, $4=3+1$ 등과 같이 적기로 한다.

---

[1] 쥐 페아노, 산수원리, 보까, 뚜린, 1889.

* 이장우, 《현대집합론의 원리》, 경문사, 1998, 157－163쪽 및 Charles C. Pinter, 《Set Theory》, 경문사, 1998, 181-187쪽 참조.

** 독자 누구나 무심코 써왔을 자연수에 관한 공리조차 이처럼 만만치 않다. 이 공리 모두를 익힌 뒤 유기수에 관한 규칙 C_1-C_4 중 특히 C_3(241쪽)을 곱씹어 봄 직하다.

*** 사실 수학적 귀납법(61쪽)은 바로 여기 다섯 번째인 이 공리에 바탕을 두고 있다.

Addition of natural numbers is defined inductively by:

(1) $$1+y=y^+$$

(2) $$x^++y=(x+y)^+$$

Using (1), (2), and mathematical induction, the following basic properties of addition of natural numbers can be proved.

(3) $x+(y+z)=(x+y)+z$ (associative law)

(4) $x+y=y+x$(commutative law)

(5) $x+z=y+z$ implies $x=y$(cancellation law)

We shall prove only (3) here and leave (4) and (5) to the reader as exercises.

[Proof] (3) Let y and z be any two arbitrary members of N that are fixed through-out this proof. In order to prove (3) by mathematical induction, we first show that

$$1+(y+z)=(1+y)+z$$

This is accomplished by the following step-by-step derivation:

$$\begin{aligned} 1+(y+z) &= (y+z)^+ && \text{by (1)} \\ &= y^++z && \text{by (2)} \\ &= (1+y)+z && \text{by (1)} \end{aligned}$$

Thus, the associative law (3) is valid for $x=1$ and the first requirement for mathematical induction is satisfied. Our next task is to show that the validity of

$$k+(y+z)=(k+y)+z \qquad \text{(induction hypothesis)}$$

implies the validity of

$$k^++(y+z)=(k^++y)+z$$

We have

$$\begin{aligned} k^++(y+z) &= [k+(y+z)]^+ && \text{by (2)} \\ &= [(k+y)+z]^+ && \text{by induction hypothesis} \\ &= (k+y)^++z && \text{by (2)} \\ &= (k^++y)+z && \text{by (2)} \end{aligned}$$

which was to be proved. The proof of (3) is now complete, by mathematical induction.

자연수의 덧셈은 다음과 같이 귀납적으로 정의한다.

(1) $1+y=y^+$ (2) $x^+ + y=(x+y)^+$

자연수의 덧셈에는 다음과 같은 기본성질이 있다. 즉 임의의 자연수 x, y, z에 대하여

(3) **결합법칙** $x+(y+z)=(x+y)+z$
(4) **교환법칙** $x+y=y+x$
(5) **소거법칙** $x+y=y+z \Rightarrow x=y$

여기서는 (3)을 증명하기로 하고 (4), (5)는 독자의 연습문제로 돌린다.

[증명] (3) 이 증명을 통하여 y와 z는 N의 임의의 원소지만 지금은 일단 고정된 것으로 생각한다. (3)을 수학적 귀납법으로 증명하기 위하여 먼저 $x=1$인 경우, 즉

$$1+(y+z)=(1+y)+z$$

임을 밝혀야겠지만 그것은 다음과 같이 유도할 수 있다.

$$\begin{aligned} 1+(y+z) &= (y+z)^+ && (1) \\ &= y^+ + z && (2) \\ &= (1+y)+z && (1) \end{aligned}$$

그러므로 $1+(y+z)=(1+y)+z$

따라서 결합법칙 (3)은 $x=1$에 대하여 성립한다. 이로써 수학적 귀납법에 대한 첫 번째 가정을 알아 보았다. 다음 단계는 귀납적 가정

$$k+(y+z)=(k+y)+z$$

으로부터 등식 $k^+ + (y+z)=(k^+ + y)+z$

이 성립함을 밝히는 일이다. 여기서

$$\begin{aligned} k^+ + (y+z) &= [k+(y+z)]^+ && (2) \\ &= [(k+y)+z]^+ && \text{귀납적 가정} \\ &= (k+y)^+ + z && (2) \\ &= (k^+ + y)+z && (2) \end{aligned}$$

따라서 $k^+ + (y+z)=(k^+ + y)+z$

마침내 수학적 귀납법에 따라 (3), 즉 결합법칙이 성립한다.

Multiplication of natural numbers is defined inductively by:

(6) $$1y=y$$

(7) $$x^+y=xy+y$$

The following basic properties of multiplication can be derived from (6) and (7) by mathematical induction:

(8) $x(yz)=(xy)z$ (associative law)

(9) $xy=yz$ (commutative alw)

(10) $xz=yz$ implies $x=y$ (cancellation law)

(11) $x(y+z)=xy=xz$ (distributive law)

Properties (8), (9), and (10) for multiplication resemble properties (3), (4), and (5), respectively, for addition. The proofs for (8), (9), and (10) are similar to the proofs for (3), (4), and (5), and hence the proofs of theses are again left to the reader. We prove the distributive law here.

[Proof] (11) Let y and z be any two arbitrary members of N that are fixed through out the proof. We shall prove (11) by mathematical induction on x. For $x=1$, by (6), it is clearly true that

$$1(y+z)=1y+1z$$

Thus, the first part of mathematical induction is satisfied. To complete the proof, we assume that

$$k(y+z)=ky+kz$$

is true for some $k \in \mathrm{N}$. Then

$k^+(y+z)=k(y+z)+(y+z)$	by (7)
$=(ky+kz)+(y+z)$	by induction hypothesis
$=ky+\{kz+(y+z)\}$	by (3)
$=ky+\{(kz+y)+z\}$	by (3)
$=ky+\{(y+kz)+z\}$	by (4)
$=ky+\{y+(kz+z)\}$	by (3)
$=(ky+y)+(kz+z)$	by (3)
$=k^+y+k^+z$	by (7)

Therefore, by mathematical induction, (11) is valid.

자연수의 곱셈은 다음과 같이 귀납적으로 정의한다.

(6) $1y=y$
(7) $x^{+}y=xy+y$

다음과 같은 곱셈에 관한 기본성질도 (6), (7)과 수학적 귀납법에 따라 유도할 수 있다.

(8) 결합법칙 $x(yz)=(xy)z$
(9) 교환법칙 $xy=yx$
(10) 소거법칙 $xz=yz \Longrightarrow x=y$
(11) 분배법칙 $x(y+z)=xy+xz$

곱셈에 관한 성질 (8), (9), (10)은 차례로 덧셈에 관한 성질 (3), (4), (5)와 비슷하다. 그뿐만 아니라 (8), (9), (10)의 증명은 (3), (4), (5)의 증명과 유사하므로 그 증명들도 독자에게 맡기고 여기서는 (11)만을 증명하여 보자.

[증명] 이 증명에서도 집합 $\mathbb{N}$의 원소 y, z는 일단 고정된 것으로 하고 x에 대한 수학적 귀납법을 적용하기로 한다.

먼저 $x=1$인 경우 다음은 (6)에 따라 분명히 성립한다.

$$1(y+z)=y+z=1y+1z, \quad \text{즉} \quad 1(y+z)=1y+1z$$

그러므로 수학적 귀납법의 첫 번째 가정을 만족한다. 이제 적당한 $k\in\mathbb{N}$에 대하여 다음이 성립한다고 가정하자.

$$k(y+z)=ky+kz \qquad \text{귀납적 가정}$$

그러면

$$\begin{aligned} k^{+}(y+z) &= k(y+z)+(y+z) && (7)\\ &=(ky+kz)+(y+z) && \text{귀납적 가정}\\ &=ky+\{kz+(y+z)\} && (3)\\ &=ky+\{(kz+y)+z\} && (3)\\ &=ky+\{(y+kz)+z\} && (4)\\ &=ky+\{y+(kz+z)\} && (3)\\ &=(ky+y)+(kz+z) && (3)\\ &=k^{+}y+k^{+}z && (7) \end{aligned}$$

따라서 $k^{+}(y+z)=k^{+}y+k^{+}z$

마침내 수학적 귀납법에 따라 (11), 즉 분배법칙이 성립한다.

The concept of order in N can be introduced as follows :

(12) $x>y$ if and only if there exists $z \in \mathrm{N}$ such that $x=y+z$.

From this definition of order, it follows that

(13) $x>y$ and $y>z$ imply $x>z$ (transitive law)

and

(14) for any x and y in N, one and only one of the followings holds:
$x>y,\ x=y,\ y>x$ (trichotomy law)

In fact, all known properties of natural numbers may be dervied from the Peano axioms. In concluding this appendix, we state the following theorem, which asserts that the natural number system is unique.

Theorem Let N and N′ be two sets satisfying the peano axioms 1 through 5. Then there exists a one-to-one correspondence (called an isomorphism) $f : \mathrm{N} \rightarrow \mathrm{N}'$ such that

$$f(1)=1' \text{ and } f(n+1)=f(n)+1'$$

where $1'$ denotes the special element of N′ satisfying 1−5.

자연수의 집합 ℕ에서의 순서에 관한 정의를 다음과 같이 내린다.

(12) 적당한 $z \in \mathbb{N}$가 존재함으로써 $x=y+z \Leftrightarrow x>y$.

우리는 위 정의로부터 다음을 얻는다.

(13) **추이율** $x>y \wedge y>z \Longrightarrow x>z$

(14) **삼분법*** ℕ의 임의의 원소 x, y에 대하여 다음 중 한 경우만 꼭 성립한다.

$$x>y, \qquad x=y, \qquad y<x$$

사실 자연수에 관하여 이미 알려진 모든 성질들은 지금까지 알아본 페아노의 공리로부터 유도된다. 여기서 자연수계는 다음 정리를 감안할 때 유일하다는 주장을 증명 없이 소개하고 이 부록을 마치고자 한다.

정리** 페아노의 공리 1부터 5까지를 만족하는 두 집합 ℕ, ℕ′에 대하여 다음과 같은[**동형사상**, 사실은 (ℕ, ≦)에서 (ℕ, ≦′)으로의 순서동형사상이라고 하는] 일대일 대응 $f: \mathbb{N} \longrightarrow \mathbb{N}'$이 존재한다.

$$f(1)=1', \quad f(n+1)=f(n)+1'$$

여기서 1′은 집합 ℕ′의 특별한 원소로서 페아노의 공리 1~5를 모두 만족한다.

* 이 시점에서 페아노의 공리를 돌이켜 보건데 공리 1에 따라 "1"이 자연수임이 규정되고 그 외의 자연수는 공리 2로 하여금 구성할 수 있음을 명시하고 있다.

한편 공리 3은 집합 ℕ이 무한집합임을 말해주고 있고 공리 4를 함수의 시각으로 생각할 때 그것은 단사임을 알 수 있다. 마지막 공리 5는 바로 수학적 귀납법의 바탕이 된다.

우리는 이같은 체계를 갖춘 수들을 1,2,3,⋯으로 나타냄으로써 자연수라고 부르고 있다.

** 끝내 이 책을 성의껏 공부해온 독자로 하여금 마지막 이 정리의 증명을 시도하여 유종의 미를 거둔다면 옮긴이에게 큰 보람이 아닐 수 없다!

해답

연습문제 1.1

1. (S) **2.** (S) **3.** (N) **4.** (S) **5.** (N)
6. (N) **7.** (N) **8.** (N) **9.** (S) **10.** (S)

20. 차례로 8, 16, 2^n

21.

p	q	r	
T	T	T	
T	T	F	
T	F	T	
T	F	F	
F	T	T	
F	T	F	
F	F	T	
F	F	F	

연습문제 1.2

13. 그렇다. **14.** 그렇다.

15. (4)와 (5), (7)과 (8), (9)와 (10), (11)과 (12)

16. (a) F : "나는 내 동생에게 음식을 나눠준다."라고 놓음으로써 $\sim(\sim F)$.*
(b) $A \to W$ (c) $\sim S \leftrightarrow P$ (d) $(F \vee \sim F) \to G$ (e) $(E \vee R) \to P$

연습문제 1.3

8. 함수 $f(x)$의 도함수가 존재하지 않고 나의 키는 175cm가 아니다.

---

* 이것은 명제 F의 이중부정이다[13쪽 문제 12 및 29쪽 정리 2(a) 참조].
16.(b)–(d)에 대해서도 (a)에서처럼 대문자가 나타내는 명제를 적어 보아라.

10. (a) $\sim(p_1 \wedge p_2 \wedge p_3 \wedge \cdots \wedge p_n) \equiv \sim p_1 \vee \sim p_2 \vee \sim p_3 \vee \cdots \vee \sim p_n$

(b) $\sim(p_1 \vee p_2 \vee p_3 \vee \cdots \vee p_n) \equiv \sim p_1 \wedge \sim p_2 \wedge \sim p_3 \wedge \cdots \wedge \sim p_n$

연습문제 1.4

3. 먼저 다음과 같은 진리표를 작성한다.

	p	$\wedge$	$\sim q$	$\rightarrow$	c	$\leftrightarrow$	p	$\rightarrow$	q
	T	F	F	T	F	T	T	T	T
	T	T	T	F	F	T	T	F	F
	F	F	F	T	F	T	F	T	T
	F	F	T	T	F	T	F	T	F
단계	1	2	1	3	4	5	1	2	1

이 표에서 단계 5를 보면 $p\wedge\sim q\rightarrow c$와 $p\rightarrow q$는 동치임을 알 수 있다.

연습문제 1.5

1. $p\wedge(p\rightarrow q)\equiv p\wedge\sim(p\wedge\sim q)$ $\rightarrow$의 정의(정의 4)

$\equiv p\wedge(\sim p\vee q)$ 드 모르간의 법칙, 이중부정법칙

$\equiv (p\wedge\sim p)\vee(p\wedge q)$ 분배법칙

$\equiv c\vee(p\wedge q)$ $p\wedge\sim p\equiv c$

$\equiv p\wedge q$ 교환법칙, 정리 7(b)

$\Rightarrow q$ 단순화 법칙

그러므로 $p\wedge(p\rightarrow q)\Rightarrow q$

3. $(p\wedge\sim q\rightarrow c)\Leftrightarrow\sim[(p\wedge\sim q)\wedge\sim c]$ $\rightarrow$의 정의(정의 4)

$\Leftrightarrow\sim(p\wedge\sim q)\vee c$ 드 모르간의 법칙, 이중부정법칙

$\Leftrightarrow\sim(p\wedge\sim q)$ 정리 7(b)

$\Leftrightarrow(p\rightarrow q)$ $\rightarrow$의 정의(정의 4)

그러므로 $(p\wedge\sim q\rightarrow c)\Leftrightarrow(p\rightarrow q)$

4. $(p\vee q)\wedge\sim p\equiv\sim p\wedge(p\vee q)$ 교환법칙

$\equiv(\sim p\wedge p)\vee(\sim p\wedge q)$ 분배법칙

$\equiv c\vee(\sim p\wedge q)$ $\sim p\wedge p\equiv c$

$\equiv\sim p\wedge q$ 교환법칙, 정리 7(b)

$\Rightarrow q$ 단순화 법칙

그러므로 $(p \vee q) \wedge \sim p \Rightarrow q$

5. $(c \to p) \equiv \sim(c \wedge \sim p)$ $\to$의 정의(정의 4)

$\equiv \sim c \vee p$ 드 모르간의 법칙, 이중부정법칙

$\equiv t \vee p$ $\sim c \equiv t$

$\equiv t$ 교환법칙, 정리 7(a)

그러므로 $(c \to p) \equiv t$ 따라서 $c \Rightarrow p$

7. $(p \vee q \to q) \Longleftrightarrow \sim[(p \vee q) \wedge \sim q]$ $\to$의 정의(정의 4)

$\Longleftrightarrow \sim(p \vee q) \vee q$ 드 모르간의 법칙, 이중부정법칙

$\Longleftrightarrow q \vee (\sim p \wedge \sim q)$ 드 모르간의 법칙, 교환법칙

$\Longleftrightarrow (q \vee \sim p) \wedge (q \vee \sim q)$ 분배법칙

$\Longleftrightarrow (\sim p \vee q) \wedge t$ 교환법칙, $q \vee \sim q \equiv t$

$\Longleftrightarrow \sim p \vee q$ 정리 7(a)

$\Longleftrightarrow \sim(p \wedge \sim q)$ 드 모르간의 법칙, 이중부정법칙

$\Longleftrightarrow (p \to q)$ $\to$의 정의(정의 4)

그러므로 $(p \vee q \to q) \Longleftrightarrow (p \to q)$

9. $(p \to r) \wedge (q \to r) \Longleftrightarrow (\sim p \vee r) \wedge (\sim q \vee r)$ 문제 8

$\Longleftrightarrow (r \vee \sim p) \wedge (r \vee \sim q)$ 교환법칙

$\Longleftrightarrow r \vee (\sim p \wedge \sim q)$ 분배법칙

$\Longleftrightarrow \sim(p \vee q) \vee r$ 드 모르간의 법칙, 교환법칙

$\Longleftrightarrow (p \vee q \to r)$ 문제 8

그러므로 $(p \to r) \wedge (q \to r) \Longleftrightarrow (p \vee q \to r)$

11. $(p \to q) \wedge (p \to \sim q) \Longleftrightarrow (\sim p \vee q) \wedge (\sim p \vee \sim q)$ 문제 8

$\Longleftrightarrow \sim p \vee (q \wedge \sim q)$ 배분법칙

$\Longleftrightarrow \sim p \vee c$ $q \wedge \sim q \equiv c$

$\Longleftrightarrow \sim p$ 정리 7(b)

그러므로 $(p \to q) \wedge (p \to \sim q) \Longleftrightarrow \sim p$

12. $(p \to q) \vee (p \to r) \Longleftrightarrow (\sim p \vee q) \vee (\sim p \vee r)$ 문제 8

$\Longleftrightarrow (\sim p \vee \sim p) \vee (q \vee r)$ 결합법칙, 교환법칙

$\Longleftrightarrow \sim p \vee (q \vee r)$ 멱등법칙

$\Longleftrightarrow (p \to q \vee r)$ 문제 8

그러므로 $(p \to q) \vee (p \to r) \Longleftrightarrow (p \to q \vee r)$ $\Longleftrightarrow$에 관한 추이율*

*문제 12의 결론을 얻기까지의 증명과정에 있어서 $\Longleftrightarrow$에 관한 추이율이 몇 차례 생략되어 있다. 문제 4·5·7·9·11에 대해서도 일일이 같은 맥락으로 생각하여 보아라.

연습문제 1.6

4.*	$\sim\exists x\, p(x) \equiv \sim\exists x \sim(\sim p(x))$	이중부정
	$\equiv \forall x \sim p(x)$	한정기호 부정법
	그러므로 $\exists x\, p(x) \equiv \forall x \sim p(x)$	≡에 대한 추이율
6.	$\sim\forall x \sim q(x) \equiv \exists x \sim(\sim q(x))$	한정기호 부정규칙
	$\equiv \exists x\, q(x)$	이중부정
	그러므로 $\sim\forall x \sim q(x) \equiv \exists x\, q(x)$	왜?

14.* ② 거짓 반례 :

p : $\triangle ABC \equiv \triangle DEF$, q : $\triangle ABC$, $\triangle DEF$의 대응하는 각의 크기가 제각기로 놓으면 ①이 참이다. 그러나 $\triangle ABC \backsim \triangle DEF$의 경우 q는 참이지만 p는 거짓일 수 있다. 따라서 ①의 역 $q \to p$는 거짓이다.

③, ④에 대해서는 독자에게 미룬다.

15.* (a) ⑤ (참)

증명 :	$a=b$	……①
	$\Rightarrow aa=ba$	등식의 성질(그것을 말하여라!)
	$\Rightarrow aa=bb$	①의 우변의 a에 b를 대입함
	$\Rightarrow a^2=b^2$	문자 a, b 각각에 대한 거듭제곱
그러므로	$a=b \Rightarrow a^2=b^2$	⇒의 추이율

역(거짓)

반례 : $a=3$ $b=-3$일 때 $3^2=(-3)^2$ 그러나 $3\neq-3$

이(거짓)

반례 이 : $a\neq b \to a^2\neq b^2$의 대우

$a^2=b^2 \to a=b$

는 거짓이다(위 15*(a) 역 참조).

그런데 이, 그 것의 대우는 동치[29쪽 정리 2(d)]이므로 주어진 이는 거짓이다.

(b) 독자에게 미룬다.

---

* (1) 이 문제 풀이에 있어서 저자는 "대입법"을 인용하고 있다. 그러나 지금껏 "대입"에 대한 용어의 정의없이 여기서 그것을 인용한다는 것은 천부당만부당하다.

(2) 동치관계 ≡의 "좌변", "우변"이란 용어는 그대로 받아들이지만 "이항"이란 언급 대신 "≡에 관한 대칭률"이라 함이 마땅하다고 본다.

연습문제 1.7

1. (a) 직접증명

1. $A \vee (B \wedge C)$	
2. $B \to D$	
3. $C \to E$	
4. $D \wedge E \to A \vee C$	
5. $\sim A$/그러므로 C	
6. $B \wedge C$	1, 5, 논리합의 삼단법
7. $B \wedge C \to D \wedge E$	2, 3, 구성적 양도논법
8. $D \wedge E$	7, 6, 삼단긍정법
9. $A \vee C$	4, 8, 삼단긍정법
10. C	9, 5, 논리합의 삼단법

(b) 간접증명

6. $\sim C$	간접증명
7. $B \wedge C$	1, 5, 논리합의 삼단법
8. C	7, 단순화법칙
9. $C \wedge \sim C$	6, 8, 논리곱
10. $C \wedge \sim C \equiv c$	9는 모순과 동치

7.

1. $P \wedge C \to R$	
2. $R \to G$	
3. $H \to \sim I$	
4. H	
5. $\sim G \vee I$/그러므로 $\sim(P \wedge C)$	
6. $\sim I$	3·4에 대한 삼단긍정법
7. $\sim G$	5·6에 대한 논리합의 삼단법
8. $\sim R$	2·7에 대한 삼단부정법
9. $\sim(P \wedge C)$	1·8에 대한 삼단부정법

8.

1. $W \vee H \to L \wedge S$	
2. $\sim S$/그러므로 $\sim H$	
3. $\sim S \vee \sim L$	2, 합의 법칙
4. $\sim(L \wedge S)$	3·교환법칙·드 모르간의 법칙
5. $\sim(W \vee H)$	1·4에 대한 삼단부정법
6. $\sim W \wedge \sim H$	5에 대한 드 모르간의 법칙
7. $\sim H$	6에 대한 단순화 법칙

9. 1. $E \wedge S \rightarrow G$
2. $G \rightarrow H$
3. $\sim H$/그러므로 $\sim E \vee \sim S$
4. $\sim G$ 2·3에 대한 삼단부정법
5. $\sim(E \wedge S)$ 1·4에 대한 삼단부정법
6. $\sim E \vee \sim S$ 5에 대한 드 모르간의 법칙

연습문제 1.8

1. $n=1$일 때 등식 ③이 성립함을 알 수 있다. 다음에 귀납적 가정으로서 $2 \leqq k$인 정수 k와 $0 \leqq r \leqq k$인 모든 정수 r에 대하여 다음이 성립한다고 가정한다.

$$C(k, r)=\frac{k!}{r!(k-r)!}$$

그러면 정의 7과 이 가정에 따라

$$\begin{aligned} C(k+1, r) &= C(k, r)+C(k, r-1) \\ &= \frac{k!}{r!(k-r)!}+\frac{k!}{(r-1)!(k-r+1)!} \\ &= \frac{k!(k-r+1)}{r!(k-r)!(k-r+1)}+\frac{k!r}{r(r-1)!(k-r+1)!} \\ &= \frac{k!(k+1)}{r!(k-r+1)!} \\ &= \frac{(k+1)!}{r!(k+1-r)!} \end{aligned}$$

그러므로 $\quad C(k+1, r)=\dfrac{(k+1)!}{r!(k+1-r)!}$

따라서 위 등식은 $k+1$일 때에도 성립하므로 수학적 귀납법의 결론에 따라 증명이 끝난다.

2. 정리 8과 수학적 귀납법을 이용하여 이 문제를 풀 수 있으되 그러려면 먼저 다음 보조정리부터 수학적 귀납법으로 증명할 필요가 있다.

보조정리 음수가 아닌 정수 n에 대하여 정수 r이 0보다 작거나 n보다 크면

$$C(n, r)=0$$

연습문제 2.1

2. $D \subseteq A \subseteq B \subseteq C$

5. $A \subseteq \phi$라고 가정하면 $\forall x[(x \in A) \to (x \in \phi)]$는 참인 명제이다. 그런데 $x \in \phi$는 거짓이므로 $(x \in A) \to (x \in \phi)$이 참이려면 모든 원소 x에 대하여 $x \in A$는 거짓이어야 한다. 따라서 $A = \phi$

6. (a) 다음 논증의 타당성을 확인하도록 한다.

1. $\forall x[(x \in A) \Rightarrow (x \in B)]$	가정 $A \subset B$
2. $\exists x[(y \in B) \wedge (y \notin A)]$	
3. $\forall x[(x \in B) \Rightarrow (x \in C)]$	가정 $B \subseteq C$
/ 그러므로 $A \subset C$	결론
4. $\forall x[(x \in A) \Rightarrow (x \in C)]$	1, 3, 추이율
5. $y \in B$	2, 단순화법칙
6. $y \notin A$	2, 단순화법칙
7. $y \in C$	3, 5, 삼단긍정법
8. $\exists y[(y \in C) \wedge (y \notin A)]$	7, 8, 논리곱
9. $A \subset C$	4, 8, $\subset$의 정의

8. (a) 거짓 (b) 거짓 (c) 거짓 (d) 거짓 (e) 거짓 (f) 참

9. 이것을 n에 관한 수학적 귀납법에 따라 증명한다.

$$C(1, r) = \begin{cases} 1, & r=0 \text{ 또는 } 1\text{인 경우} \\ 0, & \text{그 외의 경우} \end{cases}$$

이므로 주어진 주장은 $n=1$일 때 참이다. 여기서 $k(k \geq 2)$개의 원소를 지닌 집합 S가 주어질 때 임의의 정수 r에 대하여 S의 부분집합으로써 r개의 원소를 지닌 것이 $C(k, r)$개 존재한다고 가정한다(귀납적 가정). 지금 $k+1$개의 원소를 지니는 임의의 집합

$$A = \{a_1, a_2, \cdots, a_k, a_{k+1}\}$$

를 생각할 때 A의 부분집합으로써 r개의 원소를 지닌 것은 집합

$$A - \{a_{k+1}\} \text{ 곧 } \{a_1, a_2, \cdots, a_k\}$$

의 부분집합으로써 r개의 원소를 지닌 것이며 이런 것은 $A - \{a_{k+1}\}$의 부분집합으로써 $r-1$개의 원소를 지닌 것의 각각에 원소 a_{k+1}를 덧붙여 얻어진다. 그러므로 귀납적 가정과 정의 7에 따라 A의 부분집합으로써 r개의 원소를 지닌 것은 꼭

$$C(k, r) + C(k, r-1) = C(k+1, r)$$

개 존재한다. 이로써 수학적 귀납법에 의하여 증명이 끝난다.

연습문제 2.2

3. (b) $\phi, \{x\}, \{\{y, z\}\}, \{x, \{y, z\}\}$

5. 우선 다음 주장을 염두에 두자.

집합 $Y\in\mathscr{P}(A-B)$가 존재함으로써 $X=B\cup Y \Longleftrightarrow X\in\mathscr{P}(A:B)$

(a) $\mathscr{P}(A:B)$의 원소의 수는 멱집합 $\mathscr{P}(A-B)$의 원소의 수, 즉 2^{n-m}과 같다.

(b) $B=\phi$이면 $\mathscr{P}(A:B)=\mathscr{P}(A)$이므로 $\mathscr{P}(A)$의 원소의 수는 2^{n-0}, 즉 2^n이다.

연습문제 2.3

1. $A\cup B\supseteq B$인 것은 분명하므로 증명을 끝내려면 $A\cup B\subseteq B$임을 보여야 한다.

$x\in A\cup B\equiv(x\in A)\vee(x\in B)$	$\cup$의 정의
$\Rightarrow(x\in B)\vee(x\in B)$	$A\subseteq B$
$\equiv x\in B$	멱등법칙

그러므로 $A\cup B\subseteq B$

한편 $A\cup B\supseteq B$ 따라서 $A\cup B=B$

6. (a)

1. $x\in A\Rightarrow x\in C$	가정 $A\subseteq C$
2. $x\in B\Rightarrow x\in C$/그러므로 $A\cup B\subseteq C$	가정 $B\subseteq C$/결론
3. $x\in A\vee x\in B\Rightarrow x\in C\vee x\in C$	1, 2, 구성적 양도논법
4. $x\in A\vee x\in B\Rightarrow x\in C$	3, 멱등법칙
5. $x\in A\cup B\Rightarrow x\in C$	4, $\cup$의 정의
6. $A\cup B\subseteq C$	5, $\subseteq$의 정의

7. (i)

$C\subseteq(A\cap B)\cup C$	
$=A\cap(B\cup C)$	가정
$\subseteq A$	

그러므로 $(A\cap B)\cup C=A\cap(B\cup C)$

$\Rightarrow C\subseteq A$

(ii) $C\subseteq A$로 놓으면

$(A\cap B)\cup C=(A\cup C)\cap(B\cup C)$	분배법칙, 교환법칙
$=A\cap(B\cup C)$	문제 1

그러므로 $(A\cap B)\cup C=A\cap(B\cup C)$

따라서 $C\subseteq A\Rightarrow(A\cap B)\cup C=A\cap(B\cup C)$

(i), (ii)에 따라 $(A\cap B)\cup C=A\cap(B\cup C)\Longleftrightarrow A\subseteq C$

11. $A\subseteq C$이므로 문제 10에 따라 $A\cup B\subseteq C\cup B$. 또 $B\subseteq D$이므로 문제 10에 따라 $C\cup B=B\cup C\subseteq D\cup C=C\cup D$. 즉 $C\cup B\subseteq C\cup D$ 그러므로 $A\cup B\subseteq C\cup D$

연습문제 2.4

1. $A-(B\cap A)=A\cap(B\cap A)^c$ 예제 5

$=A\cap(B^c\cup A^c)$ 드 모르간의 정리

$=(A\cap B^c)\cup(A\cap A^c)$ 분배법칙

$=(A\cap B^c)\cup\phi$ 정리 5(c)

$=A-B$ 예제 5, 정리 4(a)

그러므로 $A-(B\cap A)=A-B$

3. $B\subseteq A^c$라고 가정하면 임의의 원소 x에 대하여 $x\in B\Rightarrow x\in A^c$, 즉 $x\in B\Rightarrow x\notin A$. 그러므로 $x\notin A\cap B$. 따라서 $A\cap B=\phi$.

역으로 $A\cap B=\phi$이면 $x\in B\Rightarrow x\notin A$, 즉 $x\in B\Rightarrow x\in A^c$. 그러므로 $B\subseteq A^c$.

4. (i) $B\subseteq A$이면

$(A-B)\cup B=(A\cap B^c)\cup B$ 예제 5

$=B\cup(A\cap B^c)$ 교환법칙

$=(B\cup A)\cap(B\cup B^c)$ 분배법칙

$=A\cap(B\cup B^c)$ 가정 $B\subseteq A$, 연습문제 2.3 문제 1

$=A\cap U$ 정리 5(c)

$=A$ $A\subseteq U$

그러므로 $B\subseteq A$이면 $(A-B)\cup B=A$

(ii) $(A-B)\cup B=A$이면 $B\subseteq(A-B)\cup B=A$, 즉 $B\subseteq A$

(i), (ii)에 따라 $(A-B)\cup B=A\Longleftrightarrow B\subseteq A$

6. (a) $(A-C)\cup(B-C)=(A\cap C^c)\cup(B\cap C^c)$ 예제 5

$=(C^c\cap A)\cup(C^c\cap B)$ 교환법칙

$=C^c\cap(A\cup B)$ 분배법칙

$=(A\cup B)-C$ 교환법칙, 예제 5

그러므로 $(A-C)\cup(B-C)=(A\cup B)-C$

10. $C-B=(A\cup B)-B$ 가정 $C=A\cup B$

$=(A\cup B)\cap B^c$ 예제 5

$=(A\cap B^c)\cup(B\cap B^c)$ 교환법칙, 분배법칙

$=(A-B)\cup\phi$ 예제 5, 정리 5(c)

$=A$ 가정 $A\cap B=\phi$

그러므로 $C-B=A$

따라서 $A=C-B$

11. $(A\cup B)-(A\cap B)$

$=(A\cup B)\cap(A\cap B)^c$	예제 5
$=(A\cup B)\cap(A^c\cup B^c)$	드 모르간의 정리
$=[(A\cup B)\cap A^c]\cup[(A\cup B)\cap B^c]$	분배법칙
$=[(A\cap A^c)\cup(B\cap A^c)]\cup[(A\cap B^c)\cup(B\cap B^c)]$	교환법칙, 분배법칙
$=[\phi\cup(B\cap A^c)]\cup[(A\cap B^c)\cup\phi]$	정리 5(c)
$=(B-A)\cup(A-B)$	예제 5
$=(A-B)\cup(B-A)$	교환법칙

그러므로 $(A\cup B)-(A\cap B)=(A-B)\cup(B-A)$

연습문제 2.6

2. (a) $\{0\}$ (b) $[0,1]$ (c) $[0,1/99]$

4. (a) $U\mathscr{P}(\mathscr{P}(\phi))=U\{\phi,\{\phi\}\}=\phi\cup\{\phi\}=\{\phi\}$

그러므로 $U\mathscr{P}(\mathscr{P}(\phi))=\{\phi\}$

(b) $U\mathscr{P}(\mathscr{P}(\mathscr{P}((\phi))))=U\{\phi,\{\phi\},\{\phi\}\},\{\phi,\{\phi\}\}$

$=\phi\cup\{\phi\}=\{\{\phi\}\}\cup\{\phi,\{\phi\}\}=\{\phi,\{\phi\}\}$

그러므로 $U\mathscr{P}(\mathscr{P}(\mathscr{P}(\phi)))=\{\phi,\{\phi\}\}$

6. (a)

$\left(\bigcup_{i=1}^{m}A_i\right)\cap\left(\bigcup_{j=1}^{n}B_j\right)=\bigcup_{j=1}^{n}\left[\left(\bigcup_{i=1}^{m}A_i\right)\cap B_j\right]$	정리 9
$=\bigcup_{j=1}^{n}\left[B_j\cap\left(\bigcup_{i=1}^{m}A_i\right)\right]$	교환법칙
$=\bigcup_{j=1}^{n}\left[\bigcup_{i=1}^{m}(B_j\cap A_i)\right]$	정리 9
$=\bigcup_{j=1}^{n}\left[\bigcup_{i=1}^{m}(A_i\cap B_j)\right]$	교환법칙

그러므로 $\left(\bigcup_{i=1}^{m}A_i\right)\cap\left(\bigcup_{j=1}^{n}B_j\right)=\bigcup_{j=1}^{n}\left[\bigcup_{i=1}^{m}(A_i\cap B_j)\right]$

9.* 가정에 따라 $A=\{a_1, a_2, a_3, \cdots, a_n\}$으로 놓으면

$U\{\{a\}\mid a\in A\}=\bigcup_{1}^{n}\{a_i\}=A$ 그러므로 $U\{\{a\}\mid a\in A\}=A$

연습문제 3.1

2. $A=\phi$ 또는 $B=\phi$ 또는 $A=B$

9. (a) $(x, y)\in(A\times C)\cap(B\times D)$

$=[(x, y)\in A\times C]\wedge[(x, y)\in B\times D]$	$\cap$의 정의
$=[(x\in A)\wedge(y\in C)]\wedge[(x\in B)\wedge(y\in D)]$	$\times$의 정의
$=[(x\in A)\wedge(x\in B)]\wedge[(y\in C)\wedge(y\in D)]$	결합법칙, 교환법칙
$=(x\in A\cap B)\wedge(y\in C\wedge D)$	$\cap$의 정의
$=(x, y)\in(A\cap B)\times(C\cap D)$	$\times$의 정의

그러므로 $(A\times C)\cap(B\times D)=(A\cap B)\times(C\cap D)$

11. (i) $a=c$이고 $b=d$이면 $\{a\}=\{c\}$이고, $\{a, b\}=\{c, d\}$

그러므로 $\{\{a\},\{a, b\}\}=\{\{c\},\{c, d\}\}$ 또는 $(a, b)=(c, d)$

(ii) $\{\{a\}, \{a, b\}\}=\{\{c\}, \{c, d\}\}$라고 가정한다. $a=b$이면 순서쌍 (a, b)는 한 원소 집합 $\{\{a\}\}$와 같다. 지금 $(a, b)=(c, d)$라고 가정하면 $\{\{c\}, \{c, a\}\}=\{\{a\}\}$. 따라서 $\{c\}=\{c, a\}=\{a\}$이므로 $a=b=c=d$. 한편 $a\neq b$이면 (a, b)와 (c, d)는 각각 한원소 집합 $\{a\}$, $\{c\}$ 하나씩을 지닌다. 그러므로 $a=c$. 집합 $\{a, b\}$, $\{c, d\}$는 또한 각각 "두 원소 집합" $\{a, b\}$, $\{c, d\}$ 하나씩을 지니므로 $\{a, b\}=\{c, d\}$. 따라서 $b\in\{c, d\}$. 그러므로 $b=d$.

왜냐하면 $b=c$이면 $a=c$. 그러므로 $a=b$. 이것은 가정 $a\neq b$에 모순이기 때문이다.

지금까지를 요약하면 $(a, b)=(c, d)\Longrightarrow a=c\wedge b=d$

한편 $a=c\wedge b=d\Rightarrow(a, b)=(c, d)$.

이상으로 $(a, b)=(c, d)\Longleftrightarrow a=c\wedge b=d$

연습문제 3.2

3.* (a)

$y\in\mathrm{Im}(\mathscr{R})\Longleftrightarrow$ 적당한 $x\in A$에 대하여 $(x, y)\in\mathscr{R}$	치역의 정의
$\Longleftrightarrow$ 적당한 $x\in A$에 대하여 $(y, x)\in\mathscr{R}^{-1}$	$\mathscr{R}^{-1}$의 정의
$\Longleftrightarrow y\in\mathrm{Dom}(\mathscr{R}^{-1})$	정의역의 정의

그러므로 $y\in\mathrm{Im}(\mathscr{R})\Longleftrightarrow y\in\mathrm{Dom}(\mathscr{R}^{-1})$

따라서 $y\in\mathrm{Dom}(\mathscr{R}^{-1})\Longleftrightarrow y\in\mathrm{Im}(\mathscr{R})$

즉 $\mathrm{Dom}(\mathscr{R}^{-1})=\mathrm{Im}(\mathscr{R})$

연습문제 3.3

1. $(x, y)\in X/\mathscr{T} \Longleftrightarrow$ 적당한 $A\in\mathscr{T}$에 대하여 $x\in A \wedge y\in A$ — $X/\mathscr{T}$의 정의

$\Longleftrightarrow$ 적당한 $A\in\mathscr{T}$에 대하여 $(x, y)\in A\times A$ — $\times$의 정의

$\Longleftrightarrow (x, y)\in \bigcup_{A\in\mathscr{T}} A\times A$ — $\cup$의 정의

그러므로 $X/\mathscr{T}=\bigcup_{A\in\mathscr{T}} A\times A$

6. $(x, y)\in X/(X/\mathscr{E})$

$\Longleftrightarrow$ 적당한 $A\in X/\mathscr{E}$에 대하여 $(x, y)\in A\times A$ — 문제 1

$\Longleftrightarrow$ 적당한 $c\in X$에 대하여 $(x, y)\in(c/\mathscr{E})\times(c/\mathscr{E})$ — $X/\mathscr{E}$의 정의

$\Longleftrightarrow (x, y)\in\mathscr{E}$ — 정리 3(b)

그러므로 $X/(X/\mathscr{E})=\mathscr{E}$

연습문제 3.4

6. (a) $[5, \infty)$ (b) $[-1, 1]$ (c) $\mathbb{R}$

9. 2^3, n^m

10. 2, n

11. $A=\mathrm{Dom}(g)$라 하고 집합 Y의 임의의 부분집합으로써 $\mathrm{Im}(g)$를 포함하는 것을 B라고 할 때 $g: A\to B$를 보이도록 한다.

12. $(x, y)\in f$라고 가정하면 f는 반사적이므로 $(x, x)\in f$. 그런데 f는 함수이므로 $y=x$. 즉 모든 $x\in X$에 대하여 $f(x)=x$.
따라서 $f: X\to X$는 항등함수 $\mathrm{I}_X: X\to X$와 같다.

13. 모든 $x\in[0, 1]$에 대하여 $f(x)=1-x$이다.

14. 집합 X의 임의의 원소 x에 대하여 $(x, f(x))\in f\subseteq g$이므로 $(x, f(x))\in g$, 즉 모든 $x\in X$에 대하여 $g(x)=f(x)$. 따라서 정리 7(161쪽)에 의하여 $f=g$.

연습문제 3.5

4. (a) $x\in A\Rightarrow f(x)\in f(A)$ — 정의 9(a)

$\Rightarrow x\in f^{-1}(f(A))$ — 정의 9(b)

그러므로 $A\subseteq f^{-1}(f(A))$

(b) $y\in f(f^{-1}(B))$이라면 적당한 $x\in f^{-1}(B)$가 존재함으로써 $y=f(x)$.

그런데 $x \in f^{-1}(B)$이므로 $y=f(x) \in B$. 따라서 $f(f^{-1}(B)) \subseteq B$.

8. $x \in p_X(\mathscr{R}) \Longleftrightarrow$ 적당한 $(x, y) \in \mathscr{R}$에 대하여 $x=p_X(x, y)$

$\Longleftrightarrow x \in \mathrm{Dom}(\mathscr{R})$

그러므로 $p_X(\mathscr{R})=\mathrm{Dom}(\mathscr{R})$. 마찬가지로 $p_Y(\mathscr{R})=\mathrm{Im}(\mathscr{R})$ (과연?)

9. (a) $y \in f(A \cap f^{-1}(B))$

$\Longleftrightarrow$ 적당한 $x \in A \cap f^{-1}(B)$에 대하여 $y=f(x)$ 정의 9(a)

$\Longleftrightarrow$ 적당한 $x \in A$와 $x \in f^{-1}(B)$에 대하여 $y=f(x)$ $\cap$의 정의

$\Longleftrightarrow y \in f(A)$이고 $y \in B$ 정의 9

$\Longleftrightarrow y \in f(A) \cap B$ $\cap$의 정의

그러므로 $f(A \cap f^{-1}(B))=f(A) \cap B$ ……①

(b) ①의 A에 X를 대입하면 $f(X \cap f^{-1}(B))=f(X) \cap B$.

여기서 $X \cap f^{-1}(B)=f^{-1}(B)$ (왜?)이므로 $f(f^{-1}(B))=f(X) \cap B$

연습문제 3.6

7. 집합 Y가 한 원소 집합이면 X-사영 $p_X : X \times Y \to X$는 단사이다.

11. $m!$

12. (a) $f^{-1}(f(A)) \supseteq A$ 「178쪽 문제 4(a) 참조」이므로 $f^{-1}(f(A)) \subseteq A$임을 보이기만 하면 충분하다. 여기서 임의의 $x \in f^{-1}(f(A))$에 대하여 $f(x) \in f(A)$. 따라서 적당한 $x' \in A$에 대하여 $f(x)=f(x')$. 여기서 f는 단사이므로 $x=x'$.
따라서 $x \in A$. 그러므로 $f^{-1}(f(A)) \subseteq A$

연습문제 3.7

6. 집합 X의 임의의 원소 x, x'에 대하여 $f(x)=f(x')$이라면 $g \circ f=1_X$에 따라 $x=g \circ f(x)=g \circ f(x')=x'$, 즉 $x=x'$. 따라서 f는 단사이다. 한편 $h(Y) \subseteq X$이고 $f \circ h=1_Y$이므로 다음을 얻는다.

$$f(X) \supseteq f(h(Y))=f \circ h(y)=1_Y(Y)=Y, \text{ 즉 } f(X) \supseteq Y$$

여기서 $f(X) \subseteq Y$ (왜?)이므로 $f(X)=Y$. 따라서 f는 전사이다. 그러므로 f는 전단사이다.

다음에는 문제 4, 5의 결과를 이용하여 아래 것을 얻는다.

$$g=g \circ 1_Y=g \circ (f \circ f^{-1})=(g \circ f) \circ f^{-1}=1_X \circ f^{-1}=f^{-1}, \quad \text{즉} \quad g=f^{-1}$$

또 $$h=1_X \circ h=(f^{-1} \circ f) \circ h=f^{-1} \circ (f \circ h)=f^{-1} \circ 1_Y=f^{-1}, \quad \text{즉} \quad h=f^{-1}$$

따라서 $g=h=f^{-1}$

9. (i) 집합 X의 임의의 원소 x, x'에 대하여 $g\circ f(x)=g\circ f(x')$이라면 $g(f(x))=g(f(x'))$이고 g는 단사이므로 $f(x)=f(x')$. 그런데 f도 단사이므로 $x=x'$. 이것은 $g\circ f: X\to Z$가 단사임을 뜻한다.

(ii) $g\circ f(X)=g(f(X))$ $\qquad$ $g\circ f$의 정의

$=g(Y)$ $\qquad$ f는 전사이다.

$=Z$ $\qquad$ g는 전사이다.

그러므로 $g\circ f(x)=Z$. 따라서 $g\circ f: X\to Z$는 전사이다.

이상으로 (i), (ii)에 따라 $g\circ f: X\to Z$는 전단사이다.

(iii) (ii)의 결과에 따라 $g\circ f: X\to Z$는 전단사임을 감안할 때, 다음을 얻는다.

$(g\circ f)\circ(f^{-1}\circ g^{-1})=g\circ(f\circ f^{-1})\circ g^{-1}$ $\qquad$ 정리 15

$=(g\circ 1_y)\circ g^{-1}$ $\qquad$ 문제 5, 정리 15

$=g\circ g^{-1}$ $\qquad$ 문제 4

$=1_Z$ $\qquad$ 문제 5

그러므로 $(g\circ f)\circ(f^{-1}\circ g^{-1})=1_Z$

한편

$(f^{-1}\circ g^{-1})\circ(g\circ f)=f^{-1}\circ(g^{-1}\circ g)\circ f$ $\qquad$ 정리 15

$=f^{-1}\circ(1_Y\circ f)$ $\qquad$ 문제 5, 정리 15

$=f^{-1}\circ f$ $\qquad$ 문제 4

$=1_X$ $\qquad$ 문제 5

따라서 $(f^{-1}\circ g^{-1})\circ(g\circ f)=1_X$

그러므로 문제 6(397쪽)의 결과에 따라 $f^{-1}\circ g^{-1}=(g\circ f)^{-1}$

연습문제 4.1

2. 집합 Y로 하여금 유한집합이라는 결론에 반하여 Y는 무한집합이라고 가정하면 $g^{-1}: Y\to X$는 일대일 대응이므로 정리 2(207쪽)에 따라 X는 무한집합이다. 이것은 X가 유한집합이라는 가정에 모순이다. 따라서 Y는 유한집합이다.

4. 함수 $f: A\to A$를 단사라 하고, $f(A)\neq A$라면 $g(x, y)=(f(x), f(y))$로 정의된 함수 $g: A\times A\to A\times A$ 역시 단사이고 $g(A\times A)\neq A\times A$. 따라서 $A\times A$는 무한집합이다.

연습문제 4.2

2. (i) $f\times g(x, z)=f\times g(x', z)\Rightarrow(f(x), g(z))=(f(x'), g(z'))$

$$\Rightarrow f(x)=f(x') \wedge g(z)=g(z')$$
$$\Rightarrow x=x' \wedge z=z' \text{ (왜?)}$$
$$\Rightarrow (x, z)=(x', z')$$

그러므로 $f\times g(x, z)=f\times g(x', z') \Rightarrow (x, z)=(x', z')$

따라서 $f\times g$는 단사이다.

(ii) $\forall_{(y, w)}\in Y\times W \Rightarrow y\in Y \wedge w\in W$

$\Rightarrow \exists_{x\in X},\ f(x')=y$ 마찬가지로 $\exists_{x\in Z},\ g(z)=w$

그러므로 $\exists_{(x, z)\in X\times Z},\ f\times g(x, z)=(f(x),\ g(z))=(y,\ w)$

즉 $f\times g(x, z)=(y, w)$

따라서 $f\times g$는 전사이다.

마침내 (i), (ii)에 의하여 $f\times g$는 일대일 대응이다.

4. 지금 $f : X-Y\sim Y-X$라 놓고 함수 $g : X\rightarrow Y$를 다음과 같이 정의한다.

$$g(x)=\begin{cases} f(x), & x\in X-Y\text{일 때} \\ x, & x\in X\cap Y\text{일 때} \end{cases}$$

그러면 $X-Y=X-(X\cap Y),\ \ Y-X=Y-(X\cap Y)$

따라서 g는 전단사이다.

그러므로 $X\sim Y$.

5. 각 $\gamma\in\Gamma$에 대하여 $f_\gamma : X_\gamma\sim Y_\gamma$라 하고 함수 $f : \bigcup_{\gamma\in\Gamma} X_\gamma\rightarrow \bigcup_{\gamma\in\Gamma} Y_\gamma$를 모든 $x\in X_\gamma$에 대하여 $f(x)=f_\gamma(x)$로 정의한다. 그러면 $\{X_\gamma \mid \gamma\in\Gamma\}$, $\{Y_\gamma \mid \gamma\subset\Gamma\}$는 각각 서로소인 집합들의 족이므로 f는 잘 정의된 전단사이다.

9. 집합 A의 각 부분집합 B에 B의 특성함수 $\chi_B : A\rightarrow\{0, 1\}$이 대응하는 함수 $f : \mathscr{P}(A)\rightarrow 2^A$는 분명히 일대일 대응이다(과연?). 따라서 $\mathscr{P}(4)\sim 2^4$.

연습문제 4.3

1. $z=\{1, 2, 3, \cdots\}\cup\{0\}\cup\{-1, -2, -3, \cdots\}$으로 놓고 생각하여라.

4. $f : A\rightarrow\mathbb{N}$, $g : B\sim\mathbb{N}$이라면 $h(x, y)=(f(x), g(x))$로 정의된 함수 $h : A\times B\rightarrow\mathbb{N}\times\mathbb{N}$는 일대일 대응이므로 $A\times B\sim\mathbb{N}\times\mathbb{N}$. 따라서 정리 10에 의하여 $A\times B$는 가부번집합이다.

5. $A_1=B_1$이고 각 $k\in\mathbb{N}$에 대하여 $A_{k+1}=B_{k+1}-\bigcup_{j=1}^{k} A_j$라면 $\{A_k \mid k\in\mathbb{N}\}$는 서로소인 가산집합의 가부번족이다. 더구나 $\bigcup_{k\in\mathbb{N}} A_k=\bigcup_{k\in\mathbb{N}} B_k$이고 A_1, B_1는 각각 가부번집합이다. 그러므로 정리 10의 따름정리를 이용하여 $\bigcup_{k\in\mathbb{N}} A_k$가 가부번집합인 것을 보일 수 있다(그리하여 보아라!). 따라서 $\bigcup_{k\in\mathbb{N}} B$도 가부번집합이다.

6. 좌표평면의 원으로서 중심의 좌표와 반지름이 모두 유리수인 것의 집합 $\mathscr{C}$에 대하여 $f(c)=(x, y, z)$로 정의된 함수 $f: \mathscr{C} \to \mathbb{Q}\times\mathbb{Q}\times\mathbb{Q}$에 있어서 점 (x, y), z는 각각 원 $c\in\mathscr{C}$의 중심, 반지름이라고 할 때 분명히 f는 단사이다(과연?). 한편 예제 5(227쪽)에 따라 $\mathbb{Q}\sim\mathbb{N}$이고 정리 10(227쪽)에 따라

$$\mathbb{Q}\times\mathbb{Q}\times\mathbb{Q}\sim\mathbb{N}\times\mathbb{N}\times\mathbb{N}, \quad \mathbb{N}\times\mathbb{N}\times\mathbb{N}\sim\mathbb{N}\times\mathbb{N}, \quad \mathbb{N}\times\mathbb{N}\sim\mathbb{N}$$

그러므로 $\mathbb{Q}\times\mathbb{Q}\times\mathbb{Q}\sim\mathbb{N}$. 따라서 $\mathbb{Q}\times\mathbb{Q}\times\mathbb{Q}$는 가부번집합이다. 여기서 $f(\mathscr{C})$는 가부번집합 $\mathbb{Q}\times\mathbb{Q}\times\mathbb{Q}$의 무한부분집합이므로 정리 8에 따라 그것은 가부번집합이다. 따라서 $\mathscr{C}$도 가부번집합이다.

7. 각 유리수를 $p/q(q\neq0)$와 같이 한 가지로 나타낸다. 여기서 $p\in\mathbb{Z}$, $q\in\mathbb{N}$이고 p와 q의 최대공약수는 1이다. 그러면 $f(p/q)=(p, q)$로 정의된 함수 $f: \mathbb{Q}\to\mathbb{Z}\times\mathbb{N}$는 단사이다. 분명히 $\mathbb{Z}\times\{1\}\subseteq f(\mathbb{Q})\subseteq\mathbb{N}\times\mathbb{N}$이므로 정리 8(221쪽)에 따라 $f(\mathbb{Q})$는 가부번집합이다. 따라서 $\mathbb{Q}$도 가부번집합이다.

연습문제 4.4

4. 이 문제의 주장의 결론에 반하여 0과 1 사이의 무리수의 집합이 가부번집합이라고 가정하자. 그런데 0과 1 사이의 유리수의 집합이 가부번집합이므로 이 두 집합의 합집합 역시 가부번집합이다. 그러므로 0과 1 사이의 실수의 집합은 가부번집합이다. 이것은 정리 12에 모순된다. 따라서 0과 1 사이의 무리수의 집합은 비가부번집합이다.

5. 평면 p에서의 원의 집합을

$$C=\{(a, b, r)\mid(x-a)^2+(y-b)^2=r^2\}$$

라고 놓자. 여기서 함수 $f: C\to R\times R\times R$을 $f(a, b, r)=(a, b, r)$로 정의하면 f는 항등함수로서 일대일대응이므로 $C\sim R\times R\times R$. 그런데 R이 비가산집합이므로 C 또한 비가산집합이다(사실 여기서 $C=R\times R\times R$).

연습문제 5.1

3. 유한집합에 대한 기수의 규칙 C_{-3}에 따라 당연하다.

5. $0=\text{card}\,\phi$, $\text{card}\,\mathbb{N}$, $\text{card}\,\mathbb{R}$

연습문제 5.2

1. 유한기수 n에 대한 유한집합을 A, 즉 $\text{card}\,A=n$이라 하자.

지금 문제의 결론을 부정하여 card $N \leqq n$이라면 card $N \leqq$ card A.
그러므로 $N \subseteq A$. 이것은 모순!(왜?) 따라서 결론을 얻는다.

2. a를 초한기수라 하고 집합 A에 대하여 card $A = a$라고 하면 무한집합 A는 제 4장 정리 11에 의하여 가부번부분집합 이를테면 B를 지닌다. 그런데 $N \sim B$이므로 card N=card B. 한편 $B \subseteq A$이므로 card $B \leqq$ card A. 따라서 card $\mathbb{N} \leqq a$.

6. 집합 B에 대하여 분명히 $B \sim B$. 한편 가정에 따라 $B \subseteq C$이므로 B는 C의 부분집합 B와 대등하다. 또 가정 $A \sim C$으로부터 $C \sim A$. 그런데 또 가정에 따라 $A \subseteq B$이므로 C는 B의 부분집합 A와 대등하다. 그러므로 정리 1(245쪽 칸토어-베른슈타인 정리)에 따라 $C \sim B$. 또한 가정에 의하여 $A \sim C$. 따라서 $A \sim B$.

연습문제 5.3

1. 주어진 주장을 부정함으로써 모순을 유도하여라.

2. 지금 $f : A \sim B$이라면 함수 $f : A \to B$로 하여금 각 $X \in \mathscr{P}(A)$에 대하여 $f^*(X) = f(X)$로 정의된 함수 $f^* : \mathscr{P}(A) \to \mathscr{P}(B)$를 얻는다. 여기서 f는 전단사이므로 f^*도 전단사이다(과연?). 따라서 $\mathscr{P}(A) \sim \mathscr{P}(B)$.

3. 이 문제에서 주장하는 결론에 반하여 가부번집합 A가 존재함으로써 $\mathscr{P}(A)$가 가부번집합이라면 card A=card $\mathscr{P}(A)$. 이것은 정리 2(271쪽 칸토어의 정리)에 모순이다. 그러므로 $\mathscr{P}(A)$는 비가부번집합이다.

연습문제 5.4

3. (c) $\mathbb{R} \sim (0, 1) \sim (1, 2)$이고 $(0, 1) \subseteq (0, 1) \cup (1, 2) \subseteq \mathbb{R}$이므로 정리 1(249쪽)에 따라 $(0, 1) \cup (1, 2) \sim \mathbb{R}$. 그런데

$$\text{card}\,(0, 1) = \text{card}\,(1, 2) = \text{card}\,\mathbb{R} = c.$$

따라서 $c + c = c$

4.* 259쪽 각주 * 참조.

연습문제 5. 5

1. (a) 기수에 관한 규칙 C_1에 따라 기수 x, y에 각각 대응하는 집합, 이를테면 X, Y가 존재한다. 그런데 정의 3에 따라

$$xy = \text{card}\,(X \times Y),\quad yx = \text{card}\,(Y \times X)$$

여기서 임의의 $(x_i, y_i)\in X\times Y$에 $(y_i, x_i)\in Y\times X$를 대응시키는 함수는 일대일대응이다(과연?). 그러므로 $X\times Y\sim Y\times X$. 따라서 $xy=yx$.

(b) $x=\text{card } X$, $y=\text{card } Y$, $Z=\text{card } Z$로 놓고 다음을 유도하여라.
$(xy)Z=\text{card}[(X\times Y)\times Z]=\text{card}(X\times Y\times Z)=\text{card}[X\times(Y\times Z)]=x(yz)$

(c) 집합 X, Y, Z에 대하여 다음을 주목하여라.

$$X\times(Y\cup Z)=(X\times Y)\cup(X\times Z) \quad \text{여기서} \quad Y\cap Z=\phi$$

3. 반례: $x=1$, $y=z=\chi_0$에 대하여 $x<y$이지만 $xz=yz=\chi_0$[예제 5(c) 참조]

5. (a) 지금 A, B를 각각 $x=\text{card } A$, $y=\text{card } B$인 집합이라 하자. 여기서
$xy=\text{card}(A\times B)=0=\text{card}\,\phi$ 그러므로 $A\times B=\phi$
따라서 $A=\phi$ 또는 $B=\phi$이므로 $x=0$ 또는 $y=0$.

연습문제 5.6

2. 지금 A, $\{a\}$를 각각 $\text{card } A=a$, $\text{card}\{a\}=1$인 집합이라고 할 때
$\{f\,|\,f:\{a\}\to A\}\sim A$ 그러므로 $\text{card } A\{a\}=a$, 즉 $a^1=a$.
마찬가지로 $\{f: A\to\{a\}\}^*\sim\{a\}$이므로 $1^a=1$.

5. 임의의 $2\leqq n$에 대하여 정리 8, 10과 등식 $\aleph_0\aleph_0=\aleph_0$에 따라 다음을 얻는다.

$$c=2^{\aleph_0}\leqq n^{\aleph_0}\leqq n_0{}^{\aleph_0}\leqq(2^{\aleph_0})^{\aleph_0}=2^{\aleph_0\aleph_0}=2^{\aleph_0}=c$$

그러므로 $n^{\aleph_0}=c=\aleph_0{}^{\aleph_0}$.

8. 문제 6에 따라 $c\leqq\aleph_0 c\leqq cc=c$. 그러므로 $\aleph_0 c=c$.

11. $Z^{Y\times X}$의 임의의 원소 f, g에 대하여 $\psi(f)=\psi(g)$라고 가정하면
$e_f=e_g$[이 두 함수는 $(Z^Y)^X$의 원소] 따라서

$$\begin{aligned}\forall a\in X\ f^a=g^a &\Rightarrow \forall a\in X\,[\forall_{b\in Y}\ f^a(\mathrm{b})=g^a(\mathrm{b})]\\ &\Rightarrow \forall a\in X\,[\forall_{b\in Y}\ f(\mathrm{b}, a)=g(\mathrm{b}, a)]\\ &\Rightarrow \forall(b, a)\in Y\times X\ f(\mathrm{b}, a)=g(\mathrm{b}, a)\\ &\Rightarrow f=g\end{aligned}$$

그러므로 $\psi(f)=\psi(g)\Rightarrow f=g$, 즉 ψ는 단사이다.
한편 $\forall h\in(Z^Y)^X$에 대하여 함수 $f\in Z^{Y\times X}$를 $f(b, a)=h(a)(b)$로 정의하면
$\psi(f)=e_f$ 여기서 $e_f: X\to Z^Y$, $h: X\to Z^Y$이므로

$$\forall x\in X,\ \forall b\in Y\ \psi(f)(a)(b)=e_f(a, b)=f^a(b)=f(b, a)=h(a)(b)$$

따라서 $\forall a\in X\ \psi(f)(a)=h(a)$ 곧 $\psi(f)=h$이므로 ψ는 전사이다.
이상으로 ψ는 전단사이다.

---

* 이 함수 f에 대하여 $f(A)$를 말하여라.

연습문제 5.7

2. 클래식 힐베르트 공간이 H라면 정리 8과 10에 따라 다음을 얻는다.

$$c \leqq \text{card } \mathbb{H} \leqq c^{\aleph_0} = (2^{\aleph_0})^{\aleph_0} = 2^{\aleph_0\aleph_0} = 2^{\aleph_0} = c$$

그러므로 $\text{card}\,\mathbb{H} = c = \text{card}\,\mathbb{R}$.

3. 연습문제 5.6 문제 5와 문제 2에서의 $\text{card}\,\mathbb{R}^{\aleph_0} = c^{\aleph_0} = c$에 따라 $\mathbb{R}^{\aleph_0}$의 원소 중 격자점의 집합의 기수는 $\aleph_0^{\aleph_0}$, 즉 c와 같다.

4. 함숫값이 다만 0, 1인 모든 일실변수함수의 집합의 기수는 2^c와 같다. 한편 n실변수인 실수값의 함수의 집합의 기수는 c^{c^n}와 같다. 그런데 $c^n = c$, $\aleph_0 c = c$이므로 다음을 얻는다.

$$c^{c^n} = c^c = (2^{\aleph_0})^c = 2^{\aleph_0 c} = 2^c \quad \text{그러므로} \quad c^{c^n} = 2^c$$

5. 문제 4의 결과에 따라 $\mathfrak{f} = 2^c$. 그러므로 다음을 얻는다.

$$\mathfrak{f} \leqq \mathfrak{f}^n \leqq \mathfrak{f}^{\aleph_0} \leqq \mathfrak{f}^c = (2^c)^c = 2^{cc} = 2^c = \mathfrak{f}$$

따라서 $$\mathfrak{f}^n = \mathfrak{f}^{\aleph_0} = \mathfrak{f}^c = \mathfrak{f} \qquad \cdots\cdots ①^{**}$$

연습문제 6.1

2. (a) 지금 $B \neq \phi$라고 가정해도 무방하다(왜냐하면 $B = \phi \Longrightarrow C = \phi$이기 때문이다). 집합 $\{f^{-1}(y) \mid y \in B\}$는 A의 분할을 이루며 선택공리에 따라 대표집합 C를 지님으로써 각 $y \in B$에 대한 $C \cap f^{-1}(y)$는 한 원소집합이다. 그러므로 함수 f의 C로의 제한 $f \mid C : C \to B$는 전단사이다. 따라서 $\text{card}\,A \geqq \text{card}\,C = \text{card}\,B$. 곧 $\text{card}\,A \geqq \text{card}\,B$.

4.*** (a) 집합 $\mathbb{R}$에 속하는 임의의 a, b에 대하여

$a = a$	반사율
$a = b \wedge b = c \Longrightarrow a = c$	추이율
$a = b \wedge b = a \Longrightarrow a = b$	반대칭률

그러므로 $\mathbb{R}$ 위의 상등관계 $=$는 부분순서이다.

8. 선택공리에 따라 $g : \{A_\gamma \mid \gamma \in \Gamma\} \to \bigcup_{\gamma \in \Gamma} A_\gamma$를 선택함수라고 하면

$$\forall \gamma \in \Gamma \; f(\gamma) = g(A_\gamma) \in A_\gamma$$

로 정의된 함수 $f : \Gamma \to \bigcup_{\gamma \in \Gamma} A_\gamma$가 존재한다. 따라서 $\Gamma \neq \phi$이면 $\mathop{\mathbb{P}}_{\gamma \in \Gamma} A_\gamma \neq \phi$

** ①은 또 하나의 놀라운 사실이다. 그 배경을 곰곰히 음미해보아라.

*** $\forall a \in \mathbb{R}$ $a - a = 0$이면 그리고 그 때에만 $a = a$(반사율).
이 정의를 참고로 독자는 추이율, 반대칭률을 각각 밝혀라.

11. 적당한 함수 $g : B \to A$에 대하여 $f \circ g = I_B$라고 가정하면 임의의 $b \in B$에 대하여 $(f \circ g)(b) = f(g(b)) = I_B(b) = b$. 여기서 $g : B \to A$는 함수이므로 $g(b) \in A$. 지금 $g(b) = a$라고 놓으면 $f(a) = b$이고 $a \in A$ 따라서 $b \in f(A)$. 여기서 b는 B의 임의의 원소이므로 $B \subseteq f(A)$. 그러므로 f는 전사이다.

한편 함수 f는 전사라고 가정하면 각 $y \in B$에 대하여 $f^{-1}(y) \neq \phi$이고 $f^{-1}(y) \in A$. 여기서 A에 대한 선택함수의 하나를 h라고 할 때 함수 $g : B \to A$를 각 $y \in B$에 대하여

$$g(y) = h(f^{-1}(y))$$

로 정의한다. 즉 각 $y \in B$에 대한 $g(y)$로써 $f^{-1}(B)$의 임의의 원소 곧 $x = g(y)$로 놓을 수 있다. 그러면 $x \in f^{-1}(B)$, $f(x) = y$.

따라서 각 $y \in B$에 대하여

$$(f \circ g)(y) = f(g(y)) = f(x) = y = I_B(y)$$

그러므로 $\qquad f \circ g = I_B$

연습문제 6.2

2.* 오른쪽 그림과 같은 부분 순순서집합

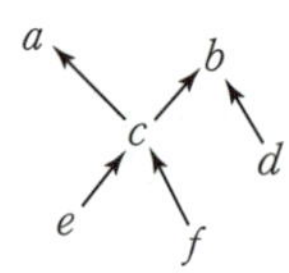

$$A = \{a, b, c, d, e, f\}$$

는 화살표 "→"로써 반순서를 나타내고 있다.

이 경우 $B = \{a, b, c, e, f\}$에는 최소상계, 최대하계 모두가 존재하지 않는다. 왜냐하면 a, b 및 e, f 사이에 각각 순서가 주어져 있지 않기 때문이다.

3. 다음은 주어진 집합 A에 대한 물음의 답이다.

상계 $\{a, b, c\}$, 하계 $\{a\}$, 최소상계 $\{a, b, c\}$, 최대하계 $\{a\}$

4. 임의의 $B \in \mathscr{B}$에 대하여 $B \subseteq \bigcup_{B \in \mathscr{B}} B$이므로 $\bigcup_{B \in \mathscr{B}} B$는 $\mathscr{B}$의 상계이다.

지금 C를 $\mathscr{B}$의 또 다른 상계라면 $B \subseteq C$이므로 $\bigcup_{B \in \mathscr{B}} B \subseteq C$.

따라서 $\sup \mathscr{B} = \bigcup_{B \in \mathscr{B}} B$

$\inf \mathscr{B} = \bigcap_{B \in \mathscr{B}} B$의 증명은 독자에게 돌린다.

7. 문제 2에서의 보기인 부분순서집합 $(A, \leqq)$의 그림에 따르면 A에는 2개의 극대원 a, b 및 3개의 극소원 d, e, f가 존재한다.

9. 대소관계 $\leqq$를 순서로하는 전순서집합 Q의 전순서부분집합

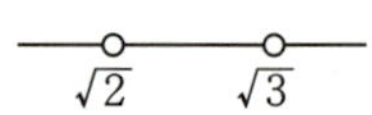

$$A = \{x \in \mathrm{Q} \mid \sqrt{2} < x < \sqrt{3}\}$$

* $\{a, c, f\}$는 부분순서집합 A의 하나의 사슬이다. 또 다른 사슬 이를테면 원소가 2개, 3개인 것을 말하여라.

에 대하여 유리수의 조밀성에 따라 A의 극대원소, 극소원소 모두가 존재하지 않는다(다만 $\sup A=\sqrt{3}$, $\inf A=\sqrt{2}$).

12. 집합 A의 전순서부분집합이 원소인 족 $\mathscr{T}$에 대하여 $B\in\mathscr{T}$라고 하자. 그러면 $\mathscr{T}$는 포함관계 ⊆에 따른 부분순서집합이다. 여기서 정리 2에서의 증명과 같은 방법을 적용함으로써 $\mathscr{T}$에는 $B\subseteq C$인 극대원소 C가 존재한다.

13. 집합 $X=\{a, b, c\}$에 대한 반순서집합 $(\mathscr{T}(X), \subseteq)$의 반사슬로서

$$C_1=\{\{a\}, \{b\}, \{c\}\},\quad C_2=\{\{a,b\}, \{b,c\}, \{a,c\}\}$$

가 존재한다. 또 다른 반사슬로써 원소가 2개인 것이 존재하면 지적하여라. 이상을 참고로 문제의 결론을 증명하여라.

연습문제 6.3

1. (a) B의 상계 $a(a\notin B)$, $b(b\notin B)$, $c(c\in B)$
B의 하계 $f(f\notin B)$
$\sup B=c(c\notin B)$, $\inf B=f(f\notin B)$
(여기서 $g\notin B$가 B의 하계가 아님은 $g\notin B$에서가 아니라 g와 $d(d\in B)$ 사이의 순서가 정해져 있지 않기 때문이다.)

2. 여기서 $(A, \leq)$를 부분순서집합이라 하고 그것의 전순서부분집합이 원소인 족 $\mathscr{T}$ 위에 부분순서 ⊆가 정의된 $(\mathscr{T}, \subseteq)$에 조른의 보조정리를 적용코자 한다. 지금 $(\mathscr{T}, \subseteq)$의 하나의 사실 $\mathscr{C}$에 대하여 $K=\bigcup_{C\in\mathscr{C}} C$로 놓는다. 여기서 $K\in\mathscr{T}$임을 보임으로써 K는 $\mathscr{C}$의 상계인 것을 밝히기로 한다. 사실 $x, y\in K$이면 적당한 $D\in\mathscr{C}$, $E\in\mathscr{C}$에 대하여 $x\in D$, $y\in E$. 그러나 $\mathscr{C}$는 $(\mathscr{T}, \subseteq)$의 사슬이므로 $D\subseteq E$ 아니면 $E\subseteq D$. 지금 $E\subseteq D$라면, $x, y\in D$. 한편 D는 전순서집합이므로 $x\leq y$ 아니면 $y\leq x$. 따라서 K는 $(A, \leq)$의 전순서부분집합이다. 그러므로 $K\in\mathscr{T}$. 따라서 조른의 보조정리에 따라 $(\mathscr{T}, \subseteq)$는 극대원소를 지닌다.

3. 지금 R을 단위원 1을 지닌 환이라 하고 그것의 진아이디얼*의 집합 $\mathscr{A}$를 ⊆에 따른 부분순서집합이라고 하자. 여기서 $\mathscr{C}$를 $(\mathscr{A}, \subseteq)$의 임의의 사슬이라면 $1\notin\bigcup_{I\in\mathscr{C}} I$이므로 $\bigcup_{I\in\mathscr{C}} I$는 R의 진아이디얼이다. 따라서 $\bigcup_{I\in\mathscr{C}} I\in\mathscr{A}$는 $\mathscr{C}$의 하나의 상계이다. 그러므로 조른의 보조정리에 따라 $(\mathscr{A}, \subseteq)$에는 극대원소가 존재한다. 따라서 R의 진극대아이디얼이 존재한다.

4. 벡터공간을 V로 놓고 그것의 일차독립인 것의 부분집합이 원소인 집합 $\mathscr{A}$를 ⊆에 따른 부분순서집합이라 하자. $\mathscr{C}$를 $(\mathscr{A}, \subseteq)$의 사슬이라면 $K=\bigcup_{C\in\mathscr{C}} C(K\in\mathscr{A})$는 일차독립이므로 $\mathscr{C}$의 상계이다. 따라서 조른의 보조정리에 따라 $(\mathscr{A}, \subseteq)$에는 극대

* 하나의 집합으로서의 아이디얼을 말할 때 진아이디얼은 그 집합의 진부분집합을 뜻한다.

원소(이것은 V의 기저를 이룬다)가 존재한다.

6. 격자는 그 원소의 수 n에 따라 다음과 같은 형을 지닌다. 다만 순서동형**인 것은 모두 묶어 하나로 본다.

(a) $n=1, 2, 3$인 경우 선형순서(107쪽 각주 * 참조)일 때 뿐이다.

(b) $n=4$인 경우 선형순서, 부분순서 두 가지 있다.

(c) $n=5$인 경우 5가지 있다.

(d) 오른쪽 그림과 같은 부분순서집합은 격자가 아니다. 왜냐하면 cvd가 존재하지 않기 때문이다.

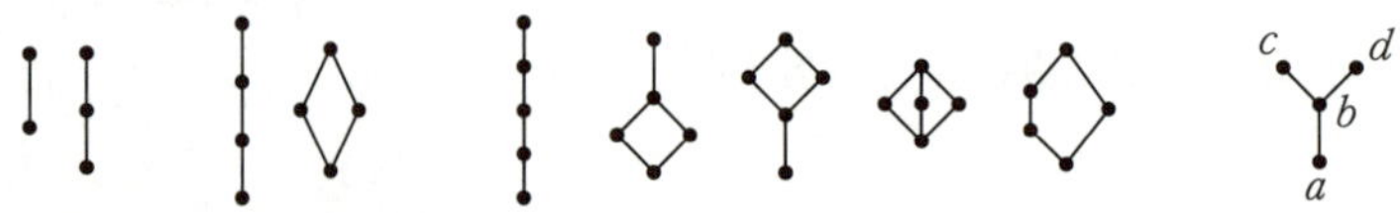

(a) (b) (c) (d)

이상을 참고로 문제의 결론을 증명하여라.

8. $\mathscr{F}$를 ⊆에 따른 부분순서집합, $\mathscr{C}$를 $(\mathscr{F}, \subseteq)$의 임의의 사슬이라고 할 때 $\mathscr{C}$에는 상계가 존재함을 보이도록 한다. 이 경우 $K=\bigcup_{C\in\mathscr{C}} C$는 그 상계의 후보의 자격이 있다. 지금 $\{x_1, x_2, \cdots, x_n\}$을 K의 임의의 유한부분집합이라면 적당한 $C_i\in\mathscr{C}$, $i=1, 2, \cdots, n$에 대하여 $x_i\in C_i$. 그런데 $\mathscr{C}$는 사슬이므로 적당한 $C\in\mathscr{C}$가 존재함으로써 $i=1, 2, \cdots, n$에 대하여 $C_i\subseteq C$.

한편 $\{x_1, x_2, \cdots, x_n\}\subseteq C$이므로 $\{x_1, x_2, \cdots, x_n\}\in\mathscr{F}$.

따라서 $K\in\mathscr{F}$이므로 K는 $\mathscr{C}$의 하나의 상계이다. 그러므로 조른의 보조정리에 따라 $(\mathscr{F}, \subseteq)$에 극대원소가 존재한다.

연습문제 6.4

1.*** 집합 Q는 가부번집합이므로 전단사 말하면 $f : \mathbb{Q}\sim\mathbb{N}$가 존재한다. 지금 임의의 $p, q\in\mathbb{Q}$에 대하여 N에 있어서

$$f(p)\leq f(q)\Leftrightarrow p\leq q$$

인 것으로 Q 위의 관계 ≦를 정의한다. 여기서 ≦는 자연수에 대한 대소관계를 나타낸다. 이런 때 $(\mathbb{N}, \leq)$이 정렬집합이므로 $(\mathbb{Q}, \leq)$도 정렬집합이다.

2. 결론부터 부정하고 보아라.

7. $(A, \leq)$가 정렬집합이면 이 집합은 엄밀한 뜻에서 감소하는 무한수열은 포함하지 않는다. 왜냐하면 그러한 수열은 A의 부분집합으로서 최소원소를 지니지 않기 때

** 이 책의 333쪽 및 이장우, 《현대집합론의 원리》, 경문사, 1998년, 정의 6.7(197−198쪽) 참조.

*** 여기서 정의된 전단사 f는 일종의 순서동형사상이다(333쪽 정의 1 및 두 각주) 참조.

문이다. 역으로 전순서집합 $(A, \leqq)$가 정렬집합이 아니라면 A의 부분집합으로서 최소원소를 지니지 않는 B가 존재한다.

지금 임의의 $a_1 \in B$를 택할 때 a_1은 B의 최소원소가 아니다. 그러므로 $a_1 > a_2$인 $a_2 \in B$를 택할 수 있다. 마찬가지로 $a_2 > a_3$인 $a_3 \in B$도 택할 수 있다. 이러한 과정을 반복함으로써 엄밀한 뜻에서 감소하는 무한수열 $a_1, a_2, a_3, \cdots$을 얻는다.

연습문제 6.5

1. 정렬집합 $(N, \leqq)$의 절편 중 $\mathbb{N}_k = \{1, 2, 3, \cdots, k\}$와 같은 형태의 것을 지적하자면 $S_k = \{x \in \mathbb{N} \mid x < k\} = \{1, 2, 3, \cdots, k-1\}$, 다만 $S_1 = \phi$.

3. (a) 정렬집합 $(A, \leqq)$의 임의의 절편 $S(S \neq A)$에 대하여 $S = S_a$인 $a \in A$가 존재한다[319쪽 정리 7(c) 참조. 사실 이 정리는 다음 문제 4에서의 증명을 거친 후 인용할 일이다]. 여기서 $S_a == \{x \in A \mid x < a\}$.

지금 $\mathscr{F}$의 임의의 절편을 각각 S_a, S_b라고 놓으면

$$S_a \cap S_b = \begin{cases} S_a & a < b \\ S_a = S_b & a = b \\ S_b & a > b \end{cases}$$

이상으로 $\mathscr{F}$의 임의의 절편의 교집합은 절편이다(같은 방법으로 합집합 $S_a \cup S_b$도 A의 절편임을 보일 수 있다).

연습문제 7.1

1. 가정에 따라 A, B를 다음과 같이 나타낼 수 있다.

$$A = \{a_1, a_2, a_3, \cdots, a_n\} \qquad B = \{b_1, b_2, b_3, \cdots, b_n\}$$

여기서 함수 $f: A \to B$를 $f(a_i) = b_i$로 정의하면 분명히 f는 순서동형사상이다. 따라서 A, B는 순서동형이다.

3. $(N, \leqq) = \{1, 2, 3, \cdots\}$ ……①

$(N, \leqq') = \{1, 3, 5, \cdots, 2, 4, 6, \cdots\}$ ……②

에 있어서 정의역이 ①인 순서동형사상으로서는 말하자면

$$f: \{1, 2, 3, \cdots\} \to \{1, 3, 5, \cdots\} \qquad g: \{1, 2, 3\} \to \{2, 4, 6, \cdots\}$$

뿐이다. 그런데 f, g는 모두 ①에서 ②로의 순서동형사상이 아니다(왜?) 따라서 A, B는 순서동형이 아니다.

연습문제 7.2

1. 정렬집합 $\{1, 2, 3, \cdots\}$로부터 정열집합 $\{1, 3, 5, \cdots, 2, 4, 6, \cdots\}$의 진절편 $\{1, 3, 5, \cdots\}$에로의 순서동형사상 f가 존재한다[연습문제 7.1 문제 3의 풀이(409쪽)]. 그러므로 정리 4(341쪽)에 따라 $\text{ord}\{1, 2, 3, \cdots\} < \text{ord}\{1, 3, 5, \cdots, 2, 4, 6, \cdots\}$.

7. ω

연습문제 7.3

4. $\alpha=\omega$, $\beta=0$, $\gamma=1$에 대하여 $\beta+\alpha=\omega=\gamma+\alpha$이므로 $\beta+\alpha=\gamma+\alpha$. 그러나 $\beta\neq\gamma$.

6. 문제 4에서와 같은 반례에 대하여 $\beta=0<1=\gamma$, 즉 $\beta<\gamma$이지만 $\beta+\alpha=\gamma+\alpha$.

11. 존재. 왜냐하면 최대순서수 말하자면 α가 존재한다면 이 α에 대하여 $\alpha<\alpha+1$ (왜?). 이것은 α가 최대순서수란 가정에 모순이다. 그러므로 최대순서수는 존재하지 않는다.

연습문제 7.4

2. (a) 순서수에 대한 규칙 O=3(335쪽)에 따라 $\text{ord}(\phi, \leq)=0$.

지금 $\alpha=\text{ord}(A, \leq')$라고 놓으면

$$0\alpha=\text{ord}(A\times\phi, \leq^*)=\text{ord}(\phi, \leq^*)=0$$

또한 $\alpha 0=0$(과연?)

이상으로 $0\alpha=0=\alpha 0$

4. $\alpha=\omega$, $\beta=\gamma=1$에 대하여 $(\beta+\gamma)\alpha=2\omega=\omega$. 그러나 $\beta\alpha+\gamma\alpha=\omega+\omega=\omega 2$. 여기서 $2\omega\neq\omega 2$이므로 $(\beta+\gamma)\alpha\neq\beta\alpha+\gamma\alpha$.

6. 지금 $\alpha=\text{ord}(A, \leq)$, $\beta=\text{ord}(B, \leq')$, $\gamma=(C, \leq'')$라고 놓았을 때

$$A\times(B\times C)\approx(A\times B)\times C$$

임을 밝히면 된다. 그러려면 각 $x\in A$, $y\in B$, $z\in C$에 대하여

$$f(x, (y, z))=((x, y), z) \qquad \cdots\cdots ①$$

로 정의된 함수 f로 하여금 $A\times(B\times C)$에서 $(A\times B)\times C$로의 순서동형사상인 것을 유도하여야 한다. 그런데 ①은 단사(과연?)임을 먼저 염두에 두자. 여기서

$$x_1=(x, (y, z)), \quad x_2=((x', y'), z') \qquad \cdots\cdots ②$$

와 같이 각각 놓고

$$y_i=f(x_i)(i=1, 2) \qquad \cdots\cdots ③$$

라고 하자.

이제 $A\times(B\times C)$에서의 $x_1<x_2$이란 가정하에 그 전제(충분조건)을 살펴보면 다음 세 경우만이 존재한다.

(a) A에서의 $x<x'$, B, C 각각의 임의의 원소 y, z

(b) A에서의 $x=x'$, B에서의 $y<y'$, C의 임의의 원소 z

(c) A에서의 $x=x'$, B에서의 $y=y'$, C에서의 $z<z'$

이렇듯 $x_1<x_2$의 전제 (a), (b), (c) 각 경우에 대하여 $A\times(B\times C)$에서의 $y_1<y_2$일 경우는 다음과 같다.

(a)′ $A\times B$에서 $(x, y)<(x', y')$

즉 A에서의 $x<x'$, B의 임의의 원소 y, y', C의 임의의 원소 z

(b)′ $A\times B$에서의 $(x, y)<(x', y')$

즉 A에서의 $x=x'$, B에서의 $y<y'$, C의 임의의 원소 z

(c)′ $A\times B$에서의 $(x, y)=(x', y')$, C에서의 $z<z'$

지금까지를 종합하면 $x_1<x_2 \Rightarrow y_1<y_2$

②, ③에 따라

$$(x, (y, z))<((x', y'), z') \Rightarrow f(x_1)<f(x_2)$$

②에 따라

$$\Rightarrow f(x, (y, z))<f((x', y'), z')$$

그러므로 $(x, (y, z))<((x', y'), z') \Rightarrow f(x, (y, z))<f((x', y'), z')$

이상으로 f는 순서동형사상이다.

따라서 $A\times(B\times C)\approx(A\times B)\times C$ 곧 $(\gamma\beta)\alpha=\gamma(\beta\alpha)$

7. $\alpha<\alpha+1$. 가정에서 $\alpha+1\leqslant\beta$이므로 $\alpha<\beta$

$\alpha<\beta\Rightarrow\alpha+1\leqslant\beta$의 증명은 독자에게 미룬다.

8. $\alpha=1$, $\beta=2$, $\gamma=\omega$에 대하여 $\alpha<\beta$, $0<\gamma$. 그러나 $\alpha\gamma=1\omega=\omega=2\omega=\beta\gamma$.

즉 $\alpha\gamma=\beta\gamma$.

9. $2\omega=3\omega$, 그러나 $2<3$이므로 $2\neq3$.

연습문제 7.5

3. 임의의 집합 X가 주어지는데로 정렬원리에 따라 정렬집합 $(X, \leq)$를 얻는다. 여기서 $\leq$는 X 위의 정렬순서이다. 지금 $\mathscr{S}=\{\mathrm{ord}(X, \leq)\}$로 놓으면 정리 12(361쪽)에 따라 $\mathscr{S}$는 정렬집합이다. 그러므로 $\mathscr{S}$에는 최소원소, 즉 X에 대한 초기순서수가 단 하나 존재한다.

기호, 약자의 해설

$\sim$	… 아니다
$\wedge$	이고
$\vee$	또는
$\rightarrow$	… 이면 … 이다
$\leftrightarrow$	… 이면, 그리고 그때에만 … 이다
$\equiv(\Longleftrightarrow)$	… 와 동치이다
$\Rightarrow$	… 이 …를 함의한다
$\forall$	모든(각) … 에 대하여
$\exists$	… 가 존재함으로써 (적당한 … 에 대하여)
$a \in A$	a는 A의 원소이다
$a \notin A$	a는 A의 원소가 아니다
$\{x, y, \cdots, z\}$	원소 x, y, $\cdots$, z로 이루어진 집합
$\{x \mid p(x)\}$	$p(x)$가 참인 x의 집합
$\mathbb{N}$	자연수의 집합
$\mathbb{Z}$	정수의 집합
$\mathbb{Q}$	유리수의 집합
$\mathbb{R}$	실수의 집합
$\mathbb{R}_+$	양의 실수의 집합
$A \subseteq B$	A는 B의 부분집합이다
$A \nsubseteq B$	A는 B의 부분집합이 아니다
$A \subset B$	A는 B의 진부분집합이다
$A \supseteq B$	A는 B의 초집합이다
ϕ	공집합
$A \cap B$	집합 A와 B의 교집합
$A \cup B$	집합 A와 B의 합집합
$\mathscr{P}(A)$	집합 A의 멱집합(A의 모든 부분집합의 집합)
$\bigcap_{\gamma \in \Gamma} C_\gamma$	집합 C_γ들의 교집합, 여기서 $\gamma \in \Gamma$
$\bigcup_{\gamma \in \Gamma} C_\gamma$	집합 C_γ들의 합집합, 여기서 $\gamma \in \Gamma$

$\bigcap_{k=1}^{n} C_k$	집합 $C_1, C_2, \cdots, C_n$의 교집합
$\bigcup_{k=1}^{n} C_k$	집합 $C_1, C_2, \cdots, C_n$의 합집합
$\bigcap_{A \in \mathscr{F}} A$	집합 A의 족 $\mathscr{F}$의 교집합
$\bigcup_{A \in \mathscr{F}} A$	집합 A의 족 $\mathscr{F}$의 합집합
(a, b)	원소 a, b의 순서쌍
$A \times B$	집합 A와 B의 데카르트 곱
$\mathrm{Dom}(\mathscr{R})$	관계 $\mathscr{R}$의 정의역
$\mathrm{Im}(\mathscr{R})$	관계 $\mathscr{R}$의 상
$\mathscr{R}^{-1}$	관계 $\mathscr{R}$의 역관계
Δ_X	X 위의 항등관계
$\mathscr{E}$	동치관계
$x/\mathscr{E}$	x와 $\mathscr{E}$로서 정해진 동치류
$X/\mathscr{E}$	동치류 $x/\mathscr{E}$의 집합, 여기서 $x \in X$
$\mathscr{T}$	집합의 분할
$X/\mathscr{T}$	X의 분할 $\mathscr{T}$에 따른 동치관계
$c(n, r)$	이항계수
$f: X \to Y$	f는 X에서 Y로의 함수이다
$1_X: X \to X$	X 위의 항등함수
$\chi_A: X \to \{1, 2\}$	특성함수, 여기서 $A \subseteq X$
$f: X \sim Y$	함수 $f: X \to Y$는 일대일 대응이다
2^A	A에서 $\{1,2\}$로의 함수의 집합
B^A	A에서 B로의 함수의 집합
$A \sim B$	A와 B는 대등하다
Card A	집합 A의 기수
$\aleph_0$	자연수의 집합의 기수
c	실수의 집합의 기수
$\mathfrak{f}$	$\mathbb{R}$에서 $\mathbb{R}$로의 함수의 집합의 기수
A_x	A의 절편 $\{a \in A \mid a < x\}$, 여기서 $(A, \leqq)$는 정렬집합이다
$(A, \leqq) \approx (B, \leqq')$	정렬집합 $(A, \leqq)$, $(B, \leqq')$은 같은 꼴의 순서이다
$\mathrm{ord}(A, \leqq)$	정렬집합 $(A, \leqq)$의 서수
ω	정렬집합 $(\mathbb{N}, \leqq)$의 서수
Add.	합의 법칙: $p \Rightarrow p \vee q$
Simp.	단순화법칙: $p \wedge q \Rightarrow p$, $p \wedge q \Rightarrow q$
D.S.	논리합의 삼단논법 : $(p \vee q) \wedge \sim p \Rightarrow q$

D.N.	이중부정법칙 : $\sim(\sim p)\equiv p$
Com.	교환법칙 : $p\wedge q\equiv q\wedge p$, $p\vee q\equiv q\vee p$
Idemp.	멱등법칙 : $p\wedge p\equiv p$, $p\vee p\equiv p$
Contrap.	대우법칙 : $p\rightarrow q\equiv\sim q\rightarrow\sim p$
De M.	드 모르간의 법칙 : $\sim(p\wedge q)\equiv\sim p\vee\sim q$, $\sim(p\vee q)\equiv\sim p\wedge\sim q$
Assoc.	결합법칙 : $(p\wedge q)\wedge r\equiv p\wedge(q\wedge r)$, $(p\vee q)\vee r\equiv p\vee(q\vee r)$
Dist.	분배법칙 : $p\wedge(q\vee r)\equiv(p\wedge q)\vee(p\wedge r)$, $p\vee(q\wedge r)\equiv(p\vee q)\wedge(p\vee r)$
Trans.	추이율 : $(p\rightarrow q)\wedge(q\rightarrow r)\Rightarrow(p\Rightarrow r)$
C.D.	구성적 양도논법 : $(p\rightarrow q)\wedge(r\rightarrow s)\Rightarrow(p\vee r\rightarrow q\vee s)$, $(p\rightarrow q)\wedge(r\rightarrow s)\Rightarrow(p\wedge r\rightarrow q\wedge s)$
D.D.	파괴적 양도논법 : $(p\rightarrow q)\wedge(r\rightarrow s)\Rightarrow(\sim q\vee\sim s\rightarrow\sim p\vee\sim r)$, $(p\rightarrow q)\wedge(r\rightarrow s)\Rightarrow(\sim q\wedge\sim s\rightarrow\sim p\wedge\sim r)$
M.P.	삼단긍정법 : $(p\rightarrow q)\wedge p\Rightarrow q$
M.T.	삼단부정법 : $(p\rightarrow q)\wedge\sim q\Rightarrow\sim p$
R.A.	배리법 : $(p\rightarrow q)\Longleftrightarrow(p\wedge\sim q\rightarrow q\wedge\sim q)$
Q.N.	전칭명제, 특칭 명제의 부정규칙 : $\sim[\forall x\, p(x)]\equiv\exists x\sim p(x)$, $\sim[\exists x\, p(x)]\equiv\forall x\sim p(x)$
Def.	정의
I.P.	간접증명법
Th.	정리
Hyp.	가정(전제)
Concl.	결론
t	항진명제
c	모순

참고문헌*

집합론을 더욱 공부하려는 학생을 위하여 다음 책들을 추천한다.

I. 역사, 철학에 관련된 책

Benaceraf, Paul, and Hilary Putnam, eds. *Philosophy of Mathematics.* Prentice-Hall, Inc., Englewood Cliffs. N.J., 1964.

Van Heijenoort, Jean, ed. *From Frege to Gödel.* Harvard University Press, Cambridge, Mass., 1967.

Wilder, Raymond L. *Introduction to the Foundations of Mathematics.* 2nd ed. John Wiley & Sons, Inc., New York, 1965.

II. 논리학과 관련된 책

Copi, Irving M. *Symbolic Logic.* 3d ed. The Macmillan Company, 1967.

Kleene, Stephen C. *Mathematical Logic.* John Wiley & Sons, Inc., New York, 1967.

**K.G. Binmore, *Logic, Sets and Numbers.* Cambridge University press, 1980.

III. 집합론과 관련된 책

Frankel, Abraham A. *Set Theory and Logic.* Addison-Wesley Publishing Company, Inc., Reading, Mass., 1966.

Halmos, Paul R. *Naive Set Theory.* D. Van Nostrand Company, Inc., Princeton, N.J., 1960.

Hayden, Seymour, and John F. Kennison. *Zermelo-Fraenkel Set Theory.* Charles E. Merrill Publishing Company, Columbus, Ohio, 1968.

Monk, J. Donald. *Introduction to Set Theory.* McGraw-Hill, Inc., New York, 1969.

Charles C. Pinter *Set Theory.* Kyung Moon Publishers(경문사) 1998.

Suppes, Patrick. *Axiomatic Set Theory.* D. Van Nostrand Company, Inc., Princeton, N.J., 1960.

**Paul J. Cohen. *Sets Theory and The Continuum Hypothesis.* W.A. Benjamin, Inc., New York, Amsterdam, 1966.

**Paul Bernays. *Axiomatic Set Theory With a Historical Introduction* by Abraham A. Fraenkel. 2nd ed. North-Holland Publishing Company, Amsterdam, 1968.

**A. Kaufmann. *Introduction To the Theory of Fuzzy Subsets. Volume I., -undamental theoretical elements.* Academic Press, New York, San Francisco, London, 1975.

**Alexander Abian. *The Theory of Sets and Transfinite Arithemetic,* W.B. Saunders Co., 1965.

* 끝까지 열심히 공부해온 독자라면 여기 소개된 어느 책이든 누구나 혼자 읽어나갈 수 있는 실력이 쌓였을 것으로 믿어 책 제목을 원서 그대로 소개한다.

** 옮긴이가 추천한 것이다.

찾아보기

《ㅈ》

《 ㅊ 》

《 ㅋ 》

《 ㅌ 》

《 ㅍ 》

《 ㅎ 》